Lecture Notes in Computer Science

Edited by G. Goos and J. Hartmanis

57

Portability
of Numerical Software

Workshop, Oak Brook, Illinois,
June 21–23, 1976

Edited by Wayne Cowell

Springer-Verlag
Berlin Heidelberg New York 1977

Library of Congress Cataloging in Publication Data

Workshop on the Portability of Numerical Software, Oak
   Brook, Ill., 1976.
   Portability of numerical software workshop, Oak Brook,
Illinois, June 21-23, 1976.

   (Lecture notes in computer sciences ; 57)
   Workshop organized by Argonne National Laboratory.
   Bibliography: p.
   Includes index.
   1. Numerical analysis--Computer programs--Congresses.
2. Software compatibility--Congresses. I. Cowell,
Wayne, 1926-   II. United States. Argonne National
Laboratory, Lemont, Ill. III. Title. IV. Series.
QA297.W65 1976   001.6'4   77-13623

AMS Subject Classifications (1970): 00 A 10, 68 A 20
CR Subject Classifications (1974): 4.22, 4.6, 5.11, 5.25

ISBN 3-540-08446-0 Springer-Verlag Berlin Heidelberg New York
ISBN 0-387-08446-0 Springer-Verlag New York Heidelberg Berlin

Printing and binding: Beltz Offsetdruck, Hemsbach/Bergstr.
2145/3140-543210

INTRODUCTION

Those who cannot remember the past are condemned to repeat it - Santayana

The man who doesn't write portable software is condemned to rewrite it -

J. F. Traub (1971)

A computer program represents a body of problem-solving experience, often preceded by painstaking analysis, algorithm selection, and programming. The value of a program is magnified many-fold when it can be moved and effectively used in computing environments other than the one where it originated; but computing environments differ in both obvious and subtle ways and every successful computer program has evolved as an adaptation to a particular environment.

The benefits and difficulties of sharing programs are widely appreciated in the computing community, nowhere more so than among the developers and users of mathematical software, the programs that perform the basic numerical calculations of science and engineering. The finiteness of computers demands mathematical compromises that may be reached in different ways in different programs on different systems. At the same time, numerical calculation is so fundamental that there is a significant movement to create high quality mathematical software for distribution and widespread use in libraries. For some of those involved in this activity the portability problem is primarily a research challenge. Others have a business interest, seeing portability as a way to reach a wider market for their software products. Still others work in computing centers and are continually faced with the task of adapting a program for use in their home environment. Whether its primary motivation is research, business, or services, each group concerned with mathematical software recognizes the breadth of the portability problem and the need to exchange information and collaborate toward improving the current state of affairs.

To facilitate such information exchange, a Workshop on the Portability of Numerical Software was organized by Argonne National Laboratory and held in Oak Brook, Illinois on June 21-23, 1976. About three dozen numerical analysts and software developers from seven countries participated. The papers in this volume are based on oral presentations at the workshop but they were written _after_ the give and take of the discussion that surrounded each presentation; thus we may assume that the impact of that discussion, as perceived by each author, is distilled in his or her paper. The collection is offered to the computing community as a point of reference in the continuing effort to produce portable numerical software.

The authors and the editor are indebted to many individuals and groups who contributed to the workshop and to this volume. Especially deserving of mention are:

- The IFIP Working Group on Numerical Software (WG 2.5), whose members first suggested the workshop and then participated fully;
- The National Science Foundation and the U.S. Energy Research and Development Administration who provided financial support;
- Burton Garbow who made many helpful comments on the manuscripts;
- Patricia Witkowski who skillfully retyped the original manuscripts to achieve uniformity of format throughout the book.

Wayne Cowell
Argonne, Illinois
June, 1977

T. J. Aird
International Mathematical and
    Statistical Libraries, Inc.
Houston, Texas

James M. Boyle
Argonne National Laboratory
Argonne, Illinois

W. S. Brown
Bell Telephone Laboratories
Murray Hill, New Jersey

George D. Byrne
Univ. of Pittsburgh
Pittsburgh, Pennsylvania

W. J. Cody
Argonne National Laboratory
Argonne, Illinois

Wayne Cowell
Argonne National Laboratory
Argonne, Illinois

Ingemar Dahlstrand
Lund University
Lund, Sweden

Carl de Boor
Mathematics Research Center
Madison, Wisconsin

T. J. Dekker
University of Amsterdam
Amsterdam, The Netherlands

L. M. Delves
University of Liverpool
Liverpool, England

Jack Dongarra
Argonne National Laboratory
Argonne, Illinois

Kenneth W. Dritz
Argonne National Laboratory
Argonne, Illinois

J. J. Du Croz
The NAG Central Office
Oxford, England

Bo Einarsson
National Defense Research Institute
Tumba, Sweden

John Gabriel
Argonne National Laboratory
Argonne, Illinois

B. Ford
The NAG Central Office
Oxford, England

Lloyd D. Fosdick
University of Colorado
Boulder, Colorado

P. A. Fox
Bell Telephone Laboratories
Murray Hill, New Jersey

L. Wayne Fullerton
Los Alamos Scientific Laboratory
Los Alamos, New Mexico

W. Morven Gentleman
University of Waterloo
Waterloo, Ontario

A. D. Hall
Bell Telephone Laboratories
Murray Hill, New Jersey

Dennis Harms
Hewlett-Packard Co.
Cupertino, California

Pieter W. Hemker
Stichting Mathematisch Centrum
Amsterdam, The Netherlands

Frances Holberton
National Bureau of Standards
Gaithersburg, Maryland

T. E. Hull
University of Toronto
Toronto, Ontario

William M. Kahan
University of California
Berkeley, California

P. Kemp
University of Cambridge
Cambridge, England

Fred T. Krogh
Jet Propulsion Laboratory
Pasadena, California

C. L. Lawson
Jet Propulsion Laboratory
Pasadena, California

Dan Lozier
National Bureau of Standards
Gaithersburg, Maryland

Paul Messina
Argonne National Laboratory
Argonne, Illinois

Cleve Moler
University of New Mexico
Albuquerque, New Mexico

James C. T. Pool
Argonne National Laboratory
Argonne, Illinois

J. K. Reid
AERE Harwell
Didcot, England

Christian Reinsch
Leibniz-Rechenzentrum
Munich, Germany

John Rice
Purdue University
Lafayette, Indiana

J. L. Schonfelder
The University of Birmingham
Birmingham, England

Brian T. Smith
Argonne National Laboratory
Argonne, Illinois

Hans J. Stetter
Technische Hochschule Wien
Vienna, Austria

M. Sund
Gesellschaft für Strahlen und
   Umweltforschung
Munich, Germany

Christoph Überhuber
Technische Hochschule Wien
Vienna, Austria

N. Victor
University of Giessen
Giessen, Germany

W. M. Waite
University of Colorado
Boulder, Colorado

TABLE OF CONTENTS

WHAT IS PORTABILITY?

"Then you should say what you mean," the March Hare went on.
"I do," Alice hastily replied; "at least - at least I mean what
I say - that's the same thing, you know."
                                                    - Lewis Carroll

At first thought one might expect *portability* to be well defined as
a concept, whatever the difficulty of its achievement. In fact, however,
several concepts emerge when computer scientists are asked to be precise
about their use of the term and we find more than one definition in the
pages of this volume. The Numerical Algorithms Group (NAG) demands that
portable software be movable from one environment to another <u>without</u>
<u>change</u>. If such a move requires only changes that are capable of
mechanical implementation via a preprocessor, the NAG people say that
the software is *transportable*. Boyle and Cody explicitly adopt the NAG
usage. On the other hand, Brown, Hall, and Fox say that a program is
portable if the effort to move it to a new environment is much less
than the effort to rewrite it for the new environment. Reinsch suggests
that software is *portable* if successful running in a new environment is
achieved by no more than manual adjustment of machine dependent parameters
intentionally provided for that purpose. Then he exposes the folly of
trying to be too precise by reminding us that, in the current state of
the art, any definition depends on implicit assumptions which are left
undefined.

We shall take the position that the meaning of portability is contained in the collective descriptions of what is actually being done to transplant computer programs. Each "definition" serves a group of workers in their local setting. More generally, what we seek is a sense of the relationship between portability investigations and other mathematical software endeavors. In this vein we begin with Christian Reinsch's paper, Some Side Effects of Striving for Portability.

Reinsch provides perspective by discussing seven "side-effects" of our efforts to achieve portability. These are side benefits really -- desirable results of taking portability very seriously in our approach to numerical computation. Reinsch gives special treatment to the implications of a keen awareness of machine arithmetic on the part of programmers intent upon producing portable software. It is both refreshing and sobering to see what might happen if we "mean what we say."

SOME SIDE EFFECTS OF STRIVING FOR PORTABILITY

Christian Reinsch

Portability and *Numerical Software* are concepts whose meaning and contents may vary. Hence, at the beginning of this Workshop, it might be appropriate to work out some delineation of their scopes so that different interpretations and applications of these terms are avoided. This speaker is neither authorized nor otherwise entitled to foist his own and probably one-sided definitions upon the meeting so that a clarification of these points is left open and recommended for subsequent discussion. Here are some possibilities.

NUMERICAL SOFTWARE. From a first point of view, numerical software consists of individual routines, independent of each other and usually of modest length. They are the realizations of numerical algorithms as described in our textbooks and journals. Editors of the latter have a tendency to ask for test examples and numerical confirmation when authors submit papers which propose new algorithms. Therefore, many numerical analysts are inclined to program their methods and to submit for publication these products by which they achieved their results along with the corresponding theory. Among the more conspicuous examples are the program collections and algorithms sections of the journals *Communications of ACM* (since February, 1962, but now in TOMS), *The Computer Journal*, *Numerische Mathematik, BIT, Computing,* etc. Also, it is now not unfashionable for textbooks of numerical analysis to include some appendix with FORTRAN or ALGOL programs.

This view might seem too narrow or even inappropriate for many. Indeed, one could have the impression that occasionally printed programs are not fully acknowledged as software. And if one accepts as software only what is stated in FORTRAN and punched on cards or written on magnetic tape, then from this point of view, numerical software would essentially be the standard program libraries which come with every computer or the service libraries like *IMSL*, *NAG*, etc.

One step further in the definition of numerical software would include the systematic combination of such algorithms, called "multi-algorithms," which usually consist of a control program and a family of algorithms associated with a given problem. The former acts as a driver, selecting from among the latter the most promising routine. The decision rules, however, might well depend on the machine characteristics and this adds a new flavour to the portability problem. Most of the function approximation routines are indeed multi-algorithms in this sense, say mini-multi-algorithms (e.g., a Chebyshev expansion for one range of arguments and an asymptotic expansion for another range, etc.). Multi-algorithms, in particular, have been tried for numerical quadrature; another example with great reputation is the EISPACK system for algebraic eigenvalue problems.

The most comprehensive interpretation of "numerical software" would have to include all kinds of application packages, e.g., partial differential equation solvers and finite element methods in structure analysis (possibly including some interactive facilities for the geometry definition and element enumeration).

To sum it up, there are different definitions possible of what is
to be included in our notion of "numerical software" and the discussion
of portability questions might be heavily influenced by the chosen
delineation.

PORTABILITY.   Intuitively, our notion of portable numerical software
is quite clear: the attribute "portable" is to be awarded to those programs
which successfully run not only on one computer species but rather a variety
of them; at least this should hold true after the manual adjustment of some
machine dependent parameters intentionally provided for that purpose.   However,
the latter can only make up for quantitative differences rather than qualitative
ones as provided by the diversity of structures of the more advanced computers.
Examples of different computer categories are:

(i)   single processor with uniform random access memory,

(ii)   processors with paging and virtual memory facilities,

(iii)   pipeline or vector computers,

(iv)   parallel computers.

A few words will characterize the last two of these categories.   Pipeline processors possess special machine instructions and devices for "vector
operations" where one type of floating point arithmetic is executed on many
independent pairs of operands which are equidistantly stored in memory, i.e.,
a hardware analogy of the much discussed basic linear algebra modules.   To
increase speed, each operation is decomposed into a sequence of more elementary
steps such as sign handling, exponent removal, mantissa arithmetic, rounding,
etc., which are performed at different stations along an assembly line so
that a flow of operand pairs in various states of processing is maintained.
In the stationary phase of the instruction one pair of operands is processed

per machine cycle. Note that this approach to increased speed is promising not only for the super machines like the CDC STAR-100, the Texas Instruments ASC, etc., but could have a future for the small micro-programmable computers as well.

Parallel computers, on the other hand, have a fixed number of identical though independent processors connected through a network of data buses permitting the interprocessor communication. They need not perform the same instruction at the same time.

In a recent report Don Heller {4} discusses the consequences for software in the linear algebra area. To quote from the abstract: *The existence of parallel and pipeline computers has inspired a new approach to algorithmic analysis. Classical numerical methods are generally unable to exploit multiple processors and powerful vector-oriented hardware. Efficient parallel algorithms can be created by reformulating familiar algorithms or by discovering new ones, and the results are often surprising.*

Here, again, we need a delineation (which is necessarily arbitrary) of the scope of one of our basic terms. Without further discussion and more definite conclusions, the meaning of portability would remain obscure because it depends on implicit assumptions which are then left undefined. Most likely, one would aim at a definition of portability only within the range of a given computer category, and, at the moment, it is probably best to concentrate on just the first target group, viz., the familiar computer type with single processor and a uniform memory.

THE SIDE EFFECTS. Whatever definitions we agree upon, our striving for portability of numerical software has certain consequences, especially since portability has become a major issue for the professional production of numerical software. Some of the noteworthy side effects are as follows.

1.   A revived interest in the standardization of programming languages, in particular, FORTRAN. The necessity of having rigorous standards for both syntax and semantics of the underlying programming language is clear.

2.   Lots of proposals for FORTRAN extensions. Nearly any FORTRAN compiler will provide some additional features which are not part of the official language. No doubt these individual or local extensions of the possibilities a programmer has at his disposal have greatly relieved the pressure for official extensions. However, if programmers were willing to forgo such non-official extensions for the sake of portability, it is only natural for them to press for some extensions of the standards. Most common is the wish for allocation of working storage though this might cause a conflict with the general FORTRAN philosophy. Not so difficult to achieve and quite beneficial for the portability issue would be a COMMON which is not too common, i.e., the possibility for subroutines to share variables which, in turn, are protected and inaccessible from the rest of the world. Another feature, more useful for the development than for the application of numerical software, would be a data type where the programmer specifies the number of storage units per item and has the responsibility of providing the subroutines for the arithmetic operations and relations.

This would not be the right place to dwell on these features and their potential. The intention is only to demonstrate that each of us could provide his own suggestions and many of us, indeed, have tried it.

3.   An increased awareness of a potentially varying computational environment, in particular the following:

- Strict discipline when dealing with machine dependent parameters (e.g., machine precision) which, in turn, results in increased clarity.

- Greater effort toward intercepting exponent spill where most likely, leading to enhanced safety for all users. In particular, measures against underflow are often included just for portability reasons.

- A reluctance to exploit either special facilities of the hardware, extra features of the compilers or services of the system if they are non-standard, all leading to an inevitable sacrifice in performance. Two common examples are the double precision accumulation of inner products and the quick data transport to and from back-up storage.

- An attempt to avoid assembler routines even where it would be most appropriate.

4.   A more positive attitude towards printed programs which now and then seem to be not the worst starting point for portable software.

5.   Greater acceptance by programmers of rigorous and stringent test routines, including performance evaluations on different types of computers.

6.   Reluctance to correct hardware flaws with software measures. Many computers are said to have flaws and pitfalls in their arithmetic and we learn that counter-measures in the software are required to overcome these

snags. Different computers show different shortcomings which, in turn, need different remedies, and this could be one major source of the portability problem. Some programmers apparently are proud of their capacity to debug a machine; however, our interest in portability should convince us that we ought to oppose a zealous and obedient readiness to elude and circumvent the arithmetic flaws in our machines by all sorts of dodges and tricks in our programs. The reason for this is not so much a potential loss of speed or storage economy, rather it is the inevitable derogation of clarity which should worry us. This is because our comprehension of devices which are opposite to algebraic common sense always requires a disproportionate amount of effort and time from both the writer and reader of our programs. The reader is in an especially unfavourable situation since he usually lacks the pertinent background knowledge of the nature of the shortcoming. Incidentally, we can add that reduced clarity means also reduced versatility, i.e., reduced portability on a higher level. To prevent misunderstanding, it should be remembered that these remarks concern portable software or, more generally, software which is intended for publication and distribution. We are not discussing programs which are to be used at one installation only.

There is another good reason to abstain from making additions to otherwise pure algorithms. The practice inhibits feedback to the machine designers, or more precisely, it suppresses reaction to their creations. To make this clear, let us assume for a moment that there were heavy penalties for algorithm polluters. Then, since nobody would dare add the mentioned extra measures, programs would run safely only on clean computers, and, by Darwin's law of evolution, only such computers (and their manufacturers) would survive. Vice versa, we have to admit that

the present hardware misery is at least partly caused by too much
ingenuity and obedience on the programmer's side. Is it not a
questionable attitude that we leave the hardware designers alone
and without our patient guidance, that we lazily wait for their
products and become busy only afterwards when we must make the
best ot it? Should we not wage another crusade of *Unsafe-At-Any-Speed?*

    7.   Greater awareness of machine parameters and arithmetic
prerequisites. This last item in the list of side effects is so
important that we shall expand upon it at some length. It is the
fact that programmers and users finally become interested in the
arithmetic quality of their computers and alert to their peculiarities.
For example, this interest has been demonstrated in the discussions of
the *IFIP Working Group 2.5* which is soliciting independent views on
the minimum prerequisites for machine arithmetic.

    The arithmetic quality of a computer is determined by certain
machine parameters (e.g., precision, exponent range), the set of
arithmetic operations with their rounding algorithm, and the way
arithmetic exceptions (e.g., division by zero, exponent overflow)
are handled.

    MACHINE PARAMETERS. Numerical algorithms operate on data
structures with components in infinite domains: the natural numbers
N, the integers Z, and the real field R, while programs must use
the finite subsets of these domains provided by the computer or,
more precisely, by the numbers which are exactly representable in
its main memory. Typical are

(a) The system of integers or fixed point numbers:

$F_o := \{ -m, \ldots, n \}$ with m and n from N.

(b) The system of normalized, single precision floating point numbers:

$F_1 := \{ M \cdot B^{E - t} : M = 0 \text{ or } B^{t-1} \leq |M| < B^t, \alpha \leq E \leq \beta \}$

with mantissa M, radix B, and exponent E from Z. t is the number of significant digits and there is no reason to assume that it is an integer. B must not be from $\{-1, 0, 1\}$ but is actually always from $\{2, 8, 10, 16\}$. $M = -B^t$ might be feasible on computers with the two's complement representation of a negative mantissa; in this case F_1 is enlarged by just one number, $-B^\beta$, at its extreme left end and all other floating point numbers with this mantissa have two different codings.

(c) The system of normalized, double precision floating point numbers, F_2, similar to F_1 in its structure but with enlarged t. Note that it does not matter whether the operations on $F_1 \times F_2$ are provided by hardware or software.

It has often been observed that concise information on these subsets should be available at runtime and that some conventions are necessary for portability reasons. Cody {3} gives a thorough discussion and many references. A modest solution would be to provide three parameters, viz.,

$I := \max \{ x : x \text{ and } -x \text{ in } F_o \}$ to characterize F_o,

$\Omega := \max \{ x : x, -x, \frac{1}{x}, -\frac{1}{x} \text{ all in } F_1 \}$ and B to

characterize F_1 and F_2,

so that within these limits the operations $x \rightarrow -x$ in F_o and $x \rightarrow -x, x \rightarrow \frac{1}{x}$ in F_1 or F_2 cannot cause overflow. Intercepting overflow or underflow is the main application of the parameters I and Ω so that more accurate bounds of the range are of little value. The parameter t is not included here because the density of the floating point number system alone is not an indication of the achievable accuracy. Thus, if one prefers to have only one parameter to specify the machine precision for the floating point system, ε_1 (or ε_2), one should include the accuracy of the basic arithmetic operations in its definition. We postpone therefore the precise definition of the fourth parameter to the next section.

From a numerical point of view, the list of machine parameters need not be longer. In particular, there is no room, need, or use for parameters whose values depend on the computer coding and our prejudice on its interpretation. A notorious example is the so-called bias on the exponent which is undefined unless we agree on the position of the radix point within the mantissa (why always on the left for IBM machinery but on the right for CDC equipment?) and a rule for assigning values to the bit pattern in the exponent field of the floating point coding. Another example is the coding of negative numbers in sign-magnitude, one's complement, two's complement, etc., which in itself is immaterial to any numerical computation. It may be that this is obscured by the fact that the sign coding almost automatically influences the rounding algorithm of the arithmetic operations and that the latter, of course, is decisive.

ARITHMETIC OPERATIONS. We consider the precision of the fixed point operations (on $\mathbb{F}_o \times \mathbb{F}_o$) and the floating point operations (on $\mathbb{F}_1 \times \mathbb{F}_1$). Fixed point operations have application in multi-precision packages. To be useful they must be exact, i.e.,

- multiplication must yield the double length product,
- division must yield both quotient and remainder in the Euclidean sense: a/b yields q and r such that $a = b \cdot q + r$ and $0 \leq r/b < 1$.

- tests on > 0 and ≥ 0, < 0 and ≤ 0 must be strictly distinguished so that 0 never passes for positive or negative (even if it has representations with positive and/or negative sign).

The most debated items of computer arithmetic are the floating point operations; see, e.g., Knuth {5}, Cody {2}, Shriver {6}, Atkins and Garner {1}, Sterbenz {7}. At the moment, we are only concerned with the precision of these operations, i.e., the deviation of the computed result, $fl(a * b)$, from the exact result, $a * b$, with operands a and b from F_1 and * from $\{+, -, \times, /\}$. The precision is determined by the computer's rounding algorithm.

In a comprehensive definition, rounding in F_1 is any function $\tilde{\rho}$: $[-\Omega, \Omega] \to F_1$ so that $\tilde{\rho}(x)$ is one of the two neighbors of x in F_1 . <u>Proper</u> rounding, ρ, delivers the <u>closest</u> neighbor and the one away from zero in case of a tie. The latter convention assures a symmetric rounding, i.e., $\rho(-x) = -\rho(x)$ holds true for all arguments. Note that $\rho(x)$ can be computed from incomplete information on x: one needs the sign and the first t+1 digits of the representation of $|x|$ in the radix B system. Note also that $\frac{1}{2} B^{1-t}$ is a bound for the relative error caused by proper rounding in F_1 .

An appraisal of the computer's rounding algorithm will include the following points:

- what is the maximum error?
- what is the average error; in particular, is it locally unbiased?
- which of the algebraic laws hold and which are violated?
- how costly is its realization?
- how easily can it be completely described?

The optimum algorithm in this sense is fl(a * b) = ρ(a * b) where the computed result is the floating point number closest to the exact result.

- It gives the smallest possible rounding error for any floating point operation thereby minimizing the overall influence of rounding errors on a computation. Moreover, the relative error is always small and bounded by $\frac{1}{2} B^{1-t}$.

- The rounding is locally unbiased since ties are negligible in any statistical sample.

- The most important algebraic properties, symmetry and monotonicity, are preserved, the latter since a * b $\leq$ c * d implies fl(a * b) $\leq$ fl(c * d). In addition, one has fl(a - b) = fl(a + (-b)) and fl(a x 1) = a.

- Proper rounding is achievable because, as outlined above, one needs only the sign and the first t+1 digits of the normalized exact result. It is well known and easy to check that this can be obtained using an accumulator with 2t+1 digits in a straightforward manner, or with just t+2 digits if one is careful.

- Its description could not be simpler.

No convincing reasons to oppose proper rounding are known. Nonetheless, only a few machines, if any, give such favourable results. On the other hand, virtually all computers produce results which at least can be interpreted as the exact answers for slightly perturbed operands:

$$fl(a * b) = \{a \cdot (1+\alpha)\} * \{b \cdot (1+\beta)\}$$
$$\text{for all a, b from } \mathbb{F}_1 \text{ and * from } \{+, -, \times, /\}$$

where α and β depend on a, b, and * but are uniformly bounded in magnitude. This is just a mathematical means of interpreting the computed result since the hypothetical operands $a \cdot (1+\alpha)$ and $b \cdot (1+\beta)$ bear no relation to the manipulated operands which are actually held in the machine registers during the execution of a floating point operation.

The smallest such bound on $|\alpha|$ and $|\beta|$ is *by definition* the machine precision ε_1 which has been announced above as the fourth parameter characterizing the computational environment. From this definition, it follows that the maximum error is

$$(|a| + |b|) \cdot \varepsilon_1 \qquad \text{for addition and subtraction,}$$
$$|a \cdot b| \cdot (2\varepsilon_1 + \varepsilon_1^2) \qquad \text{for multiplication,}$$
$$(|a/b|) \cdot 2\varepsilon_1 / (1 - \varepsilon_1) \qquad \text{for division,}$$

which gives a method for determining ε_1 for any rounding algorithm by comparing the actual error in the worst case with the above bounds. Most computers are furnished with a rounding algorithm so that ε_1 is of order B^{1-t}. This is not true if unnormalized operands are admitted.

For practical purposes, it is of little importance that the relative error might be unbounded in the subtract-magnitude case. However, it is extremely important that the rounding error is locally unbiased as a simple example can demonstrate. On a CDC CYBER 175 the largest eigenvalue of the Hilbert matrix of order 100 was computed. This is certainly a very well-conditioned problem of modest size. The actual error was found to be -1.8×10^{-12} if the chopping mode was used but only 10^{-14} if the almost unbiased "round" mode was used.

Arithmetic relations can be exactly evaluated if a modified floating point subtraction is available which ignores the exponent in the post-normalization of the mantissa so that overflow and underflow are impossible.

Exponent spill would cause difficulties if merely the floating point
difference of the operands were computed and then compared with zero.

The variety of existing floating point algorithms has a very bad
influence on the portability of mathematical software. Thus, even if
disinterested in this issue, most programmers will plead for some
standard. It seems that only proper rounding is the correct candidate.

ARITHMETIC EXCEPTIONS. The reliability of mathematical software
is based on a rigorous error analysis of the employed algorithms.
Wilkinson's {8} universally accepted method makes the assumptions that

$$fl(a * b) = \{a \cdot (1+\alpha)\} * \{b \cdot (1+\beta)\}$$

with $|\alpha|$, $|\beta| \leq \epsilon_1$ for all a, b from F_1 and * from $\{+, -, \times, /\}$.

As mentioned before, virtually all existing computers with floating point
arithmetic satisfy this assumption provided the operands are normalized
and the result is in the proper range. It follows that any violation
of the above assumption must be flagged, particularly over- and underflow
of exponents. Otherwise these programs are not really reliable or must
be spoiled by an undue amount of scaling, thereby defying the reasons
floating point arithmetic was invented. Such a flag should therefore
be understood mainly as an indication (without dramatizing the point)
that the assumptions from which the quality of a given library routine
had been inferred are no longer satisfied. In the event of underflow
a program can well be continued with zero as default result rather than
terminated.

In the long run an extended error-handling facility is expected
to prevail which allows the programmer to specify the proper treatment

of the arithmetic exceptions if he is dissatisfied with the default treatment.
It seems that this is not so much a hardware problem since the commonly
provided interrupt is enough for that purpose; rather the facilities of
the programming languages should be extended in this direction. A closer
inspection shows that it is of little value to the programmer to know
what is left behind in the machine registers after the interrupt occurred:
one should not make use of it because it is bound to be machine dependent
and non-portable, because it might be undefined for pipeline computers,
and because in most cases some scaling has to be done and the computation
must be resumed at an earlier point. Nevertheless, a processor handbook
cannot be qualified unless it gives a complete description of the floating
point operations and this has to include the rounding algorithm as well
as the register contents in the exceptional cases.

CONCLUSION. It is impossible to define a minimum quality of the
hardware arithmetic and to condemn as intolerable what is below that
limit. The reason is that one can overcome any shortcomings of the
hardware by appropriate counter-measures in the software. Therefore
it is impossible to develop, and defend as mandatory, a set of standard
assumptions which would form the arithmetic prerequisites for and a
guarantee of the correct behaviour of numerical software. Of course,
this is true in particular for the proposal made above, viz., computed
results are assumed to be exact for slightly perturbed operands provided
no exponent spill occurred, setting a flag; computed arithmetic relations
are exact.

Nevertheless, this standard should provide a base which would allow
us to write and use without any risk numerical software which takes as its

guidelines only the principles of rounding error analysis and is completely free of any special features designed to combat the idiosyncrasies of a given computer. In other words, these standards are the specifications of an abstract model computer which could serve as an interface between the numerical analyst on the one hand and the programmer of mathematical software on the other hand. The portability issue should be eased by this interface.

REFERENCES

1. Atkins III, D. E. and Garner, H. L. (Eds.): Special Section on Computer Arithmetic. IEEE Transactions on Computing C-22, Number 6 (June 1973).

2. Cody, W. J.: Desirable Hardware Characteristics for Scientific Computation: A Preliminary Report for the SIGNUM Board of Directors. ACM SIGNUM Newsletter 6 (1971), 16.

3. Cody, W. J.: Machine Parameters for Numerical Analysis. This Volume, 1977.

4. Heller, D.: A Survey of Parallel Algorithms in Numerical Linear Algebra. Department of Computer Science, Carnegie-Mellon University (February 1976).

5. Knuth, D. E.: The Art of Computer Programming. Volume 2/ Seminumerical Algorithms. Reading, Massachusetts: Addison-Wesley Publ. Co. (1969).

6. Shriver, B. D.: System Design for Scientific Computation. A. Gunther et al. (Eds.): International Computing Symposium 1973, p. 211. Amsterdam: North-Holland Publ. Co. (1974).

7. Sterbenz, P. H.: Floating-Point Computation. Englewood Cliffs: Prentice-Hall, Inc. (1974).

8. Wilkinson, J. H.: Rounding Errors in Algebraic Processes. London: Her Majesty's Stationery Office (1963).

MACHINE CHARACTERISTICS AND PORTABILITY

It is often said that the use of computers for scientific work represents a small part of the market and numerical analysts have resigned themselves to accepting facilities "designed" for other purposes and making the best of them. I am not convinced that this is inevitable, and if there were sufficient unity in express- ing their demands there is no reason why they could not be met.

— J. H. Wilkinson (1971)

We have to admit a new kind of elements, Environment Enquiries, in our common programming language. These should be designed to place a carefully chosen set of information about the available equipment at the disposal of the programmer, for use in directing the control of the process.

— Peter Naur (1967)

Two points of view are expressed by the above quotes and their respective technical backgrounds are developed in the papers of this section. The first viewpoint holds that the influence of numerical analysts on machine design, so evident in the early years of computing, should be revived. The second asks for a systematic approach to the determination and use of machine-dependent quantities by programmers. These viewpoints are not in conflict. Indeed, we believe that numerical analysts would generally endorse the second (although differing sharply as to the choice and functioning of Environment Enquiries) while seeing the first as desirable and, if ultimately successful, as making the second much simpler.

If machines had well-documented, soundly-based, generally accepted arithmetic characteristics, the impact on the development of quality numerical software and on the transportability of that software would be enormous, even if the characteristics were realized in different ways (e.g., with different word lengths). Christian Reinsch, in his lead-off paper, discussed awareness of machine characteristics as one of the "side effects" of striving for portability. In this section T. J. Dekker explicates a set of machine requirements for reliable, portable software. T. E. Hull brings out the additional point that standardizing the semantics of floating point arithmetic and elementary functions would facilitate the formulation of portable proofs about programs.

The papers by Reinsch, Dekker, and Hull provide technical backbone for those dedicated to the first viewpoint above. The papers by W. J. Cody and B. Ford discuss the kinds of information about a machine that might be derived through Environment Enquiries for use by a programmer writing machine-dependent yet transportable programs.

MACHINE REQUIREMENTS FOR RELIABLE, PORTABLE SOFTWARE

T. J. Dekker

ABSTRACT

Arithmetical and relational operations in a
machine must be designed in such a way that it enables
us to make reliable, portable, efficient software.
Therefore, operations must deliver either the correct
result or an indication (trap, flag bit) that the
operation has not been correctly performed. Incorrect
results should be best possible, in order to increase
software robustness.

Some consequences are: relational operations
must always be correct; the result of integral
arithmetical operations must be exact and that of
real operations should be correctly rounded when
the exact result is within certain (machine dependent)
bounds; operations on single-precision operands yielding an exact (or correctly rounded) double-precision
result should also be available.

Here "portable software" means software which can be easily transported from one type of machine to another, only minor changes (preferably processed automatically) being necessary to get the programs running on other machines.

Several people have formulated their wishes concerning computer hardware and arithmetic; these wishes are based on either mathematical principles or practical experience. We mention the work of Cody {5} and {6}, Kahan {10}, Reinsch {13} and, in particular the discussion on the optimal base of floating-point systems by Brown & Richman {3}, and Brent {1}.

IFIP WG 2.5 organised a discussion, chaired by C. H. Reinsch, on machine arithmetic on June 23, 1976, immediately after the NSF/ERDA Workshop on Portability of Numerical Software at Oak Brook (Illinois), and has plans to issue an official document containing its proposal on a set of desirable hardware features for numerical calculation.

In this paper, we propose an incomplete set of requirements and wishes mainly based on mathematical principles. This paper was presented (just) prior to the discussion mentioned above; in the written version presented here, we have made only some minor changes in view of that discussion.

In section 2, we introduce some notations and formulate some general principles; in sections 3 and 4, we consider integral and real arithmetic respectively in more detail; in section 5, we mention some suggestions on internal representation.

Acknowledgement. The author is grateful to Dr. P. van Emde Boas for his valuable comments on the manuscript of this paper.

2. Notations and General Principles

For a certain machine, we consider an implemented system I_m of integers and an implemented system R_m of real numbers. (For instance,

R_m may be the machine's system S_m of single-precision real numbers or its system D_m of double-precision real numbers).

Let a and b be either elements of I_m or elements of R_m. We consider integral and real machine operations, operating on elements of I_m or R_m respectively, corresponding to the following mathematical operations (which for most machines are implemented by hardware):

relational operations a ∿ b,

where ∿ stands for <, ≤, =, ≠, ≥ or >;

monadic arithmetical operation - a;

dyadic arithmetical operations a * b,

where * stands for +, -, x, div and mod when a and b are in I_m, and for +, -, x, / when a and b are in R_m.

Here a div b and a mod b denote integer division and remainder respectively which are defined, for b ≠ 0 only, as follows:
The result of a div b is an integer q such that a - b x q either equals 0 or has the same sign as a and minimal absolute value; the result of a mod b is an integer r such that a - r is a multiple of b and r either equals 0 or has the same sign as a and minimal absolute value. The other operator symbols mentioned have the usual mathematical meaning.

Remark. Some people prefer machine operations corresponding to integer division and remainder defined in a different way, cf. Meertens {12}.

The corresponding integral or real machine operations are denoted by writing the circumflex symbol ^ above the operator symbol. Thus, a $\hat{\sim}$ b, $\hat{-}$ a and a $\hat{*}$ b denote integral machine operations if a and b are in I_m, and real machine operations if a and b are in R_m.

A machine operation delivers a certain result which may or may not be correct, and, moreover, may or may not deliver an error indication. For real machine operations corresponding to dyadic arithmetical operations, "correct result" must be understood in the numerical sense, defined in section 4; for all other machine operations, the result is correct if and only if it equals the result of the corresponding mathematical operation.

An error indication may be delivered by effectuating a $\underline{\text{trap}}$, i.e., a $\underline{\text{synchronous}}$ change in the flow of control (in contrast to an $\underline{\text{interrupt}}$ which is an $\underline{\text{asynchronous}}$ change in the flow of control), cf. Tanenbaum {14, p. 130}, by setting a certain $\underline{\text{flag bit}}$ to a certain value, by delivering a certain $\underline{\text{exceptional value}}$ (such as "undefined" or "infinite") as result of the operation, or by combinations of these.

$\underline{\text{Definition}}$. A machine operation is $\underline{\text{correct}}$ if it delivers a correct result and no error indication.

It is not quite clear, what we should require, when an exceptional value occurs as operand of a machine operation. For simplicity, we leave this open, and consider only machine operations as far as they operate on non-exceptional argument values. Accordingly, we assume that $I_m(R_m)$ does not contain exceptional values but only those integers (real numbers) which are obtainable as result of a correct integral (real) machine operation.

We are now ready to propose our requirements and wishes for the machine operations considered in the form of some general principles. Our aim is to have hardware features facilitating the production of $\underline{\text{reliable}}$, $\underline{\text{portable}}$ software; as secondary, but still important, goals we mention $\underline{\text{efficiency}}$ and $\underline{\text{robustness}}$ of software.

Principles.

P 1 (reliability requirement)

A machine operation must never deliver an incorrect result without error indication.

P 2 (portability requirement)

A machine operation must be correct whenever the result of the corresponding mathematical operation is within a certain (machine dependent) range; this range must be easily definable by means of machine parameters.

P 3 (efficiency requirement)

It must be possible, at least as an option, that machine operations deliver their error indications by means of effectuating a trap, preferably combined with setting a flag bit or delivering an exceptional value as result.

P 4 (robustness wish)

Within the limitations of the machine, the ranges of correct operation mentioned in P 2 and the machine precision of R_m should be "optimally" chosen, and also incorrect results delivered should be "best possible."

Remarks.

(1) Machine operations satisfying requirement P 1 may not be easy to implement efficiently, because it entails a test for correctness of each result which may cause some loss of efficiency; this requirement is, however, necessary to obtain reliable arithmetic.

(2) As to requirement P 2, we shall below give some more detailed formulations for the various machine operations considered in the form of requirements which we claim to be easily realizable, and some wishes which are important but possibly less easy to fulfill for some (especially small) computers.

(3) As to P 3, only a trap enables us to prevent or handle errors efficiently. Without a trap facility, one has to insert the necessary instructions in the main flow of control, which leads to less efficient code, cf. Tanenbaum {14, p. 130}. We propose P 3 as a requirement, because it is not difficult to implement. Correspondingly, software should provide, at least as options, hard fail reaction (i.e., error message and trap) and soft fail reaction (i.e., error indication in a certain program variable), as is done by NAG.

A related interesting language feature is the traplabel in PASCAL, i.e., a label with asterisk; its effect is that, instead of a trap, execution continues at next traplabel. This makes hard fail less hard to the user and is much safer than soft fail.

(4) As to P 4, we shall below formulate some wishes on "optimal" ranges and "best possible" incorrect values. We propose P 4 only as a wish, because it is not always obvious how to define optimal ranges and best possible incorrect values for a certain machine.

3. Integer Arithmetic.

For a certain machine, we consider an implemented system I_m of integers. Let igiant be the largest integer such that the correct system I_c defined by

$$(3.0.1) \quad I_c = \{ i \mid abs\ (i) \leq igiant \}$$

is contained in I_m.

Requirements for Correct Operation.

Let a and b be elements of I_m. We propose the following requirements as more detailed formulations of P 2.

I 1 (relational operations)

The operation a $\hat{\underset{\sim}{}}$ b must be correct.

I 2 (sign reversion)

If a is in I_c, then the operation $\hat{}$a must be correct.

I 3 (dyadic arithmetical operations)

If a * b is in I_c, then the operation a $\hat{*}$ b must be correct.

Remarks.

3.1. Machine operations satisfying these requirements are easily, and efficiently, implementable. Only requirement I 1 may cause some difficulty, because in some machines the relational operations are (for reasons of simplicity) implemented as (a $\hat{}$ b) $\hat{\underset{\sim}{}}$ 0, i.e., first the difference of the operands is calculated and then this difference is related to 0. In favor of these machines, one might replace I 1 by the weaker requirement "if a - b is in I_c then the operation a $\hat{\underset{\sim}{}}$ b must be correct."

We propose I 1 as it stands, for the following reasons:

(1) The relational operations are fundamental and may be indispensable for the programmer to predict and prevent overflow;

(2) Equality and inequality are easily implementable by means of an "exclusive or" operation which is available as hardware instruction in many machines;

(3) For the other relational operations, one can, with a slight loss of efficiency, avoid the calculation of the difference of two operands having opposite sign, so that no overflow can occur.

3.2. For portability, one has to know only the machine parameter igiant.

3.3. A consequence of I 1 is that the operation a $\hat{\underset{\sim}{}}$ b must be correct also if a and b have the same value, but different machine representations (e.g., "+0" and "-0"). In other words, no distinction whatsoever must be made between different representations of one same value.

3.4. Further Wishes.

The optimal range of correct operation is clearly obtained when $I_c = I_m$. This is easily realizable, and therefore strongly recommended, unless I_m is represented as a subset of R_m (which representation may have other advantages, see section 5).

3.5. As best possible incorrect values we suggest: exceptional value "undefined" (if available in the system, especially to be recommended as result of division by 0 (or remainder) and of operations having an exceptional value as argument); igiant with appropriate sign (in case of overflow, especially when $I_c = I_m$); element of R_m of right order of magnitude (in case of overflow, when $I_m \subset R_m$).

4. Real Arithmetic.

For a certain machine, we consider an implemented system R_m of real numbers. Let dwarf and giant be two parameters which are certain elements of R_m, and let the safe arithmetical range R' of R_m be defined by

$$R' = \{x \mid x = 0 \text{ or } dwarf \leq abs\ (x) \leq giant\}.$$

Moreover, let the correct system R_c be defined by

$$R_c = R_m \cap R'.$$

In order to define "correct in the numerical sense," we use the notions "faithful" and "optimal" introduced by Dekker {7} and defined as follows.

Definition. An element x of R_m is a faithful representation of a real number ξ if x is either the largest element of R_m not larger than ξ or the smallest element of R_m not smaller than ξ.

In particular; if ξ is in R_m, then ξ itself is its only faithful representation; if ξ is between two successive elements x and y of R_m, then both x and y are faithful representations of ξ; if ξ is larger than the largest (smaller than the smallest) element x of R_m, then x is the only faithful representation of ξ.

Definition. An element of R_m is an optimal (or properly rounded) representation of a real number ξ if x is an element of R_m nearest to ξ.

Note that this defines x uniquely as a function of ξ, except when ξ lies halfway between two successive elements x and y of R_m, in which case both x and y are optimal representations of ξ.

Definition. The result of a machine operation corresponding to an arithmetical operation is faithful (optimal) if it is a faithful (optimal) representation of the result of the corresponding mathematical operation.

Definition. A machine operation corresponding to an arithmetical operation is faithful (optimal) if it delivers a faithful (optimal) result and no error indication.

Requirements for Correct Operation.

Let a and b be elements of R_m. We propose the following requirements as more detailed formulations of P 2.

R 1 (relational operations)

The operation a $\hat{\sim}$ b must be correct.

R 2 (sign reversion)

If a is in R_c, then the operation $\hat{-}$ a must be correct.

R 3 (dyadic arithmetical operations)

If a * b is in R', then the operation a $\circledast$ b must be optimal.

Remarks.

4.1. Machine operations satisfying these requirements are easily, and efficiently, implementable. Only requirement R 1 may cause some difficulty for the same reason as mentioned for requirement I 1 in remark 3.1. Although some people find it acceptable to replace R 1 by the weaker requirement "if a - b is in R', then the operation a $\hat{\cap}$ b must be correct," cf. Reinsch {13}, we propose R 1 as it stands, for the same reasons as mentioned for I 1 in remark 3.1.

As to requirement R 2, note that this requirement implies that the correct system R_c must be _additively symmetric_, i.e., if x is in R_c, then so is -x; in practice, this is always fulfilled.

As to requirement R 3, optimal arithmetic is easily realizable. Indeed, only two _guard digits_ (i.e., extra digits in an accumulator, used to hold the mantissa of an intermediate result temporarily during an operation) are needed to obtain properly rounded results, cf. Knuth {11, p. 183 & 194 ex. 5} and Cody {5}. One might replace R 3 by the weaker requirement that the operation be faithful, instead of optimal, so that chopping up or down is allowed. For practical applications, it is important, however, that this chopping is (statistically) unbiased. We propose R 3 as it stands, because optimal arithmetic is nearly as easy to realise as unbiased faithful arithmetic.

4.2. For portability one has to know only the machine parameters dwarf and giant mentioned above, and also the machine precision, _fleps_, and (for some applications, see also 4.6) the base of the real number representation.

4.3. Remark 3.3 similarly applied to requirement R 1. A desirable additional requirement is that the arithmetic is _deterministic_, i.e., that the rounding (or unbiased chopping) of arithmetical operations is always performed in the same way so that the results of a computation can be reproduced exactly. In other words, if a = b, then, for each function f consisting of a finite sequence of machine operations, the relation f(a) = f(b) must hold true.

Further Wishes.

4.4. The optimal range of correct operation is clearly obtained when $R_c = R_m$; i.e., giant is the largest and dwarf the smallest positive

element of R_m, and in view of the additive symmetry (see remark 4.1),
-giant is the smallest and -dwarf the largest negative element of R_m.
Moreover, it is desirable that dwarf admits a normalized representation,
so that real numbers of the order of magnitude dwarf may be represented
in a satisfactory relative precision (the machine precision fleps).
Further wishes mentioned by various authors are the following. It is
desirable that dwarf is of the order of magnitude of the inverse of
giant (multiplicative symmetry), cf. Cody {6}. This is easily
realizable and, in fact, holds true on many machines. As to the size
of giant (and dwarf) and the machine precision fleps, one may suggest
suitable values, based on practical experience; e.g., giant should be
of the order of magnitude 10^{300} (and dwarf 10^{-300}), and fleps about
10^{-9}, cf. Cody {5}, and dwarf should not be larger than the square
of fleps, cf. Brown & Hall {2}.

The Base of Real Number Systems.

4.5. In most if not all machines, a floating-point representation is
used to represent real numbers; i.e., real numbers are represented
in the form

$$\text{mantissa} \times \text{base}^{\text{exponent}},$$

where mantissa is an integer or a pure fraction and exponent an integer,
both in a certain machine dependent finite range, and base is an integral
constant larger than 1.

The choice of the base is important, not only for scaling purposes
(one can multiply and divide numbers by the base exactly, apart from
overflow and underflow), but also because the base influences the accuracy

of the arithmetic. A large base causes a "wobbling precision," i.e., a large variation in relative precision (from fleps to fleps x base), cf. Cody {6}. The best value for the base appears to be 2 or 4, cf. Brown & Richman {3} and Brent {1}. In view of communication with human beings, base 10 has obvious advantages (in particular, no conversion is needed at input and output) and is acceptable, provided one or two guard digits are used in the arithmetic unit.

4.6. As best possible incorrect values, we suggest:

Exceptional value "undefined" (if available in the system, especially to be recommended as result of division by 0 and of operations having an exceptional value as argument); giant with appropriate sign (in case of overflow); dwarf with appropriate sign or 0 (in case of underflow). A multiplication or division causing underflow should not deliver 0 as best possible result, because this introduces zero divisors (i.e., non-zero numbers yielding 0 as product) which are unattractive for mathematical as well as practical reasons.

Note that, according to our definitions of "faithful" and "optimal" results, an error indication in case of overflow or underflow is not required; these error indications are desirable, however, (at least as options) in order to signal loss of relative precision. On the other hand, one might, for special purposes, consider the implementation of an (optional) "error indication" for the situation that a * b is R' but not in R_c, so that rounding or chopping is necessary. (This error indication might be useful to implement interval arithmetic.)

4.7. Facilities for Extended Precision.

It is desirable that hardware facilities are available for transitions between systems of various precision, at least from the single precision system S_m to the double precision system D_m and vice versa, cf. Cody {5}. These transitions, and also the exact multiplication of two single precision real numbers yielding a double-precision result, should be efficiently implemented. Moreover, S_m must be a subset of D_m; i.e., each element of S_m must be exactly representable (although not in the same way) as an element of D_m. In particular, the giant of D_m must not be smaller than the giant of S_m, and the dwarf of D_m not larger than the dwarf of S_m. Then the transition from S_m to D_m is trivial and exact. The transition from D_m to S_m must yield as result an element of S_m which is an optimal (or at least a faithful unbiased) representation of the given element in D_m.

5.   Internal Representation of Numbers.

The internal representation of numbers is not very important to the user. Therefore, we here mention only some suggestions on internal representation of integers and real numbers.

5.1. Sign-magnitude representation is preferable, because of its logical simplicity and independence of the subfields (viz., the sign and the magnitude), cf. Cody {5} and {6}.

5.2. According to definition (3.0.1.), I_c is additively symmetric and we have required that R_c be additively symmetric (see remark 4.1). When also our wishes $I_c = I_m$ (3.4) and $R_c = R_m$ (4.4) are fulfilled, then also I_m and R_m are symmetric. An additively symmetric system contains an odd number of elements, whereas the number of possible bit patterns in a machine word is always even. So, there is (at least) one bit pattern left over which can be used for the exceptional value "undefined," as advised by Tanenbaum {14, p. 421}.

5.3. The Grau representation of floating-point numbers, used in the machines Burroughs B 5500 (base 8) and Electrologica X 8 (base 2) is as follows, cf. Grau {9}.

The mantissa is shifted such that the corresponding exponent has minimal absolute value; i.e., the mantissa is adjusted to the left when exponent > 0, and to the right when exponent < 0. This representation has the following interesting features:

Addition and subtraction never lead to underflow; 0 is a regular element of the system (both mantissa and exponent being 0); the integers form a natural subset of the system (viz., the subset with exponent 0). A drawback of this representation is that real numbers of the order of magnitude of the smallest positive element cannot be represented in a satisfactory relative precision; moreover, Grau's representation seems to be not quite compatible with multiplicative symmetry (see also 4.4).

5.4. The representation of double-precision floating-point numbers should contain the representation of single-precision numbers as its leading part, so that transition from one system to the other is possible in a simple and efficient way (see also 4.7).

REFERENCES

1. Brent, R. P., On the Precision Attainable With Various Floating-Point Number Systems, IEEE Trans. on Computers, pp. 601-607, (1973).

2. Brown, W. S., and Hall, A. D., FORTRAN Portability Via Models and Tools, To be published (1977).

3. Brown, W. S., and Richman, P. L., The Choice of Base, Comm. ACM 12, pp. 560-561 (1969).

4. Bus, J. C. P., and Dekker, T. J., Two Efficient Algorithms With Guaranteed Convergence for Finding a Zero of a Function, ACM Transactions on Mathematical Software 1, pp. 330-345 (1975).

5. Cody, W. J., Desirable Hardware Characteristics for Scientific Computation, SIGNUM Newsletter 6, pp. 16-31 (1971).

6. Cody, W. J., Desirable Arithmetic Design of Proposed MATHNET Computer, Letter to Gabriel, Argonne National Laboratory (1974).

7. Dekker, T. J., A Floating-Point Technique for Extending the Available Precision, Num. Math. 18, pp. 224-242 (1971).

8. Einarsson, B., IFIP Working Group on Numerical Software, SIGNUM Newsletter 9, pp. 3-4 (1974).

9. Grau, A. A., On a Floating-Point Number Representation for Use With Algorithmic Languages, Comm. ACM 5, pp. 160-161 (1962).

10. Kahan, W., Implementation of Algorithms, Part I, Technical Report 20, University of California, Berkeley (1973).

11. Knuth, D. E., The Art of Computer Programming, Vol. 2/Seminumerical Algorithms, Addison-Wesley (1969).

12. Meertens, L. G. L. T., A Note on Integral Division, ALGOL Bulletin, No. 39, pp. 30-32 (1976).

13. Reinsch, C. H., Building a Library of Numerical Algorithms: A Case Study From the Handbook "Linear Algebra," Paper presented at the first meeting of WG 2.5 at Oxford (1975).

14. Tanenbaum, A. S., Structured Computer Organisation, Prentice-Hall (1976).

SEMANTICS OF FLOATING POINT ARITHMETIC
AND ELEMENTARY FUNCTIONS

T. E. Hull

ABSTRACT

This paper begins with a discussion of properties
that one might like to have satisfied by calculations
involving floating point arithmetic and elementary
function approximations. It concludes with a summary
of the main characteristics that it is believed any
detailed specification of these operations should
provide. The point of view is based on starting
with a description of what is needed by someone who
either uses or analyzes numerical software, and then
proposing specifications that are both sufficient
to meet these needs and relatively simple. One
principal motivation is that proofs concerning
programs should be portable, so in studying software portability it is as important to concern ourselves with semantic as well as syntactic details.

1.   Introduction

We are interested in the portability of numerical software. Of
course there would be no problem if both the syntax and semantics were
completely standardized. However, they are not, and a lot of effort
is being put into coping with the problem. For example, much of
this effort is devoted to the development of automatic techniques
to aid in the transporting of programs. However, while this effort
is being made, it is also important to encourage the development of
standards in order to keep the problem itself to a minimum.

In this paper we are concerned with what sort of standards
we ought to aim for regarding the operations involving floating
point arithmetic and elementary function calculations. We will
restrict ourselves to real, double precision, etc., arithmetic and,
in particular, we will not discuss complex arithmetic (although the
points we make can be modified appropriately for this case as well).
We also do not discuss integer arithmetic; some of our points are
applicable, but there are a number of other questions about integer
arithmetic that we would like to avoid in this paper altogether, such
as whether there should be a separate integer arithmetic at all, or
whether rational arithmetic is needed, and so on.

Most of the semantics of a program should be uniquely specified.
We should know, for example, whether or not the range of a DO is
traversed at least once, or what happens if the parameter value in a
computed GO TO is out of range. However, with floating point arithmetic
and elementary functions our specifications may not need to be unique,
even under ideal conditions. For example, different number bases or

word lengths might be acceptable, as long as proofs about the accuracy
of the calculations are portable (in terms of system dependent parameters
such as a relative roundoff error bound). Or, as another example, even
with the same relative roundoff error bound u, it may be enough to
insist that sqrt(x) be within 2u of the square root of x, without
insisting that sqrt(x) be the same with all systems; once again, a
proof involving u might still be portable even though the outputs
might not be exactly the same.

We begin in the next section with an indication of some general
requirements that would have to be met if proofs about accuracy are to
be portable; this section is intended only to give an overall view of
some basic requirements. This is followed by a section which shows,
mostly by example, that some further restrictions, as well as some
further facilities, will have to be required. Then, in another section,
we finally outline what we believe to be the sort of specifications
that are needed. It is intended that they be sufficiently flexible for
practical purposes, while at the same time being as reasonably simple
as they can be and still provide a sufficient basis for portability of
proofs. The specifications are not complete; this is partly because
most of the details are not particularly important for this discussion,
but also partly because there are a few questions (one in particular)
that need further, more detailed study.

2.   Some General Requirements

Proofs about numerical algorithms depend in a critical way on the
following three kinds of requirements:

(a)   <u>Accuracy of floating point arithmetic</u>. We need to know that

$$f\ell(x.op.y) = (x.op.y)(1+e),$$

where "$f\ell$" means "floating point result of," .op. is an arithmetic
operator, and e is bounded in absolute value by some (small) relative
roundoff error bound u. A requirement like this is the key one in using
a backward error analysis to prove, for example, that a particular algorithm
will produce the exact solution of a slightly perturbed problem.

This requirement may need to be supplemented by a corresponding one
for conversion on input and output. Also, a convention about the order
in which arithmetic operations are carried out is needed to resolve
ambiguities that arise, as in expressions like x+y+z, because floating
point arithmetic is not associative or distributive.

(b)   <u>Accuracy of elementary functions</u>. We need corresponding
requirements for elementary functions, such as square root, sine, log,
etc. These will in general be of the form

$$f\ell(f(x)) = (1+n_1 e)f(x(1+n_2 e)),$$

where f is the function, the n's are numbers, and the e's are usually
distinct (but still bounded in absolute value by u). The n's can be
small integers; in fact it would also be desirable if they were, at
least mostly, the same for all functions, to keep the requirements
simple. It would also be desirable if there were very little
restriction on the ranges of x and f(x), for the same reason.

(c)   <u>Overflow and underflow, etc</u>. With both (a) and (b) above
there is the possibility of overflow or underflow, and provision must
be made for these eventualities. This would of course mean introducing
further parameters into the proofs, such as M and m for the largest and
smallest positive numbers respectively.

As for the system itself, I am convinced that the best situation would be one in which there were special bit configurations to represent overflow, underflow, undefined, etc., and that the operations were closed over the entire set of floating point numbers, augmented by these special configurations. (Although I would like to modify some of the details, I have in mind something very similar to what has been advocated by Neely {6}). I would propose that the default action when special configurations arise would be to proceed, but to output appropriate messages as well. Other options should also be available, including one that left to the user the specification of what steps were to be taken.

If we had specifications of the sort described in (a), (b) and (c), much of what we prove about numerical calculations could be transported. The theorems would be in terms of parameters, such as u, M and m, which would depend on the particular system being used, but the theorems themselves could be transported with the programs. Some of what was proposed in (c) regarding augmentation of the number system, and closure, goes somewhat beyond what one usually thinks of in connection with portability. However, something like this seems to me to be necessary in any ideal system towards which we should aim; it also ought to be quite feasible from the point of view of hardware, language and software system design.

3. Some Further Restrictions and Facilities

In the preceding section we outlined the main kind of requirements we need to support proofs about numerical calculations. Now, in this section, we want to comment on these requirements. They appear to have the right sort of flavor, but, as we shall see in this section, there are at least four ways in which they need some modification or extension:

(a)   There are situations in which the arithmetic operations ought to be more accurate. Although most error analyses require only what is specified in (a) of the preceding section, there will certainly be occasions when the analyst would insist on more. For example, he would almost certainly require e to be zero in some circumstances, such as the addition of small integers. This would always be true of course if he was using floating point numbers as subscripts, or for counting, but, even leaving these cases aside (for which integers might be used), he probably wants small integer values handled exactly when possible; even psychological reasons might compel him to want this requirement. The point is that requirement (a) of the preceding section is not quite enough.

(b)   There are other situations in which the arithmetic operations should not be allowed to be more accurate. This may seem at first to be slightly paradoxical. However, I have in mind a situation like the ridiculous one offered by PL/C, where all calculations, both single and double precision, are done in double precision. The designers of this particular monstrosity may have thought they were doing us a favor, in providing more accuracy than asked for, but they have made life difficult for intelligent users who might want to measure the effect of roundoff by comparing the results of single and double precision calculations. However, our point is that PL/C still satisfies requirement (a) of the preceding section, and we therefore have another reason for modifying that requirement.

(c)   We need more control over precision and rounding. Besides
the example just mentioned, there are many others that show the need
for more precise control over the precision of numerical calculations.
Some examples show the need to be able to carry out intermediate parts
of a calculation in higher precision. Others show the need to be able
to repeat a portion of a calculation in different precisions (the
example in (b) above is of this type). Quite a few of these examples
have been discussed in detail by Hull and Hofbauer {2, 3}, along with
proposed language facilities for providing the kind of control that
is needed.

Again our point is that the requirements of the previous section
are not quite right. Besides more precise control over precision, a
case can be made for more precise control over rounding. The best
specification would probably be one that normally provided "true"
rounding rather than chopping (not because of the greater accuracy,
but because of the lack of bias in true rounding), along with special
functions to permit rounding up, or rounding down, in special circumstances. The special functions would be needed to provide control
over convergence of interations in some cases, to provide error
bounds in others, to make interval arithmetic easy to implement,
and so on.

(d)   What about preserving properties that are satisfied in the
real number system? There are many properties that are true in the
real number system, and that one might like to preserve, as much as
possible, in the number system and operations of a computer. The
question raised here is "which, if any, should we aim for in a standard?"

We would have to consider the feasibility in each case and, if feasible, the cost. We would also have to consider whether or not the added complexity is worth the resulting advantages.

Without trying to settle this question, let us list a number of examples of what we have in mind. First, there are properties such as requiring $\log(1) = 0$, or $\cos(0) = 1$, or a somewhat more general property like $abs(\sin(x)) \leq 1$. (Should we have to program so as to protect ourselves against the possibility of $1-\sin(x)$ becoming negative, when attempting, for .example, a numerical integration of $sqrt(1-\sin(x))$?) Second, there are many identities that one might like to count on, such as $sqrt(n^2) = n$ (at least for integer values of n), or $\sin(x) = -\sin(-x)$, or $1/(1/x) = x$; and what about keeping $\sin^2 x + \cos^2 x = 1$ (at least within some guaranteed tolerance)? Finally, there are some "monotonicity" properties, such as $y \leq z$ implying $xy \leq xz$ when $x > 0$, or $y \leq z$ implying $sqrt(y) \leq sqrt(z)$.

It would often be convenient to assume that properties like those just mentioned were valid in machine computations, if only because we might tend to take them for granted anyway. But, to what extent can they be achieved and, if they can, at what cost? Which, if any, should we insist upon?

4.   Outline of Specifications

It seems to me that the requirements for floating point arithmetic that have been discussed so far can be met most reasonably in only one way, and that is to require the result of each arithmetic operation to be as exact as possible for the particular precision being used. First of all, this requirement is very simple; it is simple to remember and simple to use. (The only minor point that needs to be resolved is what to do when the rounding is exactly half in the last place.)

This requirement leads immediately to the fundamental requirement for floating point error analyses that was described in 2(a). It also meets the further restrictions regarding small integers and different precisions that were mentioned in 3(a) and 3(b), as well as providing the advantage of true rounding that was referred to briefly in 3(c).

Ideally, the number system should be augmented to provide for overflow, underflow, undefined, etc., along the lines suggested by Neely. The hardware should also be designed so that it would be easy to provide needed extensions to the programming languages. These extensions would include provision for handling overflow, etc., provision for different precisions along the lines suggested by Hull and Hofbauer, and facilities for rounding up and rounding down. (A convention regarding expressions such as x+y+z is also needed in the programming language, as mentioned earlier.)

(What has been said so far is applicable to any floating point number system, regardless of its base. I would like to state, parenthetically, that my own preference would be for base 10, and I believe that this base may be the most economical, if we take all relevant aspects of the computing environment into consideration. Incidentally, one modest advantage that would appear with base 10 is that we would not need to worry about errors on conversion during I/O or compilation, and, in particular, whether or not these errors might differ!)

One further point should be made in connection with different precisions. These precisions would be multiples of some fixed word length, possibly numbers of decimal digits. It would have to be decided whether or not the exponent range would change with the precision. My own belief is that the exponent range ought to change, perhaps at a rate of something like one decimal digit of exponent for each additional five digits of fraction part. However, it must be pointed out that this feature introduces a new possibility for overflow, when a higher precision number is converted to lower precision, and this possibility must be taken into account.

There remains the question of what specifications we ought to aim for regarding the elementary functions. In my opinion, we should seriously consider insisting on only the basic accuracy requirements of the sort described in 2(b) - and striving to keep the n's and the ranges of the variables as uncomplicated as possible.

This proposal is quite drastic, and would mean, in particular, that no restrictions of the kind discussed in 3(d) would be part of the specifications. Some subroutines do satisfy restrictions of this sort, and a number of eminent individuals, including my friends Vel Kahan and Jim Cody, have advocated much stricter requirements than I am suggesting we consider for adoption. (For related work see Kahan {4, 5} and Cody {1}.) However, I worry about the number of these requirements, and the list seems to me to be endless. And, to return to the basic motivation for this paper, I am reluctant to have the correctness of a program depend on such requirements. If a program is to be transported, proofs about what it does must be transported as well. Should we contemplate having proofs that depend on $abs(sin(x))$ always being less than or equal to 1? How many of these requirements should we insist upon, even in ideal circumstances that we might hope to attain some day?

My proposal regarding functions is only one possibility, of course, and agreement will not be easy to reach. But this problem must be faced, and, at the very least, deserves more careful study.

5. Concluding Remarks

Portability of programs means portability of proofs about programs, and a lot of effort is needed to move programs from one system to another. However, although no system conforms exactly to existing standards, the very existence of such standards tends to reduce the effort needed to transport programs between different systems.

Most of the work on standardization has tended to neglect the semantics of floating point arithmetic and elementary functions. Progress in this area would be particularly helpful in reducing the problems associated with the portability of numerical software. Besides, having well-defined, clean specifications in this area would also make programming simpler.

It is hoped that the discussion presented in this paper will help draw attention to the needs in this area and suggest some steps toward the development of useful specifications.

REFERENCES

1. Cody, W. J., Desirable Hardware Characteristics for Scientific
    Computation, ACM SIGNUM Newsletter, 6, 1 (January, 1971) 16-31.

2. Hull, T. E. and Hofbauer, J. J., Language Facilities for Multiple
    Precision Floating Point Computation, With Examples, and the
    Description of a Preprocessor, Technical Report No. 63, Department
    of Computer Science, University of Toronto (1974).

3. Hull, T. E. and Hofbauer, J. J., Language Facilities for Numerical
    Computation, Proceedings of the ACM-SIAM Conference on Mathematical
    Software II, Purdue University (1974) 1-18.

4. Kahan, W., A Survey of Error Analysis, Proceedings IFIP Congress
    1971, vol. 2, Ed. C. V. Freeman, North-Holland Publishing Co.
    (1972) 1214-1239.

5. Kahan, W., Implementation of Algorithms, Part I, Technical Report
    20, Department of Computer Science, University of California,
    Berkeley (1973); available through NTIS, AD-769 124.

6. Neely, Peter M., On Conventions for Systems of Numerical Representations,
    Proceedings ACM Annual Conference (1972) 644-651.

MACHINE PARAMETERS FOR NUMERICAL ANALYSIS

W. J. Cody*

ABSTRACT

The general acceptance of conventions for basic linear algebra modules suggests that an effort will soon be made to establish similar conventions for machine dependent parameters and constants. This paper suggests basic concepts and principles which might guide such an effort and raises questions which must be adequately addressed if such an effort is to be successful.

*Work performed under the auspices of the U.S. Energy Research and Development Administration.

> *"Having a common programming language will make*
> *sure that we understand what the other person writes.*
> *This does not necessarily mean that we possess con-*
> *venient machinery for realizing his intentions."*
>
> — *Peter Naur*

1. Introduction.

Naur's eloquent plea for programming language extensions to provide access to specific machine characteristics {16} has spawned a number of proposals as to which machine characteristics are important and how they should be made available {1, 8, 9, 17}. All of these proposals envision a set of "standard" parameters or procedures which could be used to write transportable* mathematical software. Formal standards are extremely difficult to establish, but recent experience with basic linear algebra modules, or BLAS {4, 12, 14}, suggests that informal standards (Redish and Ward {17} call them "conventions") can be established if properly formulated and pursued. The BLAS are not likely to be universally included as language elements in compilers in the near future, but they are currently used by writers of mathematical software to convey reasonably precise algorithmic details to contemporaries. Thus the BLAS fit the pattern envisioned by Naur in the quote that heads this paper.

There are many, including the author and a recently established IFIP working group {3}, who feel that the lessons of the BLAS can now be applied to finishing what Naur started. The advantages of a widely used set of

---

*We follow NAG {8} in defining portable to mean that software can be moved from one environment to another without change, and transportable to mean that some changes may be needed, and that they are capable of mechanical implementation via a preprocessor.

parameters and procedures to describe and access machine dependent
quantities are obvious to those who work with mathematical software,
but recent proposals contain minor disagreements as to which parameters
and procedures are useful, and how they should be made available.

This paper presents basic concepts and principles which can be
used to mediate these differences, and perhaps to prepare a proposal for
basic machine parameters, or BMPs, although this is not such a proposal.
These principles are finally applied to some parameters for numerical
analysis which might be candidates for BMPs.

2.   Background.

Naur {16} set the pattern for most of the work which followed. He
not only suggested the use of machine characteristics in writing portable
software, but he also recommended extensions to common programming languages
to provide access to these parameters. Redish and Ward {17} subsequently
proposed an extensive list of parameters and suggested a systematic
scheme for naming them. They further proposed procedures or subroutines
for manipulating components of a floating-point number, such as determining
or specifying the unbiased exponent. (Sterbenz {18} has made similar
suggestions.) Finally, they suggested that the values of certain mathematical constants such as pi be available, perhaps through a subprogram.

These two papers include most of the basic proposals that have
been made. All subsequent work involves implementations of one or more
of these proposals, with extension in detail dictated by practical
experience. For example, Aird and his colleagues {1} have reworked
and augmented the Redish-Ward parameters and constants and have determined
specific values for these quantities for a large number of contemporary

American computers. This information is then used in a Fortran preprocessor which automatically prepares a machine-specific version of a program from an appropriately formatted master version. Some of the parameters are used to control the preprocessing, the number of significant decimal digits acceptable to a given compiler on a specific machine determining the length of decimal constants, for example. Other parameters, such as the mathematical constants, are automatically inserted as needed into the machine-specific source program produced by the preprocessor. Basically similar schemes have been implemented in the NAG {8} and NATS {6} software projects.

In a slightly different approach, Fox, et al. {9} have compiled a similar list of machine parameters and specific values which are imbedded as data in machine-specific Fortran subroutines. These are then available to users as Redish and Ward {17} suggested.

The efforts by Aird, Fox and their colleagues represent actual implementations to meet specific needs and are not necessarily intended as exercises in standardization. But the mere existence of this material in program form is a first step towards a de facto standard or convention, and its impact should not be underestimated. The value of the work is tempered by the lack of uniformity in names of parameters, and by the fact that all information is predetermined static information.

Several attempts have been made to determine selected machine parameters dynamically. Malcolm {15} presented prototype portable Fortran subroutines which use arithmetic operations to determine the base NB of the floating-point representation, the number of (implicit and explicit) significant base-NB digits in the representation of the significand, and whether the arithmetic chops or rounds. Gentleman and

Marovich {10} pointed out that these programs malfunction when arithmetic registers retain numbers to greater significance than do storage registers, and that correct performance further depends upon compilers producing correct values for simple constants. They modified the programs to meet some of these objections, but the admonition to be cautious still applies.

Recently George {11} has used character manipulation to determine machine parameters related to the representation of integers and floating-point numbers as well as character related parameters. The technique requires formatted I/O to an external unit, a weakness that would be eliminated if the core-to-core I/O currently available in some compilers and proposed for ANSI Fortran {5} becomes standard. Again, however, his technique fails when the local I/O conventions contaminate the formatted I/O stream with carriage control information.

The only remaining work of which the author is aware is that of the IFIP working group. The effort is too new to be described here, but their working papers indicate an intention to concentrate upon static as opposed to dynamic treatment of parameters.

3. Some Principles.

Almost all of the work just summarized suggests particular machine parameters as being important, where the term machine parameter is here being used in a very general sense. But we feel that there are three different classes of machine parameters plus the class of mathematical constants, and that the general characteristics of each class govern how the class should participate in any BMP proposal.

The first class consists of parameters of the arithmetic, i.e.,
parameters related to the floating-point, fixed-point and integer
arithmetic systems, where an arithmetic system is defined to be the
appropriate combination of number representation and arithmetic mode.
Parameters which specify the radix, and tell whether the arithmetic
operations are rounded or chopped would fall into this class.

The second class consists of algorithmic parameters, i.e., parameters
which are essential to control of an algorithm. These include numerical
parameters such as the machine epsilon, the smallest positive floating-point
number eps such that

$$1.0 + eps > 1.0$$

in the machine arithmetic, as well as similar parameters for non-numerical
algorithms. All algorithmic parameters are additionally characterized as
being hardware-dependent.

The third class consists of system parameters, such as the designation
of the standard input unit. These depend upon the operating system conventions for a particular computing center, and are software-dependent
as opposed to hardware-dependent.

Finally, there are the mathematical constants such as pi. These are
the only true constants in the lot. They cannot be defined by any standard
because they already exist as separate mathematical entities.

The best treatment of parameters differs from group to group. The
only commonalities in the treatments are the usual requirements for
systematic names and precise definitions. Ford {8} avoids the problem
of the current diversity of names for similar parameters by stating that

the names are of secondary importance to the definitions, but we disagree.
Clarity of communication of algorithms is fundamental to this whole effort,
and clarity requires uniqueness of terminology. For the mathematical constants
the only things that can be "standardized" are the names, a characterization
which distinguishes that group of parameters as a whole from the other groups.

For maximum utility, names should suggest the definition of the parameter
and should default to the appropriate variable type in Fortran, the most
important language with default variable typing. Several of the authors
surveyed emulated the Fortran treatment of the intrinsic functions by
suggesting word stems for a set of parameters and prefixing the stem with
an "I" for an integer parameter, an "R" for a real parameter and a "D"
for a double precision real parameter. Similar conventions should be
used for naming subprograms.

Precise definitions are fundamental to establishing BMPs. We may
not be able to agree on the appropriateness of a particular parameter,
but the parameter can neither be used nor discussed intelligently unless
it is unambiguously defined. This is not always an easy task, and an
inability to define a parameter may be a valid reason to omit it from
the BMPs.

Several of the previous authors have defined parameters, particularly
algorithmic parameters, in terms of the parameters of the arithmetic. For
instance, the machine epsilon previously defined might be defined as

$$eps = NB ** (-NT+1)$$

instead, where NB is the radix of the floating-point representation and
NT is the minimum number of implicit and explicit base-NB digits in the

significand of a stored or active floating-point number. This value of eps is sometimes larger than the value previously defined, but it is still a practical value and

$$1.0 + eps > 1.0$$

on all current computers.

Such mathematical definitions are precise, and they naturally select certain primitive parameters which span the parameter space in some sense. Ideally, we would like to identify these primitive parameters and use them to define other parameters, but there is danger in this approach if it is not carefully presented. The definitions are not always safe for dynamic computation of the parameters because of inaccuracies of the hardware and software. A BMP proposal must distinguish between the definitions and the values for non-primitive parameters, emphasizing the importance of precise precomputed values for specific machines.

Only generally useful parameters should be singled out for inclusion in the BMPs. Aird's parameter list has been selected for its utility in his preprocessing environment. There are parameters in his list which are useful only for controlling his preprocessor and which should not be included in a general list of BMPs. There are proposed parameters in other lists whose usefulness in any environment is difficult to imagine. For example, the number of lines per output page is often determined by a carriage control tape on a line printer. Because the current tape is neither known to nor selectable by even the operating system, a parameter specifying the number of lines per page appears to be useless from a practical viewpoint. We will discuss other such parameters in the next section.

It is essential to treat parameters dynamically instead of statically when they are not constant for a given environment but change with the dynamic situation. As an example, CDC 6000/7000 series computers offer either chopped or one of two types of rounded floating-point arithmetic, the type used for a Fortran program often being a compiler option. Evidently the parameter describing the type of arithmetic being used can be given a standard name but not a static value. Some form of dynamic determination must be provided.

The other form of dynamic treatment is manipulation. Certain parameters are most useful if they can be modified and manipulated in prescribed ways. Such parameters are most frequently found among the parameters of the arithmetic, although there are arguments for dynamic treatment of many algorithmic parameters as well. We will return to this subject a little later.

4.   Parameters for Numerical Analysis.

The specific parameters we want to discuss do not comprise an exhaustive list of interesting and important parameters, even in the author's experience. They have been selected to illustrate specific points, or because they have been overlooked in other lists. We have purposely chosen bland names for the parameters because it is not our intent to make serious suggestions for names for BMPs.

We begin with parameters for the floating-point arithmetic. These include parameters which relate to the representation of a number and parameters which describe the functioning of the arithmetic unit.

Among the former is the radix NB of the floating-point representation. The parameter can be defined unambiguously and we believe it to be a constant for any particular machine. It is useful for defining algorithmic parameters,

as we have already seen, but it is also useful by itself. Because certain
common values of NB, such as 16, imply the existence of "wobbling word length"
and accompanying problems with the significance of important constants {7},
the value of NB may determine which of two computations is to be made. The
author can foresee no need to manipulate this parameter, but because programs
exist for the dynamic determination of its value the parameter could easily
be treated either statically or dynamically. Either way, NB appears to be
a legitimate candidate for a primitive BMP.

A second parameter that is often mentioned is the bias on the exponent.
We do not believe the bias is a legitimate BMP because, first, we have not
found a need for it that could not be satisfied by other means, and second,
it is not necessarily a "constant" for a particular machine. CDC machines
use one bias for numbers less than $2**47$ in magnitude, and a different bias
for larger numbers. The obvious use for the bias is in the fabrication and
dissection of a floating-point number, but the bias is immaterial for these
purposes. The important thing is to be able to access and manipulate the
unbiased signed exponent IE. Experience with the Sterbenz subprograms {18}
which extract and insert IE suggests that they satisfy all needs for exponent
manipulation tasks. They permit the synthesis of a floating-point number
with a significand obtained from one number and either a manufactured
exponent or one extracted from another number. If, for instance, the NT
defined in the previous section is available, an epsilon can be synthesized
that will augment a given number by n units in its least significant digit
provided the given number is not too small in magnitude. In particular,
the machine epsilon can be constructed from its definition in terms of NT.
It thus appears that subprograms for the dynamic extraction and insertion
of IE are candidates for the BMPs, but that the bias and probably the significand are not needed.

The above implies a need for NT. The precise definition of NT
given in the last section takes into account the possibility that the
explicit representation of the leading bit of a normalized binary
significand is suppressed, and the possibility that the number of
significant digits retained in storage differs from the number retained
in the arithmetic registers. Both of these possibilities are realized
in contemporary computers. With this definition, NT is static but can
be determined dynamically. It is certainly a candidate for a primitive
BMP.

Another parameter frequently associated with the floating-point
arithmetic system is NR which states whether the arithmetic fits the
intermediate results of an operation to the machine precision by rounding
or by chopping. As stated earlier NR must be determined dynamically, and
programs to do this are known provided "machine precision" refers to the
precision retained in the arithmetic registers. But there is additional
rounding or chopping on machines where stored numbers retain less significance
than those in the arithmetic registers. Does this require another parameter,
a different definition of NR, or should it be ignored? The answer is not
clear.

There has also been a proposal {8} to have NR designate the type of
rounding done, but this information appears to be impossible to obtain
and not very useful even if it were available. There are far too many
rounding schemes to be simply catalogued, and dynamic detection of a
particular scheme or class of schemes seems beyond our capabilities.
To have any influence at all on numerical software the numerical
behavior of each scheme must be understood. This is too much to expect
for practical purposes.

Another arithmetic parameter is NG, the number of available guard characters. If NT + NG NB-ary digits, of which some may be leading zeros, are developed in the arithmetic registers at an intermediate stage of an arithmetic operation, and the NG extra digits participate in postnormalization of the significand, then there are NG guard digits. Certain numerical algorithms ought to be modified when NB is 16 and NG is 0, so NG is of some interest. It is subject to misunderstanding and subsequent misuse, however. Consider chop mode single precision arithmetic on CDC 7000 series machines, where NB = 2, NT = 48 and NG = 0, i.e., only the first 48 bits of a result participate in the postnormalization shift. One might suspect that

$$1.0 - eps = 1.0$$

whenever $eps < 2^{**}(-47)$. In reality

$$1.0 - eps < 1.0$$

in the machine even when $2^{**}(-95) \leq eps < 2^{**}(-47)$ because all single precision arithmetic is performed in double length registers. NG has nothing to do with the significance of the arithmetic registers for intermediate results. This possibility of a misunderstanding makes the usefulness of NG as a BMP questionable.

Finally there is the parameter NN which tells whether negative floating-point numbers are represented in sign-magnitude, base complement, or base-1 complement notation. We have unsuccessfully attempted to detect differences in representation using arithmetic calculations, including those suggested by Knuth {13}. If we cannot detect any computational effect due to the representation of negative numbers, then we suspect that NN is not a BMP candidate.

This list of interesting parameters of the arithmetic is surprisingly
short in comparison with other lists. We find little of importance beyond
NB, NT, some subprograms for the manipulation of IE, and the appropriate
counterparts for double precision numbers and integers.

For the discussion of algorithmic parameters we define an <u>achievable</u>
(<u>floating-point</u> or <u>integer</u>) <u>number</u> {8} as a storable number which can be
obtained as a result of any of the standard arithmetic operations without
hardware detected error. All algorithmic parameters will be assumed to
be achievable numbers. There are certain obvious candidates for algorithmic
parameters which many of the surveyed papers suggest in one form or another.
These include the machine epsilon, the largest floating-point number XINF
such that -XINF is also an achievable number, and the smallest positive
floating-point number XMIN such that -XMIN is also achievable. Note in
this last case that there are certain floating-point numbers which are
considered to be positive on CDC equipment under addition or subtraction,
but are considered as zero for multiplication or division. XMIN must
be non-zero for all operations.

There are other useful algorithmic parameters described in various
lists. For example, XMININ and XMAXIN, the smallest and largest achievable
positive floating-point numbers whose reciprocals are also achievable (usually
XMININ is not 1.0E0/XMAXIN). But let's consider a useful parameter which
is not often suggested. We motivate it as follows. Let

$$X = XMIN * 1.25E0$$

Then, if LOGIC is a logical variable, and

$$LOGIC = (X .LE. XMIN)$$

what is the value of LOGIC? On IBM equipment it is FALSE, but on CDC
and UNIVAC equipment it is TRUE.   In both cases the results are correct
according to both the Fortran standards {2} and the proposed revised
Fortran standards {5}! The standards are ambiguous, saying that the
value of the logical expression is true or false depending upon whether
the relation is true or false, but that the decision can be based upon
a comparison of the arithmetic difference between the two variables
and zero.   These alternatives can give conflicting results when the
difference underflows.   The ambiguity is resolved in favor of what
common sense says the result should be if both X and XMIN are scaled
by SCALE, defined by

$$SCALE = NB**NT,$$

because

$$Y = X * SCALE$$
$$LOGIC = (Y .LE. XMIN*SCALE)$$

is free of underflow.   This is only one example of the utility of the
parameter SCALE which we submit as a candidate BMP.

Strong cases can be made for other parameters as well, but the
supporting evidence should always show the utility of the candidate.
The inclusion of a parameter simply because it can be defined will
weaken a proposal for BMPs to the point where it will not receive
serious consideration.

5.   Another Possibility.

There is one other possibility which we have ignored to this
point: we could require that all references to machine dependencies
be through function subprograms. Naur {16}, for example, suggested
the function EPSILON(U) whose value is defined to be the smallest
positive floating-point number such that

(U+EPSILON(U) .EQ. U)   .AND.   (U-EPSILON(U)   .EQ. U)

has the value FALSE, but he did not consider his proposal to be
realistic. This approach is very attractive to computer scientists
working on automated programming aids, however, because it permits
conceptually clean source code which can be moved from one computer
environment to another without modification. The imbedding of a
machine-specific value of the machine epsilon in a data statement,
for example, renders a program only transportable and requires that
the program author know the value for his machine, while the invocation
of the function value EPSILON(1.0E0) is completely portable and requires
no special machine knowledge.

Attractive as this approach appears, it is a very difficult one.
Someone must provide the necessary functions for each computer/compiler
combination, or programs using these functions are less transportable
than before. Specifications for the function programs must be very
carefully drawn, or ambiguities such as exist in the Fortran standards
will occur. The function programs must be written by experts in
systems programming as well as the numerical behavior of the machine.
It is important, for example, to know that

(X   .GT.   0.0E0)   .AND.   (SCALE*X   .EQ.   0.0E0)

is TRUE for certain floating-point numbers on CDC equipment if a
function defined by

SCALEF(X)   = X * (NB**NT)

is to be properly implemented. How many other such pitfalls exist
and who knows enough about them to design function specifications

and implementations? How can such valuable people be enticed into helping in this effort? There are many such questions and problems to be resolved if this approach is to be viable.

We do not feel that this approach is without merit, but we do feel that it will be much more difficult to accomplish than a more traditional approach. Part of the success of the BLAS lies in their being described in (sometimes ambiguous) portable Fortran source as well as prose. The Fortran implementations are not always efficient, and users are urged to substitute hand-coded programs or even inline Fortran meeting the same design specifications whenever that approach is expedient. But at least programs using the BLAS can rely on some implementation being readily available. That is not the case here. It is not clear how some of the interesting functions for machine dependencies could be written in portable Fortran. We don't even know how to write a Fortran implementation for the logical test for

$$(A \quad .LE. \quad B)$$

that gives the sensible result in every case. How can we hope to write a portable Fortran exponent manipulation subprogram, for example, or determine XINF without triggering an arithmetic exception?

6. Conclusion.

In summary, we believe there are three classes of machine parameters aside from the mathematical constants: parameters of the arithmetic, algorithmic parameters and system parameters. These classes will enter into any proposal for BMPs in somewhat different ways. Because some of the parameters of the arithmetic are fixed for a given type of machine they can be treated statically by giving names and definitions and assigning machine-dependent values. But others in this group can assume a variety of values and are best treated dynamically. Depending upon the parameter and the intended use the dynamic treatment might be limited to procedures for determination of the current value, or it might include procedures for manipulation of values. Certain of these parameters are primitives which span the parameter space in some sense. In any case, standard names, definitions and calling sequences are necessary.

Algorithmic parameters are almost exclusively static for a given machine, and can be assigned names, definitions, and machine-dependent values as part of the BMP proposal. System parameters, on the other hand, depend upon the operating system. Names and definitions can be specified in the BMP proposal, but values must be locally assigned. Finally, only names can be provided for the mathematical constants because values are already prescribed.

There is an alternate way of treating algorithmic and systems parameters by defining them in terms of functions. Each approach has advantages and disadvantages, and the proper choice between these alternatives may require a great deal of study.

The effort to establish BMPs will perhaps only be made once. Success will establish a usage that will not be easily changed, while failure will discourage further attempts. We believe that the basic concepts and principles presented here could be the basis for a successful first effort.

REFERENCES

1. Aird, T. J., The Fortran Converter User's Guide, IMSL, Houston, 1976.

2. American National Standard Fortran, American National Standards Institute, New York, 1966.

3. Anonymous, IFIP Working Group 2.5 on Numerical Software, SIGNUM Newsletter 10, Nos. 2 & 3, November, 1975, p. 16.

4. Anonymous, SIGNUM Basic Linear Algebra Project, SIGNUM Newsletter 10, Nos. 2 & 3, November, 1975, pp. 3-4.

5. Anonymous, X3J3/76 dpANS FORTRAN, February, 1976.

6. Boyle, J. M., Portability Problems and Solutions in NATS, Proceedings of the Software Certification Workshop, W. Cowell, (Ed.), Argonne National Laboratory, 1973, pp. 80-89.

7. Cody, W. J., An Overview of Software Development for Special Functions, Lecture Notes in Mathematics 506, Numerical Analysis Dundee 1975, G. A. Watson, ed., Springer-Verlag, Berlin, 1976, pp. 38-48.

8. Ford, B., Portability of Numerical Software in the NAG Project, to appear.

9. Fox, P. A., Hall, D., and Schryer, N. L., Machine Constants for Portable Fortran Libraries, in Computer Science Technical Report No. 37, Bell Laboratories, Murray Hill, N. J., 1975.

10. Gentleman, W. M., and Marovich, S. B., More on Algorithms That Reveal Properties of Floating-Point Arithmetic Units, CACM 17, 1974, pp. 276-277.

11. George, J. E., Algorithms to Reveal the Representation of Characters, Integers and Floating-Point Numbers, ACM TOMS 1, 1975, pp. 210-216.

12. Hanson, R. J., Krogh, F. T., and Lawson, C. L., A Proposal for Standard Linear Algebra Subprograms, Technical Memorandum 33-660, Jet Propulsion Laboratory, Pasadena, 1973.

13. Knuth, D. E., The Art of Computer Programming, Vol. 2, Seminumerical Algorithms, Addison Wesley, Reading, Mass., 1971.

14. Lawson, C. L., Hanson, R. J., Kincaid D., and Krogh, F. T., Basic Linear Algebra Subprograms for Fortran Usage, unpublished draft, 1976.

15. Malcolm, M. A., Algorithms to Reveal Properties of Floating-Point Arithmetic, CACM 15, 1972, pp. 949-51.

16. Naur, P., Machine Dependent Programming in Common Languages, BIT 7, 1967, pp. 123-131.

17. Redish, K. A., and Ward, W., Environment Enquiries for Numerical Analysis, SIGNUM Newsletter 6, 1971, pp. 10-15.

18. Sterbenz, P. H., Floating-Point Computation, Prentice Hall, Englewood Cliffs, N. J., 1974.

PREPARING CONVENTIONS FOR PARAMETERS FOR
TRANSPORTABLE NUMERICAL SOFTWARE

B. Ford

1.   Introduction

The vast majority of numerical software at the present time is
written in FORTRAN, although a significant amount is in Algol 60 and
Algol 68 programs. The FORTRAN and Algol 60 languages were designed
with the implicit assumption that a program written in a language
common to many machines could not take advantage of special features of
any of them {10}. Although the initial proposal by Naur for environmental
enquiries led to the inclusion of a limited enquiry mechanism in Algol 68,
the facilities offered are inadequate for numerical computation. A recent
supplement to the Algol 60 Revised Report appears to suffer a similar
deficiency {4}.

There is a pressing need to augment each of these languages with
the parameters of a model numerical computing configuration. Why such
parameters are necessary and how the set for FORTRAN prepared by IFIP
Working Group 2.5 evolved is the subject of this paper.

2.   Definitions

Portability

A program is portable between configurations a and b if it will
compute to prescribed levels of accuracy and efficiency in each computing
environment without any change.

Transportability

A program is transportable between configurations a and b if it will compute to prescribed levels of accuracy and efficiency in each computing environment with automated changes made by a computer program.

Portability is an unattainable ideal, except possibly for the most trivial programs. What we would accept is a transportable program.

3. Developing Numerical Software

The development of numerical software entails at least three distinct steps:

(i)   The design of the algorithm

(ii) Its realisation as a documented source-language subroutine or program

(iii) The certification of the compiled code on a given configuration (i.e., a machine together with its operating system, compiler and available libraries) and its detailed documentation for that configuration.

3.1 Designing an Algorithm

When designing an algorithm the primary concern is that it computes the solution of the specified problem. In addition to other qualities {5, 2} a consideration is that the algorithm effectively utilizes the computing configuration for which it is implemented. Increasingly, numerical analysts and programmers are seeking to design algorithms that can be transferred without change to different computing configurations. The vast majority of algorithms will function in their new environments. Generally, however, the accuracy of the results or the efficiency of the computation will be impaired due to the failure to make optimal use of the new environment. In a number of cases the algorithm will not function at all, or will compute meaningless numbers. Hence, to form the basis of a portable or transportable program an algorithm must be expressed in an adaptable form {6}, i.e., it can be automatically adapted to utilise effectively the particular characteristics of its present computing environment. This requires access to the features of an abstract numerical configuration, onto which all existing numerical configurations can be mapped. As with all models, the abstract numerical configuration may contain only the gross features required, and a number of existing numerical configurations may only implement a subset of the characteristics of the model. Use of the model will also ease communication of the precise form of the algorithm to other people.

3.2 Writing a Routine

With regard to the abstract numerical configuration and its effect on the writing of a routine Naur is quite specific: "The only way the use of Environmental Enquiries can cause a substantial extra effort is the case when the programmer wishes to describe a solution of a problem which will take a wider class of configurations into account. This, however, is a task which is outside the scope of present day languages." But this task is precisely our objective. And for the present, it must be performed in the languages he had in mind, particularly FORTRAN. We believe that whilst the cure suggested by Naur has the seeds of a solution to the problem, the particular approach he adopted (which has been followed by the majority of other writers {11, 1, 3} in this area to date) is too piecemeal, selecting individual requirements in an inconsistent and often uncoordinated manner. What is required is the precise specification of the abstract numerical configuration discussed above, designed for ready incorporation into the existing languages.

4. Design of the Abstract Numerical Configuration

The design of the abstract numerical configuration is dictated
by experience, by generality, by simplicity and by convenience of use.
Its features are selected for their functional independence of one
another and their significance in numerical computation. Development
and refinement of the model is essential during the next few years. It
is hoped that an abstract numerical configuration, realistic in terms of
current configurations yet complete from the viewpoint of algorithm and
software requirements, will be evolved in time for the first revision of
FORTRAN 77 and for inclusion in any major Algol, general purpose or
numerical language development.

It is anticipated that through its use analysts will discover the
relevant machine characteristics to use in the realisation of algorithms
as transportable software.

5. Parameterisation of the Abstract Numerical Configuration

When including a parameterisation of the abstract numerical
configuration in each of the various languages (e.g., FORTRAN, BASIC,
ALGOL 60, and ALGOL 68) it is essential to conform, where possible, with
the existing philosophy of the language and to preserve the effectiveness
of the parameters and their ease of use for algorithm adaptability.

The actual numerical configuration presented to a high-level language
programmer is created by the hardware-designer and the compiler-writer.
Both have felt free to indulge their own fancies and foibles; both have
frequently failed to document the details of the configuration that they
have created. (We are here talking primarily about data representation
and arithmetic operations.) Moreover since there is not yet any generally
agreed model of a numerical computing configuration, neither hardware-

designers nor compiler-writers have seen their role in terms of assigning specific values to the parameters of the model and realising the resulting configuration. Agreement on such a model and on its parameterisation in the major high level languages will be a significant step towards defined and constant interfaces with the high level language compilers. In the meantime it is often far from easy to determine the appropriate parameter values to describe a particular existing configuration, and indeed there may be no assignment of parameter values which will describe the configuration precisely: we hope however that the parameterisation will usually permit a description that is sufficiently close.

6. Preparing Conventions

The discussion so far has concentrated on the objectives of the author in setting out with others to prepare conventions for parameters for transportable numerical software. Pursuit of such objectives takes time, research, patience, and compromise. The value to the community of having generally agreed, accepted, and used conventions is obvious and requires no further elaboration.

7. Background to the Present Proposals

Consideration of parameters for transportable numerical software was initiated by Naur in his fundamental paper proposing Environmental Enquiries in 1967 {10}. The theme was taken up for numerical analysis in the paper by Redish and Ward in 1971 {11}. At the Jet Propulsion Laboratory Fortran Preprocessor Workshop in November 1973 there was a meeting of Aird, Byrne, Krogh, Pool, and the author to discuss machine parameters in FORTRAN. This resulted in a first draft of a paper called

"Machine Constants," written by the author for the inaugural meeting
of the IFIP Working Group on Numerical Software (WG 2.5) in Oxford
in January 1975. Independently Aird et al. {1} published a note in
the SIGNUM Newsletter in 1974. The response of WG 2.5 to "Machine
Constants" was to request a more precise specification of its purpose.
Written comments suggested further improvements and so a second draft
was undertaken.

To this point an attempt had been made to cover the requirements
of scientific and technical programmers, package and preprocessor
developers, and error analysts in the parameters and constants included
in the discussion. This was found to be impractical and the interests
of scientific and technical programmers above were made primary. In
addition it was recognised that it was necessary to examine in detail
those underlying characteristics which resulted in particular parameters
being required in programs. Further, the mode of implementation of
parameters required careful consideration {9, 8}. These considerations
were reflected in the second draft.

A third draft was prepared and discussed during a workshop on
transportable numerical software at Argonne in August 1975. This
discussion included Aird, Boyle, Cody, Krogh, Smith and the author.
No consensus of views emerged, although there was some agreement on
both general issues relating to purpose, objectives, and implementation,
and on particular points.

At about this time work was in hand at Bell Laboratories on the basic
utilities for a portable Fortran library {7}. This included investigation
of suitable machine constants. The emphasis of this study was on modelling
the number representation of a configuration.

8.   <u>Synopsis and Comments on Draft 4 (Appendix 1)</u>

The fourth draft "Machine Characteristics and their Parameterisation in Numerical Software" was completed in January 1976, and distributed to the numerical software community for criticism, expansion, and comment prior to the ERDA/NSF Portability Workshop at Oak Brook in June 1976. The paper starts with the background given above and then stresses the need to develop adaptable algorithms for transportable subroutines with their dependence on configuration parameters. The fundamental importance of developing the abstract numerical configuration is next examined, with emphasis being given to the two problems of selecting the relevant characteristics and specifying them as a set of parameters.

It was evident from the correspondence received before and comment at the workshop that a complete characterisation of the abstract numerical configuration was not yet possible. Hence the approach to be adopted for further discussion would be to select a restricted, rather simplistic model involving only those characteristics for which we could see an obvious, immediate need. We would try to select the characteristics so that they complemented one another without side effect or confusion. A particular effort would be made in the next draft to include every major facet of the model. Reasonable doubt about the need for a particular characteristic would result in its exclusion, at least at that stage. The aim was for a limited initial characterisation of a computing configuration, as it affects numerical software.

The proposals of the fourth draft contain two main sets of characteristics, relating to the hardware and to the software interfaces. The first set divides

into two groups, one concerned with number representation and the other
with machine arithmetic. The second set consists of logical unit numbers
and the limits on data storage and of data transfer.

The major weakness of the draft is the attempt at a detailed model
of machine arithmetic. It is patently a failure (although a number of
correspondents had requested it). Further its relevance to algorithm
design is far from clear. Only the base of the floating-point number system
and the length of the mantissa for floating-point number representation
are of obvious use. However the characteristics concerned with number
representation (relative precision, overflow and underflow) are of
fundamental importance. Some analysts prefer a number representation
in a symmetric range, rather than a range defined by overflow and underflow.

In the second set the inclusion of a standard teletype line was
considered to be impractical and the punch unit and tape unit unnecessary.
Otherwise the proposals met broad agreement.

A number of other issues require discussion. The parameterisation
of the characteristics presents some interesting problems, as does the
implementation of the parameterisation in software. Some of the information
can be collected dynamically. Certainly the difficulties and inefficiencies
of obtaining all the information dynamically are well understood. Function
calls, the use of COMMON blocks and explicit value substitution all have
their strengths and weaknesses. The draft employs a simple set of parameters.
Whilst each characteristic requires and deserves specific consideration
for its mode of parameterisation, where possible individual parameter
names are preferred.

The parameter name is selected so that it may be treated where necessary as a stem to permit assignment of a prefix letter to signify integer (I), single precision (R) and double precision (D) as required. All names contain at most 6 characters.

The presentations at the workshop by Cody and the author provoked significant interest in the topic of parameters and encouraged constructive comment and debate. This resulted in an informal meeting at the close of the workshop between some attendees and members of the IFIP Working Group 2.5 (on numerical software). Draft 4 was reviewed in detail with considerable time being spent agreeing which characteristics with what definitions and explicit names should be included in Draft 5. This invaluable input, together with the correspondence and comments from about one hundred individuals, was collated by the author and presented to the IFIP WG 2.5 meeting the following day, with the suggestion that a final draft be prepared for submission to IFIP. It was evident that there was now general agreement on suitable conventions for parameters for transportable numerical software.

Draft 5 was written by Reid and the author, assisted by Dekker, and after minor editing agreed to by IFIP WG 2.5. "Parameters for Transportable Numerical Software" is appendix 2 of this paper. The proposals rest there at the present time (1st September, 1976).

APPENDIX 1

MACHINE CHARACTERISTICS AND THEIR PARAMETERISATION
IN NUMERICAL SOFTWARE

Compiled by: B. Ford

History

An early draft of this note was used as a discussion document during
the meeting of the IFIP Working Group 2.5 (on Numerical Software) in
Oxford in January 1975. The meeting requested a precise specification
of the purpose of the note and suggested a number of other improvements.
Written comment led to further changes. A third draft was discussed
during a workshop on transportable numerical software in the Applied
Mathematics Division of the Argonne National Laboratory in August 1975.
It is hoped that the following keeps faith with all the parties involved
to date. The contents however remain proposals for discussion. The
compiler would welcome correspondence on all or any aspects of the
material, or regarding apparent omissions.

Specification of Objectives

The development of numerical software entails at least three distinct
steps: the evolvement of an algorithm or composition of a series of algorithms
to undertake a specified computation, the realisation of the algorithm
in a computer language as a routine or program and the certification of
the compiled code as successfully tested and documented software for a
given configuration. Whilst this final compiled code is configuration
specific, research being undertaken by numerical analysts and programmers
will, it is hoped, enable portable algorithms and transportable subroutines
to be developed.

The algorithm writer requires recourse to a set of carefully selected machine hardware parameters, which give details of the machine arithmetic and of the derived numerical characteristics of the hardware. Algorithms may then be designed, using these defined parameters, to incorporate the necessary elements for correct computation within the hardware constraints of existing systems.

When writing transportable routines from such algorithms the programmer needs a set of machine software parameters which give details of the standard logical units and describe aspects of the operating system, compiler and compiler library interface with his software tools and target source codes.

The set of names representing these machine parameters may then be used in programs enabling transportable codes to be written.

The fundamental aspect of the present activity is the selection of the relevant machine characteristics at the hardware and software interfaces, and their specification as sets of parameters.

Although this characterisation of a computing configuration is attempted in generality, perhaps inevitably there are known exceptions even to the simple features chosen. Whilst these cases may weaken the value of the present exercise it is believed that there remains the opportunity of a commonly recognised and used characterisation being generally adopted.

An initial study has been made of the machine hardware characteristics used regularly within the various fields of numerical analysis and of the machine software characteristics required widely in the development of numerical software. Care has been taken to include all such elements

within the parameterisations. It is hoped and intended that this first
selection will be criticised, extended and revised through discussion
with algorithm writers in the various numerical areas and with numerical
software librarians.

The names given to the ensuing parameters have been selected carefully
and it is believed that they are sensible and systematic. Nevertheless,
these names are of secondary importance and it is anticipated that groups
with existing bodies of source code will continue to use their present
names. It should be a comparatively simple matter to machine process
between various sets of names as a prelude to the exchange of source
codes.

Parameter Selection

The hardware parameters consist of two sets; one giving details of
machine arithmetic and the other giving details of derived numerical
characteristics. This second set has been divided into two groups,
the fundamental group and the particular group. The fundamental group
is made up of those elements essential to the writing of algorithms in
the majority of numerical areas. The elements of the particular group
are those required regularly in a particular numerical area. The
criterion for the inclusion of further parameters in this group is
simply the demonstrated need for them in a collection of algorithms.

The software parameters consist of three sets, the first giving details
of the standard input/output units, the second giving limits on data storage
and the third giving limits on data transfer.

Representation

The representation of an integer, real, or extended real quantity is that introduced by the statement INTEGER X, REAL X, DOUBLE PRECISION X respectively in FORTRAN as defined by the ANSI X3.9-1966 report.

For these representations some realisation of the computed and stored numbers obtained from arithmetic operations on various machines is required.

An achievable number is a stored number which can be obtained (without hardware detected error) on the computer by the primitive operations +, -, x, / between two numbers held in an arithmetic unit.

Problems of precision necessitate the inclusion of names for integer, real and extended real types for a number of the characteristics chosen. The characteristics involved are evident from the context. In these cases the proposed name is employed as a stem with the prefix I, R and D as relevant.

HARDWARE INTERFACE

Characteristic	Definition	Name
MACHINE ARITHMETIC		
Radix of the integer representation	Base of the integer representation	IRADIX
Radix of the mantissa of the floating point representation	Base of the mantissa of the floating point representation	MRADIX
Radix of the exponent of the floating point representation	Base of the exponent of the floating point representation	ERADIX
Length of integer	Number of binary digits in the integer representation including the sign if present	INTBIN
Length of mantissa	Number of binary digits in the mantissa including the sign if present	MANBIN
Length of exponent	Number of binary digits in the exponent including the sign if present	EXPBIN
Bias of exponent	Integer origin of exponent evaluation	BIASEX
Type of achievable number representation	For a sign modulus representation the value is the integer 0 For a one complement representation the value is the integer 1 For a two complement representation the value is the integer 2	NEGTYP
Method of rounding	For true rounding the value is the integer 0 For the forced rounding the value is the integer 1 For chopped rounding the value is the integer 2	MROUND

Fundamental Set		
Criterion of relative precision (Machine Precision)	The smallest positive achievable number such that $1.0 + x > 1.0$	EPS
Absolute measure of machine overflow (Overflow)	The largest positive achievable number x such that x and -x can be obtained without overflow	GIANT
Absolute measure of machine underflow (Underflow)	The smallest positive achievable number x such that x and -x can be obtained without underflow	DWARF
Particular Set		
Largest positive number	The largest positive achievable number x that can be represented upon the machine	INFP
Smallest negative number	The smallest negative achievable number x that can be represented upon the machine	INFM

SOFTWARE INTERFACE

Characteristic	Definition	Name
INPUT/OUTPUT UNITS		
Standard Input Unit	The logical unit number for the standard input unit of the system	NIN
Standard Output Unit	The logical unit number for the standard output unit of the system	NOUT
Standard Error Message Unit	The logical unit number for the standard error message unit of the system employable by the user	NERR
Standard Punch Unit	The logical unit number for the standard card punch unit of the system employable by the user	NPCH
Standard Tape Unit	The logical unit number for the standard paper tape unit of the system employable by the user	NTPE
Standard Teletype Line	The logical unit number for the standard teletype channel of the system employable by the user	NTTY
Limits on Data Storage		
Number of characters per word	Maximum number of characters that may always be stored per integer	NCHAR
Number of significant decimal digits	The maximum number of significant decimal digits used by a given compiler	NDSIG
Page size	The size of storage defined for use within the paging algorithm of a virtual storage system	PGSIZ
Limits of Data Transfer		
Number of input characters per line	The maximum number of characters per line including blanks that can be read from the input unit NIN per line	NLNIN
Number of output characters per line	The maximum number of characters including blanks that can be printed on a single line (on unit NOUT)	NLNUT
Number of output lines per page	The maximum number of lines that can be printed on a page without automatic page ejection (on unit NOUT)	NLNPG

Conclusions

This note has sought to select relevant machine characteristics of hardware and of software and to specify them as sets of parameters to assist the development of portable algorithms and transportable programs. It has been found impossible to develop a comprehensive characterisation which includes all such aspects of all machines. The note has therefore attempted to delineate a limited but safe region within which computation can be completed accurately and efficiently.

APPENDIX 2

PARAMETERS FOR TRANSPORTABLE NUMERICAL SOFTWARE

Prepared and agreed by the IFIP Working Group on
Numerical Software (WG 2.5)

July 1, 1976

1. History

An early draft of this note was used as a discussion document during
the first meeting of the IFIP Working Group on Numerical Software (WG 2.5)
in Oxford in January 1975. The meeting requested a precise specification
of the purpose of the note and suggested a number of other improvements
which led to a second draft. Written comment led to further changes. A
third draft was discussed during a workshop on transportable numerical
software in the Applied Mathematics Division of the Argonne National
Laboratory in August 1975. A fourth draft was written in January 1976
and distributed widely for comment and criticism. Discussion at the
NSF/ERDA Workshop on Portability of Numerical Software and at the second
meeting of the IFIP WG 2.5 (both in June 1976), together with correspondence
from other parties, led to the preparation of the present document. Although
some of the comment was contradictory, it is our belief that this final
document represents a consensus view.

2. Objectives

The development of numerical software entails three distinct steps:

    (i)   the design of the algorithm

    (ii) its realization as a documented source-language subroutine
          or program, and

(iii) the testing of the compiled code on a given <u>configuration</u>
(i.e., a machine together with its operating system, compiler
and available libraries) and its detailed documentation for
that configuration.

While the final code must depend on the configuration, it is very
desirable for the source-language version to be <u>transportable</u> (i.e., needing
only a small number of mechanical changes prior to compilation on a specific
configuration) and for the algorithm to be <u>adaptable</u> (i.e., expressed in
terms that permit efficient and accurate computation on any configuration
that is used later). Our first aim is to suggest parameters that allow the
algorithm designer to make his work adaptable in this sense. For example,
he may need the "relative precision" for a root-finding algorithm or
"page size" for an algorithm designed to solve linear equations efficiently
on a machine with virtual memory.

Our second aim is to provide a larger set of parameters that can be
used in the transportable source-language (step (ii)) in the expectation
that actual values will be readily available in the eventual run (step (iii)).
The set has to be larger since it should include such items as the standard
input and standard output units, neither of which are likely to be of concern
to the algorithm designer. The exact way that these values will be made
available is left open; some possibilities are as function calls, as values
in a named COMMON block and as flags for a preprocessor. In all cases we
hope that the actual names we suggest are used.

Our suggested parameters are listed in the table below. In addition
to a definition we give a <u>characteristic</u> name for the algorithm writer and

an explicit name for inclusion in the source code. These explicit names conform to FORTRAN conventions.

The parameters are intended to complement the intrinsic functions EPSLN, INTXP and SETXP which the members of IFIP WG 2.5 have proposed for inclusion in the new standard FORTRAN (see table).

3. Parameter Selection

The first group of parameters, which we have called the arithmetic set, is intended to provide necessary information about the arithmetic hardware. It should be noted that they are not intended to be sufficient to describe the operation of the arithmetic in full detail, nor is it expected that all will be wanted in any one program. One group of numerical analysts feels that the detail provided by the overflow and underflow parameters is un- necessary and that a range parameter is what is really needed; another group wants this detail. Since range cannot be computed from overflow and underflow, we decided that all three parameters should be included. Similarly, we expect that in many applications the relative precision parameter will be used in preference to the mantissa length, but a requirement for the mantissa has been expressed strongly by a number of numerical analysts.

The next set consists of basic input-output parameters likely to be required frequently.

The third set contains miscellaneous parameters. As long as the majority of FORTRAN compilers are designed to meet the ANSI X3.9-1966 FORTRAN standard there is a continuing requirement for a parameter for the maximum number of characters that can always be stored in an INTEGER storage unit. The number of decimal digits accepted by the compiler is needed for preprocessors generating specific code for a configuration. The page size may be required to ensure that an algorithm performs efficiently on a virtual memory machine. We suggest that this parameter be given the value of 1 for non-paging machines.

4. Representation and Arithmetic Operations

We use the systems of INTEGER, REAL and DOUBLE PRECISION numbers as described in the ANSI X3.9-1966 FORTRAN standard. For the system of INTEGER numbers we consider the arithmetic operation a * b, where * belongs to $\{+,-,x\}$, and the monadic operation -a. The computed and stored result can be computed exactly. For the systems of REAL and DOUBLE PRECISION numbers we consider the arithmetic operation a o b, where o belongs to $\{+,-,x,/\}$, and the monadic operation -a to be performed correctly if the computed and stored result can be expressed exactly as $a(1+E')$ o $b(1+E'')$ and $-a(1+E''')$ respectively, where $|E'|$, $|E''|$ and $|E'''|$ are at most comparable to relative precision.

Characteristic Name	Definition	Explicit Name		
		INTEGER	REAL	DOUBLE PRECISION
ARITHMETIC SET				
Radix	Base of the floating-point number system.	—	SRADIX	DRADIX
Mantissa Length	Number of base-RADIX digits in the mantissa of a stored floating-point number (including, for example, the implicit digit when the first bit of the mantissa of a normalized floating-point number is not stored).	—	SDIGIT	DDIGIT
Relative Precision	The smallest number x such that 1.0-x < 1.0 < 1.0+x where 1.0-x and 1.0+x are the stored values of the computed results.	—	SRELPR	DRELPR
Overflow	The largest integer i such that all integers in the range [-i,i] belong to the system of INTEGER numbers.	IOVFLO	—	—
	The largest number x such that both x and -x belong to the system of REAL (DOUBLE PRECISION) numbers.	—	SOVFLO	DOVFLO
Underflow	The smallest positive real number x such that both x and -x are representable as elements of the system of REAL (DOUBLE PRECISION) numbers.	—	SUNFLO	DUNFLO
Symmetric Range	The largest integer i such that the arithmetic operations * are exactly performed for all integers a, b satisfying $|a|$, $|b|$ ≤ i, provided that the exact mathematical result of a * b does not exceed i in absolute value.	IRANGE	—	—
	The largest real number x such that the arithmetic operations o are correctly performed for all elements a, b of the system of REAL (DOUBLE PRECISION) numbers, provided that a, b and the exact mathematical result of a o b do not have an absolute value outside the range [1/x, x].	—	SRANGE	DRANGE

Characteristic Name	Definition	Explicit Name
INPUT/OUTPUT SET		
Standard input unit	The logical unit number for the standard input unit of the system.	NIN
Standard output unit	The logical unit number for the standard output unit of the system.	NOUT
Standard error message unit	The logical unit number for the standard error message unit of the system.	NERR
Number of characters per input record	The maximum number of characters, including blanks, which can be read from a single record on logical unit NIN.	NCNIN
Number of characters per output record	The maximum number of characters, including blanks and carriage controls, which can be output to a single record on logical unit NOUT.	NCNOUT
MISCELLANEOUS SET		
Number of characters per word	The maximum number of characters that can always be stored in an INTEGER storage unit.	NCHAR
Page Size	The size of storage defined for use within the paging algorithm of a virtual storage system, in INTEGER storage units. On non-paging machines it has the value 1.	NIPAGE

Characteristic Name	Definition	INTEGER	REAL	DOUBLE PRECISION
Number of decimal digits	The largest number of decimal digits allowed and converted by a given compiler when compiling constants.	NIDEC	NSDEC	NDDEC

REFERENCES

1. T. J. Aird, E. L. Battiste, N. E. Bosten, H. L. Darilek and W. C. Gregory, "Name Standardization and Value Specification for Machine Dependent Constants," SIGNUM Newsletter 9, No. 4, 1974, pp. 11-13.

2. W. J. Cody, "The Construction of Numerical Subroutine Libraries," SIAM Rev. 16, 1974, pp. 36-46.

3. W. J. Cody, "Machine Parameters for Numerical Analysis," These Proceedings.

4. R. M. DeMorgan, I. D. Hill and B. A. Wichman, "A supplement to the Algol 60 Revised Report," Comp. J. 19, 1976, pp. 276-288.

5. B. Ford, "Developing a Numerical Algorithms Library," I.M.A. Bulletin, 8, 1972, pp. 332-336.

6. B. Ford and S. J. Hague, "Some Transformations of Numerical Software," In Press.

7. P. A. Fox, A. D. Hall and N. L. Schryer, "Machine Constants for Portable Fortran Libraries," in Computer Science Technical Report No. 37, Bell Laboratories, Murray Hill, N.J., 1975.

8. W. M. Gentleman and S. B. Marovich, "More on Algorithms that Reveal Properties of Floating Point Arithmetic Units," C.A.C.M. 17, 1974, pp. 276-277.

9. M. A. Malcolm, "Algorthms to Reveal Properties of Floating-Point Arithmetic," C.A.C.M. 15, 1972, pp. 949-951.

10. P. Naur, "Machine Dependent Programming in Common Languages," BIT, 7, 1967, pp. 123-131.

11. K. A. Redish and W. Ward, "Environment Enquiries for Numerical Analysis," SIGNUM Newsletter, 6, 1971, pp. 10-15.

PROGRAMMING LANGUAGES AND PORTABILITY

For, although the FORTRAN group hoped to radically change the economics of scientific computing [on the IBM 704] (by making programming much cheaper), it never gave very much thought to the implications of machine-independent source languages applied to a variety of machines.

— J. W. Backus and W. P. Heising (1964)

For a brief period it was thought that programming in Fortran would permit mathematical libraries to move from computer to computer, but this was an illusion, and, in fact, each new [library] effort started essentially from scratch.

— P. A. Fox (1976)

A 1965 Fortran standards document* listed the following reasons behind the existence of language differences which restrict the interchangeability of Fortran between processors:

(1) Expansion of the application area of the language;

(2) Simplification of the language for the user;

(3) Exploitation of a particular computing system in a more effective manner;

(4) Simplification of or speeding up the operation of a processor;

(5) Misunderstanding or disagreement as to what "Fortran" is for lack of definitive standards.

---

*Appendixes to ASA FORTRANs, Comm. of the ACM, Vol. 8, No. 5.

Since they point specifically to Fortran, these reasons suggest that writing numerical programs in a more powerful language, better suited to mathematical constructions, would reduce the need for expansion and simplification; thus programs would be more portable. In this section, L. M. Delves pursues this point by a careful examination of Algol 68 as a language for numerical software, judging the qualities of the language against Fortran and Algol 60 in terms of portability, efficiency, fluency, and availability.

Delves points out that efficiency and fluency (or language "power") often seem in conflict. He has some timing comparisons which suggest that the cost of Algol 68 in terms of efficiency may be quite acceptable in the light of the expressive power of the language. P. Kemp, in his paper, explores how the features of Algol 68 assist in the construction of (trans)portable code for a class of programs for which efficiency is essential, viz., procedures for the elementary functions.

Whatever may be the long-term prospects for Algol 68 (probably sooner to be realized in Europe than in North America) the huge investment world-wide in non-Algol 68 software drives us to search for alternatives to embracing a "new" programming language. The five contra-portability reasons cited above suggest that we can use Fortran (or, for that matter, Algol 60) as a portable programming language if we prescribe very carefully the nature of the computing environment and the acceptable use of the language. The papers by Pieter W. Hemker, W. S. Brown/A. D. Hall, and P. A. Fox have this approach in common although the Mathematical Centre (Hemker) and Bell Telephone Laboratories (Brown, Hall, Fox) are working with

different languages and environments. Hemker describes the experience
of moving a substantial Algol 60 library to a new computer system.
The library has a modular structure and the programs in it were written
with disciplined regard for the language definition. Brown and Hall
discuss a model of a computational Fortran environment characterized
by certain parameters while Fox reviews a mathematical subroutine
library written in a mechanically verifiable subset of ANS Fortran
and designed to run in the characterized environment.

Since Fortran cannot yet be abandoned, it is important that
responsible numerical analysts make the best possible use of the
language. Brian T. Smith's paper provides a set of tested remedies
for the common toxic effects of Fortran on portability. C. L. Lawson
and J. K. Reid offer constructive input to the current Fortran
standardization effort from the perspective of their experience in
writing numerical software.

The final paper in the section is a survey by W. M. Waite of
intermediate languages as an approach to portability. The technical
possibilities Waite suggests are very exciting, but he concludes with
a prognosis for portability techniques that is cautiously pessimistic,
reflecting the psychology and economics of the current situation.

ALGOL 68 AS A LANGUAGE FOR NUMERICAL SOFTWARE

L. M. Delves

ABSTRACT

The requirements of a language for numerical software are those of

(1) Portability: Programs should be as machine independent as possible.

(2) Efficiency: Compiled programs should run as fast as possible, given an "average" compiler.

(3) Fluency: The language should allow the expression of numerical algorithms, including their non-numerical components, as naturally and legibly as possible.

(4) Availability: The language should be available on as many different machines as possible.

These requirements are to some extent mutually conflicting; we discuss here the balance which should be struck between them, and the extent to which they are fulfilled by Algol 68.

1. Introduction

The requirements of a language for writing and for using numerical
software are those of:

(1) Portability: Programs should be as machine-independent
                        as possible.

(2) Efficiency: Compiled programs should run as fast as
                        possible.

(3) Fluency: The language should allow the expression
                        of numerical algorithms, including any
                        non-numerical components, as naturally
                        and legibly as possible; and should
                        allow them to be presented to the user
                        as simply and naturally as possible.

(4) Availability: The language should be available on
                        as many different machines as possible.

These requirements are not in any priority order; there is probably
no agreement between different groups on the relative weights to attach
to each. Different languages meet the various requirements with varying
success, and (2) and (3) in particular often seem mutually conflicting.
We discuss here the balance which should be struck in choosing a programming
language, and the extent to which the requirements are met by Algol 68.
The standard against which we judge it is FORTRAN, although comparison
with other languages (PL/I, Algol 60) is also made in places; and it
is hoped that the discussion can be followed without a detailed prior
knowledge of the language, although no attempt is made to teach it
here. The ordering of the discussion again implies no priorities, but
is purely pragmatic; we start with condition (3) to give some idea of
the capabilities of the language for those new to it.

2.   Fluency

Programming languages differ in fluency (power is an approximate synonym) both in respect of the features which they contain, and of the elegance or versatility of those features. Due to its age (basic design : 1954) FORTRAN suffers in both these respects; table 1 gives a short list of language features which are of undoubted advantage in numerical work, but which are not naturally provided by FORTRAN.

The Algol 60 language, designed only a little later than FORTRAN, included the first three features on this list (indeed, they represent possibly the most successful parts of Algol 60); but it was nonetheless not significantly more powerful than FORTRAN, because it lacked the second three. The dominance of FORTRAN over Algol 60 is certainly at least partly due to users weighting the presence of an additional feature much more heavily than the improvement of an existing one (we can write loops, and allocate storage, in FORTRAN even if inelegantly; one cannot do double length or complex arithmetic in Algol 60). This convenient user indifference to elegance has been exploited to the full by the designers of PL/I, possibly beyond the point at which consumer resistance eventually sets in. The designers of Algol 68, on the other hand, have attempted to provide a language which is as elegant as Algol 60 but much more powerful; at the same time, those features of Algol 60 which led in practice to run time inefficiencies (e.g., the f or loop syntax) have been modified. The result is a language which is recognisably of Algol origin, but is in no sense a direct extension of Algol 60. For those not familiar with the language, tables 2 - 4 attempt to convey an idea of its content; it is clear from these tables that, as it should be, it is considerably more powerful than FORTRAN.

This additional power helps both the user and the writer of numerical software; we illustrate this in tables 5 and 6. Table 5 gives the calling sequences of equivalent routines in the three versions of the NAG library which implement Crout's method for the solution of a set of linear equations; there is no doubt which of these routines is the easiest for the user to handle. Table 6 shows the coding required in FORTRAN and in Algol 68 to carry out two standard matrix manipulations. Here, the "naturalness" of the Algol 68 code is obvious; we must emphasize that this has not been bought at the expense of run time inefficiency or storage inefficiency. The examples given should use no significant intermediate storage in either language; we comment on the relative speed in a later section. The NAG Algol 68 library contains many other instances when the additional features of the language show to advantage; two or more FORTRAN routines collapsed in a natural manner into one, user options specified more briefly and descriptively, and facilities provided that cannot usefully be provided in FORTRAN.

TABLE 1

Language Features which are desirable for numerical work, but
lacking in either FORTRAN or Algol 60.

The relative lack of success of Algol 60 is attributable at least
partly to the omissions noted.

	FEATURE	SAMPLE Reason For Needing It	IN FORTRAN?	IN ALGOL 60?
1.	Dynamic Storage Control	Space saving with large arrays	X	✓
2.	Block structure	Aids storage allocation and program development	X	✓
3.	Flexible control features	WHILE clauses in loops; flexible IF syntax	X	✓
4.	Double Precision	Miserable IBM single precision	✓	X
5.	Complex	Many complex examples	✓	X
6.	Formatted I/O	Layout of results	✓	X
7.	Multiple (> double) precision arithmetic	To cope with ill-conditioned problems (or poor numerical methods!)	X	X
8.	List processing	Algebraic manipulation Automated Algorithm selection	X	X
9.	Extendability of data type of operators	To allow for new applications, e.g., Matrix Arithmetic internal arithmetic variable precision arithmetic	X	X

TABLE 2

ALGOL 68 MODES

Algol 68 has the following pre-defined modes:

Real, compl, int, bool, bit, byte, char, string, struct[+] format, proc[+]

All of these can be scalar, or array-valued, e.g.,

$$[1:n \ , \ m:p] \ \underline{real} \ array \ ;$$
$$[1: \quad 2 + n] \ \underline{proc}$$

Pointer variables, of any type, are available :

ref real x ;   (pointer to a real variable)

ref [,] ref [,] real P   (pointer to an array of pointers
                            to real arrays)

New modes can be declared by the user:

rational
numbers :   Mode rational = struct (int numerator, denominator) ;

rational p,q ;

block matrices: mode blockmatrix ref [,] ref [,] real;

blockmatrix a ;

+strictly speaking, these are not modes, but parts of
mode declarers.

TABLE 3

Algol 68 Operators

Apart from the usual arithmetic operators (+, -, *, etc.) Algol 68 provides operators which allow the testing of various of its language constructs. For example, upb x yields the upper bound of the vector variable x, so that dimensioning information need not be passed explicitly in argument lists. In addition, new operators can be freely defined, by the user, and existing operators extended in meaning

Example : matrix arithmetic

    mode vector = ref [ ] real ;

    op * = (vector u, v) real :

      (real a : = 0 ; for i from lwb u to upb u do

      a + : = u [i] x v [i] od ; a) ;

    op norm = (vector u) real : sqrt(u * u) ;

    usage :   a: = norm u; innerprod : = u * v ;

Useful groups of operators can be collected and made available automatically via a library prelude ; the NAG library contains a full set of matrix operators.

TABLE 4

OTHER AVAILABLE FEATURES IN ALGOL 68

1. Procedures may have values of any mode

    (e.g., vector-valued functions can be defined)

2. List processing facilities are built in

3. Multiple precision [long real, long long real, long long long ...]

4. Input-output well defined: a) simple, unformatted

                                 b) formatted

5. Mode Union for cases where we may not want to decide the mode

    in advance

    Example: a variable precision arithmetic package, using variables

               of mode v real :

               mode v real = union (real, long real, long long real,

               long long long real);

               with associated operator definitions for the usual

               arithmetic operations and for lengthening (or shortening)

               the current precision as required.

TABLE 5

Parameter Lists for Equivalent Routines

The examples shown are the NAG library routines implementing the
Crout method for the solution of the linear equations

$$Ax = b \quad \text{with} \quad M \quad \text{right hand sides}$$

FORTRAN:

FO4AAF (A, NA, B, NB, N, M, X, NX, WKSP, IFAIL)

ALGOL 60:

f04aaa (a, b, n, m, x, fail)

ALGOL 68:

f04aab (a, b, x, fail)

TABLE 6

Common Matrix Operations in FORTRAN and Algol 68

The coding in each case ensures that no intermediate storage is required

1) $\qquad A \leftarrow A + B + C$

Fortran   DO 100 I = 1, N

           DO 100 J = 1, M

100   A(I,J) = A(I,J) + B(I,J) + C(I,J)

<u>or</u>

CALL ADD (A, NA, B, NB, N, M)

CALL ADD (A, NA, C, NC, N, M)

<u>Algol 68</u> : a +:= b + c ;

2) $A \leftarrow B * C$

Fortran   DO 200 I = 1, L

           DO 200 J = 1, M

           TEMP = 0.0

        DO 100 K = 1,N

100   TEMP = TEMP + B (I,K) * C (K,J)

200   A(I,J) = TEMP

<u>or</u>   CALL PROD (B, NB, C, NC, A, NA, L, N, M)

<u>Algol 68</u>   A':='b*c;

2. Portability

Another area in which the advantages of Algol 68 are undeniable
is that of program portability. Differences between different installations
which can hinder portability, are, roughly in order of increasing difficulty:

(1) Card (tape) code differences

(2) Representational differences : begin or 'BEGIN',

'PLUS' or 'PLUSAB' or + : = , etc.

(3) Dialectic differences

(a) Use of super-language features; that is, code which

violates a restriction of the standard language. The

construct A (POINTER (J)) for example, contains a

non-standard suffix, and a name of non-standard

length, but would run happily under many FORTRAN

compilers.

(b) Language variations; that is, use of language

features which have been implemented in a

nonstandard manner in a particular compiler,

or for which the standard is ambiguous and the

interpretation non-universally agreed. Common

examples come from the use óf EQUIVALENCE and

COMMON statements, and from "empty"

DO loops : DO 100 I = 10, 9.

(4) Machine arithmetic differences

These can have effects on the code which range from quite

obvious to quite subtle, for example:

(a) Machine A provides REAL variables of ten decimal

digit accuracy; code is transported to machine B

which provides five.

Solution: All REAL → DOUBLE PRECISION

(but what do we correct DOUBLE PRECISION to?)

(b)   Machine A provides ten digit reals but truncates;

Machine B provides ten digit reals but rounds.

Routine L has tuning parameters carefully set

to ensure that convergence is not affected by

roundoff errors. How do we re-set them for

machine B?

2.1   Representational Differences

We split these areas of possible difficulty into

two kinds:

U   Those for which it is relatively straightforward

to provide a pre-processor to carry out a mechanical

conversion of programs from machine A to machine B.

Non U Those for which it is not.

Type U differences are widely and reasonably

regarded as being no real hindrance to portability.

Although the dividing line is not sharp, it is clear

that areas (1) and (2) above are of type U, since the

differences involved can be removed by a straightforward

edit: the preprocessor needed to move from machine A to

machine B has a structure which is independent of A and B,

and hence can itself be largely portable. We shall call

code U-portable between two machines if no non-U conversions

are required for the interchange.

.2 Dialectic Differences

Variations in language between machines are much more
of a hindrance to portability. Some of them are straightforward
enough that a simple pre-processor can reasonably catch them.
For example, non-standard FORTRAN suffixes can be easily
spotted and code inserted to standardise them, and could
reasonably be held therefore not to vitiate the portability
of a program. Other examples are harder to trap on a routine
basis; and the usual 'cure' for dialectic differences is to
insist, or attempt to insist, that putatively portable
programs be coded in a standard version of the language in
question: for FORTRAN, in ASA (ANSI) Fortran as defined in
reference {1}, or, perhaps in future, reference {2}. Such
a cure presupposes both that all writers have access to a
compiler which contains the standard language as a subset;
and that they can be persuaded to write within this subset.
Most major installations now have FORTRAN compilers which
contain ANSI or at least almost ANSI subsets; many have
ANSI-checkers, or ANSI-compilers which (attempt to) flag
any deviation from the ANSI standard. This state has been
achieved by constant pressure on the manufacturers, and
hard work by individuals and groups; but it has served
to bring forcibly home the fact that the ANSI standards
are very restrictive, being largely the intersection of
all the interested parties' compilers at the time the

standard was set. To stay within the standard, once having been used
to greater freedom, is extremely irksome; to learn the standard, with
its arbitrary restrictions, is itself painful. At my own institution,
the ANSI compiler used on the cafeteria service is bitterly resented
by staff and students alike; while a recent paper from a leading
centre of numerical software development {3} explicitly defends its
decision to not conform to the ANSI standard in the published algorithm.

The restrictive situation in FORTRAN seems inevitable in any
language whose original form is currently much less flexible than the
state of the art allows; individual compiler writers then cannot resist
the temptation, or feel it their duty, to enhance the language for their
own machines; and these enhancements, being uncoordinated, will differ
from those of other groups. Is Algol 68 any better off? In three major
respects, the answer must be: yes.

(i)   The formal definition of the language {4} ensures (hopefully)
        that ambiguities do not exist: compilers implementing the
        language should yield unique meanings for a given program.

(ii)   There was a determined effort during the language design to
        keep the various features of the language orthogonal. The
        omission by compiler writers of a particular feature, such
        as parallel processing, or heap variables, should therefore
        have no effect on other parts of the language: pure sublan-
        guages are relatively easy to define. Experience to date
        indicates that this effort has been successful.

(iii)   Because of its greatly increased flexibility and power over
        FORTRAN, the user is unlikely to feel unduly restricted by

writing in a specified standard form of the language, even if, as is inevitable (and has been explicitly catered for by the designers) compilers appear with super-language features. We return to these points in section 2.4.

2.3   Machine Arithmetic Differences

An increasing effort has been spent in the past several years (see, e.g., {5}) to understand and document those aspects of machine architecture which can lead to non-portable constants or tests appearing in programs; and to educate numerical software writers into coding those tests in terms of standard parameters which can be set once and for all at a particular installation. Thus, at its simplest level, it is common to encounter the FORTRAN statement

PI = 4.0 x ATAN (1.0)

which hopefully sets π to the local machine accuracy, and is more portable than[†]

PI = 3.1415926535

It is obviously preferable to provide π, and other mathematical and machine constants, in a form easily available to the user. Because of the lack of block structures, this is difficult to do elegantly in FORTRAN without PUBLIC variables; the NAG library is resorting to providing FUNCTION segments which return the values of the required constants, which, although it works, is hardly elegant or very convenient.

[†]I once traced a puzzling lack of accuracy in a research student's results to the initialising statement PI = 22/7. When I pointed out the offending line, he came back with better results, and the amended line

PI = 22.0/7.0

Machine constants ("environmental queries") in Algol 68, on the other hand, are defined as part of the language, and are hence available automatically under the standard names. Table 7 lists some of these; they exist in the "Standard Prelude," with values set at compile time rather than run time. Others which may be needed can be added in a "Library Prelude"; the mechanism is both elegant and convenient, and because the names are universal and do not depend on the particular library used, inherently more portable.

TABLE 7

Environmental Queries

Algol 68 provides a number of globally defined constants (normally referred to as "environmental queries") which can be accessed by any programmer, and whose values depend on the particular machine used. We list a few of these below. The language also contains a standard mechanism (the library prelude) by which other similar constants (or other pieces of code) can be added.

Constant	Mode	Value
pi	real	3.14159 .... to real accuracy
maxint	int	The largest value of mode int for this machine
maxreal	real	The largest value of mode real
small real	real	The smallest real value ε such that in the machine arithmetic used, $1 \pm \varepsilon \neq 1$
int width	int	The number of decimal digits in maxint
exp width	int	The number of decimal digits in the exponent of maxreal
real lengths	int	The number of different lengths of real which are implemented (e.g., if real lengths = 3, long long long real does not differ from long long real)
int length	int	The number of different lengths of int which are implemented
long pi	long real	3.14159 to long real accuracy
long maxint, long maxreal, etc.		Similar sets of constants for long (and short) values are defined

2.4   Portability: Experience with Algol 68

We illustrate the comments made by reference to the experience of my own department in

(a)   compiler development

(b)   computing and numerical methods teaching

(c)   numerical software development for the
       NAG Algol 68 library

This experience is not on the same scale as that which exists worldwide in transporting FORTRAN; but it seems to be the best there is.

We have an Algol 68 user population of approximately 250 including about 200 undergraduates taking computing and/or numerical methods amongst their courses, and about 50 postgraduate and staff members. Algol 68 programs are run on two computers:

(1)   Algol 68S on the department CTL Modular One computer

(2)   Algol 68R on the University ICL 1906S mainframe computer.

2.1 The Languages

Algol 68S is a sublanguage of Revised Algol 68 described in ref. {6}, and has been accepted by IFIP working group WG 2.1 as an (currently: the only) officially recognised sublanguage. It includes mode and operator declarations, omitting from the full language mainly constructs of interest in non-numerical fields; and remains much more powerful than FORTRAN or Algol 60. Its direct relevance to the current discussion is twofold:

(a)   sublanguage as defined in {6} treats all concepts
       introduced in the full language in the same way

as the full language, with a single exception: strings, which are composite entities (<u>flex</u> [ ] <u>char</u>) in the full language are primitives of the sublanguage. Further, the number of explicit restrictions imposed on allowable constructs in the sublanguage, is very small; and these points together indicate the success of the "orthogonal" mode of construction of the language.

(b)   They also mean that the user of Algol 68 is not unduly aware that he is using a sublanguage; it feels complete in itself. The NAG Algol 68 library has (to date) been written entirely within Algol 68S, and although there have inevitably been times when recourse to the full language would have been nice, on the whole the contraint imposed has not been irksome.

Algol 68S is used for all introductory student teaching and most program development, until either the programs get too long or too long running for the Modular One computer, <u>or</u> the full language is needed, when use is made of Algol 68R on the University's mainframe. This compiler (ref. {7}) implements essentially the full language, apart from parallel processing, with some dialectic differences which stem mainly from its delivery date: it was the first working Algol 68 compiler, and was completed some three years before the Revised Report {ref. 4}.

2.2 Transportability

We have a daily interest in the portability of Algol 68 programs:

(i) During term time, we run on our Modular One a five day moving average of approximately 660 jobs/day, of which about 580 are compile-and-go Algol 68S; our peak figures so far are: 890 jobs a day, of which 760 were Algol 68S. Most of these jobs are very small student programs; but about one fifth of them are not, and many of these eventually move onto the university mainframe, running in Algol 68R.

(ii) A significant part of the research effort of the department is directed towards developing algorithms for the NAG Algol 68 library, and coordinating the similar efforts of other contributing centres. All routines which are submitted are run both on Modular One and on the 1906S, often as part of an iterative cycle with the contributors.

As noted above, Algol 68R and Algol 68S have some syntactic differences. These are however in general "clean"; that is, a simple rule defines the transformation from one language to the other. As an example (the main one), the limits of a loop are given by writing

        do begin - ; - ; - end;   or do - ;   (Algol 68R)
        do - ; - ; - ; - od   (Algol 68, 68S)

and it is simple to transform mechanically in either direction. As standard practice, almost all programs are written in Algol 68; Conversion from one machine to another is carried out using a pre-processor which makes the dialectic transformation, and also the representational transformation. No code conversion is necessary because we have standardised on 1900 card code for our Modular One; but it would be simple to include this too. In practice about 95% (a rough estimate) of programs which work on Modular One, work first time after preprocessing on the 1906S; and the success rate is high enough that it is almost universal to make any changes in the program on the Algol 68 version, making pre-processing the first pass of a RUN ALGOL 68 macro on the 1906S.

The transportability of Algol 68 is recognised also in the Algol 68 NAG library. The FORTRAN and Algol 60 versions of this library contain first, a standard master image of each routine; and second, interlaced with these, replacement cards for each line which requires alteration for each machine range on which the library has been issued. By contrast, the Algol 68 master library contains pure Algol 68 text; and this is also the Modular One library. The Algol 68R version for ICL 1900 computers is defined by this text plus the preprocessor, and no 'replacement lines' have been required at all. The major differences in machine arithmetic which exist between Modular One and ICL 1900, are completely accounted for by use of the 'environmental queries' facility of the language.

It seems likely that this represents the first completely U-portable library on any pair of dissimilar machines; although we

cannot guarantee (and do not really believe) that non-U differences
will not arise with other implementations, they seem likely to be
much less important than with FORTRAN.

3. Program Efficiency

A great many people are willing to accept the benefits that would
be involved in using a higher level language than they previously had
available; but not the cost in run time efficiency that they fear will
follow. The first such argument arose over the introduction of FORTRAN;
many were reluctant to use it because their programs would then run
slower and take more core. Only the first of these considerations
remains important today; but it is a powerful one. We can make an analogy
with automatic gearboxes in cars; they add convenience, but add also to
the fuel consumption. It appears that users are willing to pay a
10-20% penalty for the convenience, but that over this figure they are
reluctant to purchase[†]. This reluctance was unfortunate for Algol 60,
since typical compilers for this language return program-dependent
runtimes of between 1.5-4 times those of an equivalent FORTRAN program,
even on machines where a genuine effort has been made to achieve
efficiency.

There are, however, reasons to expect that Algol 68 should do
better than Algol 60 in this respect. Algol 60 was designed with
the prime aim of advancing the theory (and practice) of programming
languages, and with little thought for the difficulties of the compiler
writer. With the experience of Algol 60 behind them, the designers

[†]This refers to an initial purchase of a new feature; most FORTRAN
compilers fail to get within 20% of machine code efficiency, but
now that everybody is hooked this does not matter.

of Algol 68 kept runtime efficiency very much in mind, both in what
they took out (relative to Algol 60) and what they put in. Thus:

(1)   for loops and parameter passing mechanisms, both
        necessarily slow in Algol 60, are much simplified
        and hence more efficient in Algol 68.

(2)   Arrays can often be handled much more efficiently
        than in FORTRAN, using the language facility to
        manipulate them as a whole:

                    Array 1 : = Array 2;

        or to slice them: Array 1 [1 : 10] : = Array 2 [1:10,4];

        or to point to them rather than to move them around:

        ref [,] real a pointer : = array 1;

        Similar comments apply to the manipulation of other large
        objects, such as structures.

(3)   The concept of "Building Block FORTRAN" has been discussed
        widely in recent years; the suggestion being that basic
        operations such as matrix multiply or scalar product, should
        always be carried out not by inline code but by a call to
        a standardized subroutine; the subroutine would be defined
        by a FORTRAN version, but would normally exist at any
        installation as an optimized machine code version, so that
        the overall code would run significantly faster as a result.
        The difficulty with proposals of this type lies in getting
        users to make the calls to routines with a number of arguments
        that they do not remember (see table (6)) where a simple little
        DO-loop would do the job. No such difficulties arise in Algol 68

with the equivalent matrix operator package, which would
also normally be machine coded or partly machine coded
at particular installations, although defined by a
standard Algol 68 text.

(4)   A related comment concerns performance in a paged environment.
If matrices are manipulated as a whole, only the matrix
operator package need be aware of and optimised for the
paging; where slices (rows or columns) of a matrix are
processed, it is a happy if minor point that all known
implementations of Algol 68 store matrices by rows
(because straightening (for transput) is defined this way)
and hence the standard linear algebra algorithms, developed
in unpaged environments and usually row-oriented, need no
changes.

Whether these qualitative comments add up to quantitative
efficiency can only be tested experimentally: so far, timings
have been carried out only with the Algol 68R compiler on
ICL 1900 machines. The first, by Wichman {ref. 8} showed
a 20% speed advantage for Algol 68 over Algol 60 on an
ICL 1907. The statement mix timed was chosen as representative of a large number of Algol 60 programs; but cannot
therefore reflect any differences which might arise from
using non-Algol 60 features of Algol 68. Tables 8 and 9
show comparative times obtained on an ICL 1906S for two
potentially time-consuming operations: the solution of sets
of linear equations, and the discrete Fast Fourier Transform of
a set of real data. We see that in both cases the Algol 68

times are very much better than the Algol 60; and that
they are in one case better than, and in the other worse
than, the FORTRAN times. It is dangerous to read too
much into a set of isolated results of this kind, as so
many factors affect individual routines while different
compilers may well have different strengths and weaknesses.
For example, it may (or may not) be significant that the
Algol 68 linear equation routine was coded from scratch
and hence (presumably) takes full advantage of the language
facilities; the FFT routine is based on a machine translation
from the FORTRAN. A much more extensive set of timings for
Algol 68R is currently in progress by Prudom and Hennell
{ref. 10}; so far we can only conclude that things look
very hopeful, and solicit further comparisons on other
machines.

4.   Availability

4.1   What is being done

Even the nicest and most efficient language is little
use if you cannot run it on your computer. Table 10 lists
those implementations known (by me) to be running and in
regular service with a non-null user population. The
list is long enough that most people now can, if they wish,
get access to a compiler which will run on a machine close
enough to them to be useful. However, it is clear that we
are far from the situation in which a typical installation
has three FORTRAN compilers each designed for a specific
purpose. The ready availability of FORTRAN is a powerful

argument in favour of its use for programs intended to be portable; for new languages, a typical chicken-and-egg situation exists as they strive for acceptance.

It would seem that Algol 68 has broken out of the egg, and that its spread at a satisfactory rate (faster than PL/I, for example) is reasonably assured. Signs of this emergence are:

(1) Table 10 is significantly longer than it would have been a year ago.

(2) Unlike PL/I, for which any reasonably complete compiler is something of a block-buster, the sparing introduction of primitive concepts in the language means that compilers can be written at a moderate cost. Best witness to this is the emergence of 68S sublanguage compilers on minis (Modular One, PDP11) giving a very attractive user system for these machines.

(3) By now enough students have been brought up on the language (in Computer Science Departments) to start to propagate it elsewhere. It has already essentially completely replaced Algol 60 at some British univer- sities, and has been adopted by a number of quite large commercial firms as their prime programming language.

(4) Government support can do wonders for a language, as witness COBOL. In the U.K., the real time language CORAL 66 has spread very rapidly since it became the

required language for all Ministry of Defence
real time contracts; and a number of new
CORAL 66 compilers have appeared as a result. The
Ministry now looks likely to nominate Algol 68 as
its prime high level language; and is giving valuable
support to the NAG 68 library.

4.2   What needs to be done

It will of course take more than the existence of
compilers to make significant inroads on the FORTRAN
community. The existence of the NAG library in Algol 68
should help ensure that future new users in the numerical
area will be as well supported in Algol 68 as in FORTRAN.
Existing users, with their own or their companies' large
and valuable collection of private utilities and other
aids (all in FORTRAN) will be much more reluctant to
change. For these, we should provide two facilities:

(1)   Automatic FORTRAN-ALGOL 68 Translation

As a conversion aid. One such translator is in use
at Liverpool {ref. 11}.

(2)   Mixed Language Programming

Is practicable on recent machines. Despite the
possible difficulties (different parameter-passing
mechanism, different ordering of arrays, etc.)
Algol 68 compiler writers would give the biggest
boost to the language if they could let the user
call FORTRAN routines from his Algol 68 program.

5. Summary

Algol 68 is a powerful and efficient language for numerical software; programs written in it are inherently portable and the language seems certain to spread widely and to rapidly replace Algol 60. In the long term, something will replace FORTRAN; the something may be Algol 68, and is unlikely to be PL/I.

Acknowledgements

I am grateful to all of my colleagues, and especially to C. Charlton, M. A. Hennell, P. Leng, J. M. Watt, and D. Yates, for their efforts in developing Algol 68 and the NAG library over the past several years; their views and experience have contributed much to the development of this paper. Thanks are also due to the U.K. Ministry of Defence for financial support for the Algol 68 NAG library and for this paper.

TABLE 8

Timings for the FORTRAN and Algol 68 version of the NAG
routines F04AA (Crout reduction) for the solution of a set of N
linear equations with 1 right hand side. For this routine, NAG
timings indicate that the Algol 60 times are approximately 1.4 x
FORTRAN. In this table and in table 9, times are in milliseconds
on an ICL 1906S computer, and the compilers are XFAT and Algol 68R
running in "production" mode: TRACE 0 (Fortran); INDEX AND OVERFLOW
not set (Algol 68).

N	4	8	16	32	64	Average ratio: FORTRAN
FORTRAN	6	20	82	408	2369	1
ALGOL 68	4	14	62	330	2085	0.8-0.9

TABLE 9

Timings for the FORTRAN and Algol 68 versions of the NAG routines C06AA (Fast Fourier Transform of N real data points, using Singleton's algorithm {ref. 9}). NAG timings indicate that the Algol 60 times are approximately 3.7 x FORTRAN.

N	16	64	256	1024	2048	4096	Average ratio: FORTRAN
FORTRAN	2	7	33	150	310	693	1
Algol 68	2	9	44	202	416	932	1.34

TABLE 10

Implementation of Algol 68 currently (June 1976) issued and in regular use.

Machine Range	Language	Relation to Revised Algol 68	Implemented at	Comments or restrictions
ICL 1900	68R	dialect of full language	R.R.E. Malvern	no parallel processing
CDC 7600	68	full language	Amsterdam	— " —
CTL Modular One	68S	sublanguage	Liverpool	currently no operators
PDP 11	68S	sublanguage	Carnegie Mellon	—
PDP 10	68C	dialect		Portable compiler. Number of language restrictions still current
IBM 360,370	68C	dialect	Cambridge	— " —
ICL 4130	68C	dialect	York	— " —
360/30	68/19	sublanguage	Brussels	No mode or operator declarations All routines voided
Univac 1100	68	full language	Paris - Sud	9-pass compiler
Tesla 200	68	full language	Prague	
CII 10070	68	full language	Rennes	Conversational Algol 68
IBM 370	68S	sublanguage	Durham	No operators

REFERENCES

1. Heising, W. P., Report of A.S.A. Committee X3, C.A.C.M. 7, P. 10; ibid, pp. 591-625; C.A.C.M. 8, pp. 287-8, 1965.

2. Revised Standard Fortran: Report of A.S.A. Committee X3, to appear eventually.

3. Marlow, S., and Powell, M. J. D., A FORTRAN Subroutine for Plotting the Part of a Conic That is Inside a Given Triangle, UKAEA Report AERE-R8336: HMSO, 1976.

4. van Wijngaarden, A., Mailloux, B. J., Peck, J. E. L, Koster, C. H. A., Sintzoff, M., Lindsey, C. H., Meesters, L. G. L. T., and Fisher, R. G., Revised Report on the Algorithmic Language Algol 68, Supplement to Algol Bulletin 36, University of Alberta, 1974.

5. Ford, B., Machine Characteristics and Their Parametrization in Numerical Software, These Proceedings.

6. Hibbard, P. G., Informal Description of an Algol 68 Sublanguage, Presented at the WG 2.1 Subcommittee on Algol 68 support, Cambridge 1974; to be published.

7. See, e.g., P. M. Woodward and S. G. Bond, Algol 68-R Users guide, H.M.S.O. London.

8. Wichman, B. A., Basic Statement Times for Algol 60, unpublished report 1972; see also B. A. Wichman, Five Algol Compilers, Computer Journal, 1972, pp. 8-12.

9. Singleton, R. C., Algorithm 338, Algol Procedures for the Fast Fourier Transform, C.A.C.M. 11, 1968, pp. 773-776.

10. Prudom, A. and Hennell, M. A., to be published.

11. Prudom, A., Ph.D. thesis, University of Liverpool, 1976.

WRITING THE ELEMENTARY FUNCTION PROCEDURES FOR THE ALGOL68C COMPILER

P. Kemp

ABSTRACT

ALGOL68C is an extended subset of Algol 68 as defined in the Revised Report {2}. The compiler written in Cambridge for ALGOL68C is portable and thus the elementary function procedures have been written in Algol 68. This paper describes how this has been done as a collection of transportable procedures and discusses how the features of the language assist in the writing of transportable numerical programs.

1. <u>Introduction</u>

Algol 68 is regarded by Computer Scientists as a "good" language. In particular some of its design aims were:

(i)   Easy to learn and implement.

The number of independent concepts has been minimised, and these concepts are applied orthogonally. As a result the language is powerful, yet there is neither too much to learn nor are there too many arbitrary rules for avoiding difficulties (of Algol 60 and Fortran).

(ii)   Compile time checking.

All mode (type) checking can be performed at compile time. Moreover most syntactic errors can be detected before they lead to calamitous results, so that the compiler can give good diagnostics.

(iii)   Efficiency.

Because so much checking can be performed at compile time, it is possible to produce efficient compiled code. Jenkins at the Royal Radar Establishment has shown that Algol 68 programs can run as rapidly as Fortran programs over a large class of problems.

(iv)   Independent compilation.

The language is designed so that separate segments of a program can be compiled independently without loss of efficiency or security. However the provision for calling user-written library routines is primitive.

The result is a language in which it is easy to write and debug powerful, efficient and, just as important, clear programs.

In Great Britain, three compilers for dialects of the language are widely available.

ALGOL 68-R {4} was written by the Royal Radar Establishment at Malvern and, being an early compiler (1970), is based on the original Report {1} rather than the Revised Report {2}. Thus it differs in a number of respects from what is currently thought of as Algol 68. It implements a subset of the language since one of its design aims was that it should be a one-pass compiler. It runs on ICL 1900 machines and has gained wide acceptance in the universities and elsewhere as a scientific programming language.

Algol 68S {5} was written by the University of Liverpool and is a subset of the language defined in the Revised Report specifically designed to be run on small machines for teaching purposes. It was originally written for the CTL Modular One but now runs on a variety of computers.

ALGOL68C {3} is being written at the University of Cambridge, partly as an exercise in producing a portable compiler. It implements most of the language of the Revised Report. The compiled code is an intermediate language known as Z-code. To implement the language on a new machine involves little more than writing a Z-code translator and Z-code is designed so that it can be implemented easily on a variety of machines without sacrificing efficiency on any particular one. The compiler is written in ALGOL68C and is currently available on IBM 360/370, PDP-11 and ICL 4130 machines.

In the following, "Algol 68" refers to the language of the Revised Report whereas "ALGOL68C" refers to the language implemented by the Cambridge compiler.

Procedures have been written for the elementary functions specified by Algol 68, that is sqrt, exp, ln, sin, cos, tan, arcsin, arccos and arctan. Since the ALGOL68C compiler may be implemented by people with little numerical analysis experience, the procedures should ideally be machine independent as well as satisfying the more usual requirement of accuracy, robustness and speed. This implies that the procedures will be written in Algol 68. As will be seen, Algol 68 allows the aim of machine independence to be virtually met, but at a cost in terms of efficiency on any particular machine which is too high. Hence the requirement will be reduced to one of transportability; i.e., the pro-cedures will be transformed mechanically to produce a version for a target machine.

The most difficult areas in writing portable special function software are the argument reduction, the choice of expansion function and testing for portability. The rest of this paper will discuss how the features of Algol 68 assist portability in these areas, and what difficulties are introduced. Little will be said about the numerical properties of the procedures, save to point out that since the prime implementation is on the IBM 360/370 series, the procedures embody code to avoid some of the problems of hexadecimal arithmetic. While this is not necessary on binary machines, it affects the speed very little and the accuracy not at all.

2.   Argument Reduction

In order to prevent cancellation, part of the code for reducing the argument to a standard range is usually carried out using extra precision. It is difficult to do this transportably in Fortran, largely because the basic precision used for numerical computation on some machines (e.g., IBM) is DOUBLE PRECISION and hence extra precision means deviating from the standard. Algol 68 implementors have at least avoided this problem by deciding that mode REAL should be implemented on IBM machines as 64 bit numbers. In theory Algol 68 allows extra precision to be used easily since it provides the mode long real, but this facility is not implemented in ALGOL68C. However the requirements of argument reduction can be met because Algol 68 provides for the definition of new modes and the re-definition of operators. Figure 1 shows how this can be done. The new mode LONGREAL consists of a structure with three fields. The number represented is top + lower and top is held to rather less than full accuracy so as to allow for exact multiplication by an integer whose absolute value is less than maxn. New definitions are given for * to multiply a LONGREAL by an INT and for - to subtract a LONGREAL from a REAL. The latter operation assumes that zeros are shifted into the lower end of a number when normalisation takes place after cancellation has occurred. The tangent procedure, which requires a reduction to the range $(0,\pi/4)$, can then begin:

```
        PROC  tan = (REAL x) REAL:
        BEGIN LONGREAL z = longpiby4 * ENTIER (ABS x/piby4);
              IF maxn OF z < 0 THEN print (newline,
                 "****WARNING:  loss of precision in tan") FI;
              REAL w = (x - z)/piby4;
```

.

.

.

```
MODE LONGREAL= STRUCT(REAL TOP,LOWER, INT MAXN):
  # MAXN*TOP MUST BE EXACTLY REPRESENTABLE AS A REALQUANTITY.
    IF MAXN<0 AN OVERFLOW HAS OCCURRED#


LONGREAL LONGPIBY4 = (REAL("40C90FDAA2216800"),REAL("34C234C4C6628B81"),256):
#REAL IS A PROCEDURE TAKING A STRING ARGUMENT AND RETURNING A REAL RESULT
 WHICH IS THE REAL VALUE THAT THE STRING (CONSIDERED AS HEX DIGITS) REPRESENTS

OP * = (LONGREAL Z, INT N)LONGREAL:   IF N=0 THEN (0.0,256) ELSE
         (N*TOP OF Z,N*LOWER OF Z,IF ABS N<MAXN OF Z THEN MAXN OF Z%ABS N
                                                      ELSE -1 FI)    FI:

OP  - = (REAL A, LONGREAL B)REAL:
 #THE ASSUMPTION IS THAT A AND B ARE COMPARABLE IN MAGNITUDE
  AND THAT ZEROS ARE SHIFTED IN AT THE LEAST SIGNIFICANT END OF A NUMBER
  WHEN NORMALISING AFTER CANCELLATION HAS OCCURRED#
      (A - TOP OF B) - LOWER OF B:
```

Figure 1

This piece of code introduces another useful feature of Algol 68.
Global variables can be defined in a piece of code called the prelude,
which effectively sets up the environment in which the program is to be
run. Thus environment enquiries can be made. Pi, maxreal and smallreal
(the smallest real value such that 1-smallreal < 1 < 1+smallreal) are
built into the language, and other constants required (such as piby4
and longpiby4) can be built into the prelude and be accessible to
programs. On re-implementation, changes to the values of constants
are restricted to the prelude code.

It would undoubtedly be more efficient to program the argument
reduction in machine code. ALGOL68C contains an extension of Algol 68
which allows Z-code to be embedded into programs. Unfortunately Z-code
contains no long precision capabilities at present so that this is not
a viable solution. At a later stage this might provide a way of improving
the speed of the argument reduction without reducing transportability
since Z-code is as machine independent as ALGOL68C is.

3.   Choice of Expansion Functions

The first approach tried was that of early marks of the NAG
library (see Schonfelder {6}). This used truncated Chebyshev series,
the point of truncation being determined by the precision of the processor
(which can be given as a constant in the prelude). This precision could
even be determined as a function of smallreal, thus making this approach
machine-independent in Algol 68. The code looks like:

```
INT serieslength = (sigbits|n_1,n_2,...);
#a CASE statement setting serieslength to n_sigbits#

[] REAL coeffs = (a_0,a_1,...a_N);
#N is large enough to cope with all reasonable machines#
                        .
                        .
fn:=(REAL b:=c:=0;   REAL twox = 2*x;
     FOR i FROM serieslength BY -1 TO 2 DO
         REAL d=twox * b - c + coeff [i]; c:=b;b:=d   OD;
         (twox * b + coeff[1])/2-c );
                        .
                        .
```

There are two defects in this approach. It takes twice as long
as other expansions such as economised polynomials or rational approximations,
which is important in the case of functions as basic as those being considered
here. Moreover this method of summing Chebyshev series can lead to loss
of precision on hexadecimal machines because of the multiplication and
division by 2.

An ideal solution, which would be clear and easily transportable,
would be to code the series expansion section of the procedures according
to the following pattern (precision is a variable whose value has been
set previously):

```
    IF precision < precision_1 THEN series_1
  ELIF precision < precision_2 THEN series_2
                        .
                        .
  ELIF precision < precision_k THEN series_k
                              ELSE series_{k+1}

    FI;
```

Since the variable "precision" would be defined by an identity declaration (which does not allow its value to be changed later) rather than an identifier declaration (which does), in principle the compiler can tell which condition will be satisfied and hence compile code for one series only. However all known compilers will compile code to perform all the tests and evaluate all the series, thus producing a large amount of inefficient code. The same is likely to be true of any future compiler.

Hence an approach similar to that adopted in the latest NAG routines was adopted. The maximum precision for which a particular expansion is applicable is embedded in a comment. A simple Algol 68 program can then perform the software transformation which tailors the procedure for a particular machine. Appendix 1 shows the full tangent procedure, which is unsuitable for running on any machine, since a suitable expansion series has not been extracted. However it does show how the control information is built into comments. The marker indicating the start of the control information is the string "###". Appendix 2 shows an ALGOL68C program which composes a suitable procedure, given the precision of the target machine.

Most of the expansions used are rational approximations taken from Hart {7}, modified where necessary to preserve accuracy on hexadecimal machines. In some cases the expansions given by Hart are so unsuitable for hexadecimal machines that the continued fraction approximations from the IBM Fortran Library {8} are used instead.

4. Testing for Portability

According to Ford and Bentley {9}, the main areas in which changes are needed when Fortran programs are transported to a new environment are

precision conversion and compiler idiosyncracies. Both these are absent
with ALGOL68C programs. However, since implementors of the language may
well not be numerical analysts, it is necessary for the procedure writer
to gain some confidence that his program will work in other environments.
This implies simulation of different arithmetics on one machine, preferably
in such a way that the program under test does not have to be changed itself.

Algol 68 appears to possess the properties required for this since
new modes and operators can be defined. Naively, therefore, it would
seem that the problem can be solved by re-defining the mode REAL (possibly
as a structure with fields of mode BITS) and the basic operators +,-,*,/,
ABS, etc. If this were done, the procedures exactly as coded for the IBM
370 could be run as if on any target machine. Unfortunately Algol 68
contains a restriction in that reserved modes (such as REAL) cannot be
re-defined. There are good reasons for this. Algol 68 defines a complex
set of rules for forcing mode changes when the syntax demands it (these
rules are called coercions) and these would be upset if any reserved
modes were re-defined. Allowing the programmer to re-define coercions
would lead to such potential pitfalls for the programmer that the
originators of the language did not introduce the facility; hence the
restriction on re-defining reserved modes.

It is possible to go some way in the right direction and two ways
will be suggested. A new mode REEL can be defined, together with the
operators +,-,*,/, etc. A simple case for a hypothetical machine is
shown in figure 2. This sort of approach towards providing different
arithmetic has been developed by Schonfelder with his MLARITHA package,
although this is in the context of providing multiple precision arithmetic. The program under test needs changing in the following respects:

(i)   All occurrences of "REAL" must be replaced by "REEL"

(ii)   Real constants must be replaced by a construct of the form

$$\text{reel}("<constant>)$$

(iii)   A procedure reel taking an argument of mode STRING and returning a result of mode REEL must be written.

A second and less clumsy approach which is nearly as satisfactory is shown in figure 3. It takes the pragmatic view that since the IBM 370 arithmetic is more accurate than any machine on which ALGOL68C is likely to be implemented in the foreseeable future, arithmetic is performed normally on that machine. However new operators which round or truncate the results of each operation to the precision of the target machine are provided and given different names. The only changes needed in the text of the program under test are the replacement of all occurrences of +,-,*,/ by +>,->,*>,/> respectively. Of course this approach is not 100 percent foolproof numerically. However, because the procedures are written in a high level language and are therefore unlikely to match the precision of carefully coded assembler routines in which care can be taken of the quirks of a given processor, it is certainly adequate in practice.

Two remarks should be made on the coding of the simulation operators. The first is that they need not themselves be transportable since they are only intended to be run on one machine, namely that of the procedure contributor. The second is that the simulated arithmetic would be more efficient if the operator definitions were coded in Z-code, and better

```
MODE REEL = STRUCT(INT EXPONENT.MANTISSA);

OP + = (REEL A.B) REEL:
BEGIN BOOL XEQUALSA:
      REEL X:=IF (XEQUALSA:=EXPONENT OF A > EXPONENT OF B) THEN A ELSE B FI;
      INT M = MANTISSA OF (IF XEQUALSA THEN B ELSE A FI);

# ADJUST EXPONENTS#

      WHILE EXPONENT OF X > EXPONENT OF A OR EXPONENT OF X >
                                              EXPONENT OF B
      DO EXPONENT OF X -:= 1;
         IF ABS MANTISSA OF X > MAXINT % 2
            THEN MANTISSA OF X := SIGN MANTISSA OF X * MAXINT
            ELSE MANTISSA OF X *:= 2
         FI OD;

#ADDITION#

      IF M>0 AND MANTISSA OF X > MAXINT-M
      OR M<0 AND MANTISSA OF X <-MAXINT-M
         THEN X:=(EXPONENT OF X + 1. MANTISSA OF X % 2 + M % 2)
         ELSE MANTISSA OF X +:= M
      FI;

#NORMALISATION#

      IF MANTISSA OF X = 0 THEN (0.0)
      ELSE WHILE MANTISSA OF X REM 2 = 0
              DO EXPONENT OF X +:= 1;
                         MANTISSA OF X %:= 2 OD;

      X
      FI

END;

OP - = (REEL A.B)REEL: A + REEL((EXPONENT OF B.-MANTISSA OF B));

BEGIN REEL X:=(0.1).Y:=(0.1);
      FOR I UPTO 10 DO
      EXPONENT OF Y -:= 1;
      X := X+Y;
      PRINT(NEWLINE.EXPONENT OF X." ".MANTISSA OF X) OD
END
```

Figure 2

```
INT MACHINEBASE = 16;


PROC SYSDECOMPOSEREAL = (REAL Y, REF INT EXPONENT,
                                REF REAL MANTISSA )VOID:
BEGIN
# Y = MANTISSA * MACHINEBASE ** EXPONENT;
  1>I MANTISSA I >= 1/ MACHINEBASE #
        EXPONENT := 0; MANTISSA := ABS Y;
        IF Y ^= 0 THEN
        WHILE MANTISSA >= 1 DO MANTISSA /:= MACHINEBASE:
                        EXPONENT +:= 1 OD;
        WHILE MANTISSA < 1/MACHINEBASE DO MANTISSA *:= MACHINEBASE ;
                        EXPONENT -:= 1 OD;
        IF Y < 0 THEN MANTISSA := -MANTISSA
        FI FI
ENDS;

PROC TRUNCATE = (REAL X)REAL:
BEGIN   INT SIGBITS=39;

    # THIS IS THE PRECISION OF THE SIMULATED ARITHMETIC #

        INT EXP; REAL MANT;
        SYSDECOMPOSEREAL(X,EXP,MANT);
        INT LEADINGZEROS =     IF MANT >= 0.5   THEN 0
                           ELIF MANT >= 0.25  THEN 1
                           ELIF MANT >= 0.125 THEN 2
                           ELSE                    3 FI:
        REAL ABSMFAC = ABS MANT*2.0**(SIGBITS+LEADINGZEROS);
        SIGN MANT*(2.0**(4*EXP-SIGBITS-LEADINGZEROS))*
        (INTPT ABSMFAC + IF FRPT ABSMFAC >= 0.5 THEN 1 ELSE 0 FI)
ENDS;

OP +> = (REAL A,B)REAL:TRUNCATE(A+B);
OP -> = (REAL A,B)REAL:TRUNCATE(A-B);
OP *> = (REAL A,B)REAL:TRUNCATE(A*B);
OP /> = (REAL A,B)REAL:TRUNCATE(A/B);
OP +> = (REAL A,  INT B)REAL:TRUNCATE(A+B);
OP -> = (REAL A,  INT B)REAL:TRUNCATE(A-B);
OP *> = (REAL A,  INT B)REAL:TRUNCATE(A*B);
OP /> = (REAL A,  INT B)REAL:TRUNCATE(A/B);
OP +> = (INT  A,REAL B)REAL:TRUNCATE(A+B);
OP -> = (INT  A,REAL B)REAL:TRUNCATE(A-B);
OP *> = (INT  A,REAL B)REAL:TRUNCATE(A*B);
OP /> = (INT  A,REAL B)REAL:TRUNCATE(A/B);
OP -> = (REAL X)REAL: -X;
OP +> = (INT A,B)INT:A+B;
OP -> = (INT A,B)INT:A-B;
OP *> = (INT A,B)INT:A*B;
OP -> = (INT X)INT:-X;
PRIO +>=6, ->=6, *>=7, />=7;
```

Figure 3

still if Z-code had a facility for allowing assembler code to be inserted.
Even so, this does not really matter since the simulation process is only
used when testing for portability and not in a production environment.

5.   Conclusion

There is no doubt that Algol 68 is a good language in which to write
programs, both numerical and non-numerical. It is now clear that compiled
Algol 68 code can approach the efficiency of compiled Fortran code and
that Algol 68 is becoming widely accepted by the scientific community
in Britain. My experience of Algol 68 vis à vis Fortran is that I can
get a Fortran program to compile first time but that debugging it is time
consuming; with Algol 68 on the other hand the problem lies in making it
compile, after which it will usually work very soon, thanks to the
expressive power of the language. Delves [10] has expanded further on
the facilities of the language. The programs in the appendix also
illustrate the power and clarity of Algol 68.

The question of whether the language features materially assist
the writing of portable numerical programs is less clear cut. The
provision for environment enquiries is very useful as it allows many
machine dependencies to be placed in one block. The major difficulties
affecting portability in Fortran, namely the need for precision conversions
and problems with compiler restrictions are avoided in ALGOL68C, the first
by a sensible choice of implementation precision on IBM machines and the
second by providing a portable compiler.

On the other hand there are a number of instances where the language
designers and implementers have gone only part way towards providing a
sophisticated facility. Examples we have seen are the lack of double

precision in the common implementations and the inability to re-define reserved modes. The former means that those parts of the program which require extra precision must obtain it by simulation - an expensive overhead when hardware to do the job is available. The latter has little effect on the average programmer, but means that it is more difficult than it might be otherwise to test for portability (although it is still much easier than in other languages).

REFERENCES

1. van Wijngaarden, A., (Editor), Mailloux, B. J., Peck, J. E. L., and Koster, C. H. A. Report on the Algorithmic Language ALGOL 68. Numerische Mathematik, Vol. 14 (1969) pp. 79-218.

2. van Wijngaarden, A., Mailloux, B. J., Peck, J. E. L., Koster, C. H. A., Sintzoff, M., Lindsey, C. H., Meertens, L. G. L. T., and Fisker, R. G. Revised Report on the Algorithmic Language Algol 68, Springer-Verlag 1976.

3. Bourne, S. R., Birrell, A. D., and Walker, I. ALGOL68C Reference Manual. University of Cambridge Computer Laboratory 1975.

4. Woodward, P. M., and Bond, S. G., ALGOL 68-R Users Guide. HMSO 1974.

5. Hibbard, P. G., A Minimum General Purpose Sublanguage of ALGOL 68. Algol Bulletin No. 35 (1973).

6. Schonfelder, J. L., The Production of Special Function Routines for a Multi-Machine Library. Software-Practice and Experience, Vol. 6 (1976) pp. 71-82.

7. Hart, J. F., et al. Computer Approximations. Wiley 1968.

8. IBM System/360 FORTRAN IV Library: Mathematical and Service Subprograms. IBM Form No. C28-6818.

9. Ford, B., and Bentley, J. On the Enhancement of Portability in the NAG Project: A Statistical Summary. These Proceedings.

10. Delves, L. M., Algol 68 as a Language for Numerical Software. These Proceedings.

Appendix 1. The Tangent Procedure

```
PROC TAN = (REAL X) REAL:
BEGIN   INT N = ENTIER(ABS X/PIBY4);
        LONGREAL Z = LONGPIBY4*N;
        IF MAXN OF Z < 0 THEN PRINT(NEWLINE.
                         "***WARNING: LOSS OF PRECISION IN TAN") FI;
        REAL W = (X-Z)/PIBY4;
        REAL W2=W*W;
        REAL P,Q;
#FOR A PARTICULAR IMPLEMENTATION, ONE OF THE FOLLOWING EXPANSIONS IS
 SELECTED.  THEY ARE ADAPTED FROM HART'S BOOK  #

### 7.85 > PRECISION. EXPANSION NO. 4282
   P:=212.4244575-12.55329742*W2;
   Q:=(W2-71.59606050)*W2+270.4672235;
## 10.66 > PRECISION. EXPANSION NO. 4283
   P:=(0.0528644455522*W2-8.87662377021)*W2+129.221035031;
   Q:=(W2-45.1320561006)*W2+164.529331810;
## 13.62 > PRECISION. EXPANSION NO. 4284
   P:=(-20.5535488072214*W2+2569.55342344114)*W2-34328.7404645351;
   Q:=((W2-336.930978141323)*W2+12258.9002686366)*W2-43708.7098803972;
## 19.74 > PRECISION. EXPANSION NO. 4285/5
   P:=(((0.67732772853543441922E-05*W2+0.68451087744820068706E-02)*W2
        -3.10137130696653275381*W2+211.19418034299063872)*W2
        -2613.6405295096513366;
   Q:=((0.2*W2-31.100663280634199338)*W2+953.15027258329673978)*W2
        -3327.7904778942380038;
#


CASE N MOD 4 + 1 IN
#N=0 MOD 4# (SIGN X * W * P/Q),
#N=1 MOD 4# (REAL COTAN = (Q-W*P)/(Q+W*P);

# THERE IS AN ASSUMPTION IN THE FOLLOWING CODE THAT
           1/SMALLREAL << MAXREAL  #

           IF COTAN < SMALLREAL*PI/2 THEN PRINT(NEWLINE."******ERROR:".
                                         " SINGULARITY IN TAN");
                                         SIGN X * MAXREAL * SIGN COTAN
                               ELSE SIGN X/COTAN FI),
#N=2 MOD 4# (REAL COTAN=-W*P/Q;
           IF COTAN >-SMALLREAL*PI/2 THEN PRINT(NEWLINE."******ERROR:".
                                         " SINGULARITY IN TAN");
                                         SIGN X * MAXREAL * SIGN COTAN
                               ELSE SIGN X/COTAN FI),
#N=3 MOD 4# (SIGN X*(W*P-Q)/(W*P+Q))
        ESAC
END;
```

Appendix 2.   The Expansion Selection Program

```
PROC ENDDATA = (REF FILE F)BOOL:
    (GOTO END;SKIP);

ONFILEEND(STANDIN.ENDDATA);


PROC READCHARNL = CHAR:

  BEGIN  CHAR CH;
          WHILE NOT SYSREADCHAR(B OF STANDIN.CH)
                ANDF ENSURELINE(STANDIN)
                DO PRINT(NEWLINE) OD;
          CH
END;

REAL PRECISION=16.8;
# THIS IS THE PRECISION OF THE REQUIRED CODE#

INT NOSHARPS := 0;   BOOL INCOMMENT := FALSE;

WHILE ^(NOSHARPS >= 3 AND INCOMMENT) DO CHAR CH=READCHARNL;
                                        PRINTCHAR(CH);
                                        IF CH = "#" THEN NOSHARPS+:=1;
                                                    INCOMMENT:=^INCOMMEN
                                        ELSE NOSHARPS := 0
                                        FI

                          OD;

#NOW IN THE EXPANSION SERIES SECTION#

REAL MAXPREC;

WHILE PRECISION > (MAXPREC:=READREAL) DO
            WHILE READCHAR /= "#" DO SKIP OD;
            IF READCHAR /= "#" THEN
                PRINT(NEWLINE."####ERROR - PRECISION TOO GREAT ####");
                                GOTO END  FI
            OD;

#NOW HAVE THE RIGHT SERIES#

PRINT(" MAXPREC= ".ENTIER MAXPREC."." .ROUND(100#FRPT MAXPREC).
                    "  #".NEWLINE);
NEWLINE(STANDIN);
NOSHARPS:=0;

WHILE NOSHARPS = 0 DO CHAR CH:=READCHARNL;
                    IF CH = "#" THEN NOSHARPS+:=1
                                ELSE PRINT(CH) FI
                OD;

#SKIP OTHER EXPANSIONS#

WHILE READCHAR = "#" DO
                WHILE READCHAR /= "#" DO SKIP OD
        OD;

#AND PROCEED TO THE END#

DO PRINT(READCHARNL) OD;
```

CRITERIA FOR TRANSPORTABLE ALGOL LIBRARIES

Pieter W. Hemker

ABSTRACT

A rather comprehensive numerical software library
(NUMAL {3}) was transported from a Philips EL-X8 computer
to a CDC CYBER system. The experiences justify the
following conclusion:

If (1) we use a well-defined language (e.g., ALGOL 60
or ALGOL 68), if (2) we construct well-programmed software
in that language, if (3) we have a good compiler and if
(4) the computer/compiler has well-designed arithmetic
properties, then the transportability problem scarcely
exists. This statement can also be put the other way.
The requirement of orthogonality of the conditions
(1) - (4) determines what can be considered as a
decent programming language, a good compiler, good
programming and well-behaved arithmetic.

For instance, good programming should not make
use (perhaps at the cost of some efficiency) of
idiosyncratic features of a language dialect, of a
particular compiler or of a particular kind of machine
arithmetic. From this abstract point of view, a
number of useful properties of a well-structured
portable software library are mentioned.

INTRODUCTION

In 1973, the Mathematical Centre in Amsterdam had to transport its numerical software library from a Philips EL-X8 computer to a CDC CYBER system. The library was written in ALGOL 60 and at that time it consisted of about 250 procedures. During the construction of the library, which started in the early sixties, a tradition in the use of ALGOL 60 was developed. This means that only correct ALGOL 60 in the sense of the Revised Report {2} was used and that features that were not clearly defined in the report were avoided as much as possible.

On the EL-X8 a reliable, efficient and rather complete ALGOL 60 compiler was available and the library was partly incorporated into the ALGOL 60-oriented running system.

When, in 1973, the library had to be adapted to the CDC CYBER system, the change-over took only a few months. In fact, it appeared that no essential changes in the code had to be made to adapt it for the CDC ALGOL 60 version 3 compiler and most of the transport work could be done automatically. Only in a few exceptional cases, some strange properties of the CDC machine arithmetic caused a procedure to fail in its new environment.

Since 1973 the numerical library NUMAL has been extended considerably and now it consists of about 450 specialist-oriented as well as general purpose routines in the field of numerical mathematics.

Restrictions of ALGOL 60

The easy transport from one machine to the other was mainly due to the strict use of ALGOL 60 and to the machine-independent way of programming. The latter means, e.g., that the relative accuracy of a floating-point number has to be mentioned explicitly in an input parameter in the calling sequence of a procedure. This kind of machine independence, however, was not possible for all procedures: in those programs where representations of floating-point constants are necessary, they are given to only about 15 decimal places.

The use of ALGOL 60 according to the Revised Report has some apparent disadvantages:

(1)   one cannot use input/output statements;

(2)   communication with mass storage is impossible;

(3)   double precision arithmetic is not available as
       a language feature.

The first two points force us to keep the library completely I/O free and to exclude all procedures that require mass storage. On the other hand double-precision arithmetic procedures have been introduced. Because of the favourable arithmetical properties of the EL-X8 computer, it was possible to write the elementary double-precision operations +, -, *, / in ALGOL 60 {1} for the EL-X8; for the CDC CYBER computer, however, these elementary procedures had to be written in machine code. These machine-dependent double-precision procedures were implemented because double precision was considered to be indispensable in several applications and the procedures only cause machine-dependence in a clearly distinguishable part of the library.

Structure

Besides the "machine independent" use of ALGOL 60, the library NUMAL has two other characteristics: it is an integrated library and it has a modular structure. By *integrated library* we mean that it is not merely a collection of tested and documented routines, but a coherent structure in which the different parts gear into each other. The main lines of interconnection between the various parts are given in Figure 1.

By *modular structure* we denote that also on a much smaller scale programs have been divided into pieces, which can be used separately in different places. Thus, in principle, in any two places where the same effect is required, it is effectuated by only a single piece of code.

Compiler Dependence

Though the library takes into account the restrictions that are imposed by the use of strict ALGOL 60, nevertheless a number of problems arise when the library is taken to a new compiler.

We give a short list of problems that may arise:

(1)　In general a different character representation and ALGOL symbol representation are used. A very simple program can take care of this conversion, but it is also possible that an incomplete character set is used; e.g., lower case letters are missing in the CDC character set.

(2)　Even good compilers have some restrictions. It is wise not to use all the ALGOL 60 features that are permitted by the Revised Report; e.g., most ALGOL 60 compilers do not handle "own dynamic arrays." However, when a reasonably complete compiler is available, the restrictions imposed by it introduce only minor problems for numerical programs and they are easily eliminated.

(3)   It is not defined by the Revised Report how independent
       compilation of procedures should be handled. In order
       to adapt the texts of the original procedures for use
       in the CDC system, all externally declared procedures
       needed to be referenced by a code declaration inside
       the procedure body.

These points all have to do with the peculiarities of a particular
compiler, whereas the library was constructed with no particular compiler
in mind. The only thing we can do - if we want the library to run on
another machine - is to find (or to insist on the construction of) a
compiler with a negligible number of anomalies.

Arithmetic

Two more points have to be kept in mind when we consider the
transportation of a library to a new environment (i.e., computer +
compiler), viz., machine arithmetic and elementary functions.

When an algorithm is coded independently of a particular machine
environment, a guarantee with respect to its performance can only be
given under certain assumptions on the machine arithmetic. Weak
arithmetic can spoil a sound algorithm. For instance, on CDC, a
program failed because, using CDC arithmetic, one can obtain real
numbers a and b such that a $\neq$ 0.0. b $\geq$ 1.0 and a x b = 0.0.   In
implementing programs on existing machines, one has to reckon with
this kind of peculiarity that makes the construction of truly portable
(machine independent) software almost impossible.

It would be expedient if a clear terminology existed to denominate
machine arithmetic characteristics, so that computer/compilers could be

classified according to their arithmetical properties. Such a classification would enable a programmer to guarantee his code for an environment
in which the arithmetical properties belonged to a certain class.

We do not intend to start such a classification here, but to make
the idea more clear we shall mention some useful requirements for floating-
point arithmetic. A *minimal set* of requirements should be

$$fl(a \circ b) = a(1+\alpha) \circ b(1+\beta)$$

where $fl(a \circ b)$ denotes the result of a floating point operation: $\circ = +, -,$
$*, /$; α and β are numbers depending on a and b respectively and on $\circ$ such
that $|\alpha| \leq \varepsilon$, $|\beta| \leq \varepsilon$ where ε is a machine parameter, the "relative machine
precision." In general these minimal requirements are not adequate. Addi-
tional requirements would be, e.g., *monotonicity*; i.e.,

$$a > b \rightarrow c + a > c + b;$$

$$c > 0, a > b \rightarrow c * a > c * b;$$

etc.

Machine arithmetic can be called *optimal* with respect to +, -, *,
/, if, as a result of any of these operations between two floating-point
numbers, the nearest representable number is delivered; if the result
lies exactly between two representable numbers one of these should be
chosen in a uniquely determined way.

Closely related to the machine arithmetic are questions with
respect to overflow and underflow (i.e., situations where the operands
a and b in the elementary operations are such that the *arithmetic
requirements cannot be satisfied*). For these cases it is expedient
if a user can make a choice among 3 options: (1) hard failure action
(i.e., after an error message the computation is stopped); (2) soft

failure action (i.e., after a message the computation goes on), or (3) no action (computation goes on without any message). If the computation is continued, the value delivered might be some kind of "undefined" or some "near" representable number.

Another question related to arithmetic is the conversion between the machine representation of a real number and its decimal representation in I/O or in a program text. A good compiler should allow all machine representable floating-point numbers to be converted to distinct decimal representations, and vice versa. To illustrate, the CDC ALGOL 60 version 3 compiler violates this requirement since floating-point numbers are represented by a 48-bit binary mantissa, whereas the compiler ignores the 15th and further digits of any decimal representation. Thus, there are certain floating-point numbers that cannot be distinguished by their decimal representation.

Like machine arithmetic, the elementary (i.e., compiler-provided) functions should satisfy certain clearly defined specifications. A *minimal requirement* (which can be imposed on any function) is

$$\text{"computed value of" function } (x,y,\ldots) =$$
$$\text{function } (x(1+\xi),\ y(1+\eta),\ldots)\ (1+\phi)$$

where $|\xi| \le e$, $|\eta| \le e$, ..., and $|\phi| \le f$; e is some number related to the relative machine accuracy and f is the relative function accuracy. For most elementary functions one can impose either e = 0 or f = 0, but f = 0 is preferable, since, e.g., it causes arcsin(sin(x)) always to deliver a value, whereas e = 0 might result in a call of arcsin(y) with y > 1.0. For monotonic functions, preservation of monotonicity could also be required.

As with machine arithmetic, a short and clear description of the properties of elementary functions (independent of the particular algorithm used for their computation) is important, so that a programmer will be able to guarantee his code under certain well defined standard conditions.

Like overflow and underflow, an improper call of an elementary function (such as sqrt(-2.0) or ln(-6.8E-8)) should result in one of 3 (optional) actions: (1) hard failure, (2) soft failure, or (3) no action. If the computation continues, the function should deliver "undefined" or rather a "near" representable number (e.g., ln(x) delivers ln($|x|$) if x < 0.0).

Portability

Although standardization and classification of machine properties has not yet reached a sufficiently developed state and compilers are not perfect, the experience with the transportation of the ALGOL library NUMAL justifies the following conclusion.

If we use (1) a well-defined language (or only a sublanguage with well-defined effects) and (2) a compiler which interprets the language correctly; if we have available (3) an environment (hardware + compiler) with well-behaved and well-defined arithmetic, and if we construct (4) well-programmed software, then portability is scarcely a problem.

We can think of these four aspects as independent of each other and the realization can be a task for different groups of people. The numerical or software specialist can raise standards for (1), (2), or (3) but his prime interest should be (4). If he cannot work independently of a particular machine he will not be able to create truly portable software.

This conclusion can be used as a starting point for and a philosophy behind the construction of software packages. In fact this idea is not new at all and it is even partly realized in some sense in the Handbook for Numerical Computation {4}. Indeed, the ALGOL texts in this book appear to be almost completely portable and they were easily implemented on the CDC CYBER-system*, except for (1) the construction of a double-length inner product, and (2) the change of some machine constants that were mentioned in the program texts. The result, however, is rather a collection of procedures than a coherent structure and the literal ALGOL text could not be made to run very efficiently.

Portability and Efficiency

The final efficiency of a code is to a high degree dependent on the compiler used. So, renouncing all special abilities of a particular environment, we will never obtain the most efficient code. On the other hand, if we exploit the special features we may not expect portability. Hence portable software will not be the most efficient on all computers. However, a great deal of the disadvantage of portable programming can be eliminated by the exploitation of the modular structure of a library. An enormous amount of work is essentially done by the very basic routines such as matrix-vector operations, polynomial evaluations, etc. This yields the possibility of speeding up the codes essentially by replacing the isolated, compiler translated, innermost pieces of a library by hand translated code. In this way a 2 to 3 times faster runtime was obtained for the NUMAL library.

---

*This was done, mainly for reference purposes, by the computing centers of the Universities of Utrecht and Groningen.

Portability and User Convenience

If ALGOL 60 routines from NUMAL are run on a particular machine, parameters such as the relative machine accuracy have to be specified. This and the complete absence of I/O are not very attractive to the average user. This disadvantage can be overcome by a user interface, i.e., a piece of software which establishes the connection between a non-specialist user and the set of numerical routines. This program (which itself could be machine independent to a certain extent), sets the machine parameters, interprets performance indicators and, possibly, selects a particular numerical procedure from among those available. By adding such a user interface on top of the numerical routines, we obtain *a structured library in 3 levels:* (1) The *user interface,* (2) The *numerical routines* consisting of algorithms coded for portability, and (3) a machine-dependent *speed-up part.*

ALGOL 60 - ALGOL 68

Although the realization of a numerical library according to the above mentioned criteria should be possible in any well-standardized language in which algorithms can be expressed, we have concentrated on languages of the ALGOL family. Our library in ALGOL 60 is available for external use; program texts and descriptions are distributed to subscribers and a version with a speed-up part, adapted to the CDC ALGOL 60 version 3 compiler, is maintained.

At this moment we are considering the possible construction of a software library, satisfying the described criteria, in the full language ALGOL 68. This language provides a number of useful features that are missing from ALGOL 60; e.g., multiple precision, file handling

and I/O routines. Other features, such as operation declarations,
allow for particularly clear and well-structured programming that is not
confused by opaque jumping or administrative details. In this way
ALGOL 68 programming could combine reasonably efficient coding with
a clear and realistic description of numerical algorithms.

Acknowledgements

I am grateful to Walter Hoffmann for a number of valuable remarks.

We are aware of the fact that we made a number of critical remarks
about the performance of the ALGOL 60 features available on CDC; however,
in our opinion, CDC offers ALGOL facilities superior to those of other
major manufacturers which have not recognized ALGOL as an important
language.

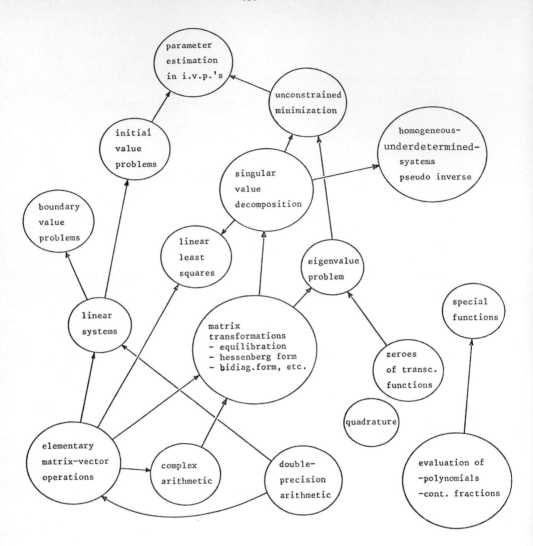

Figure 1

The main lines of interdependence
between the different parts of the
library NUMAL.

REFERENCES

1. Dekker, T. J., A Floating-Point Technique for Extending the Available Precision, Numer. Math. 18 (1971) 224-242.

2. Naur, P., Ed., Revised Report on the Algorithmic Language ALGOL 60, A/S Regnecentralen, Copenhagen, 1964.

3. NUMAL, A Library of Numerical Procedures in ALGOL 60, Mathematical Centre, 1974.

4. Wilkinson, J. H., and Reinsch, C., Handbook for Automatic Computation, Vol. 2, Linear Algebra, Springer-Verlag, Heidelberg, 1971.

FORTRAN PORTABILITY VIA MODELS AND TOOLS

W. S. Brown and A. D. Hall

ABSTRACT

   This talk is a survey of recent Bell Labs work on
FORTRAN-based software portability. Our approach is based
on our experience with the ALTRAN language and system for
symbolic algebra, and the PORT Library of state-of-the-art
procedures for numerical mathematics.

   Both ALTRAN and PORT are written in PFORT, a large and
mechanically verifiable subset of ANS FORTRAN. To make PFORT
programs easier to write and easier to read, we now use an
extension called RATFOR and a preprocessor that translates
it into PFORT. A more ambitious extension called EFL is under
development.

   From a theoretical viewpoint the key to all our work on
software portability is a model of the computing environments
in which our programs will be expected to operate. We assume
a hardware-software environment that fully supports the PFORT
language and is characterized by parameters (machine constants)
in four basic areas: (1) logical unit numbers, (2) word size,
(3) integer arithmetic, and (4) floating-point arithmetic. To
demonstrate the use of this model, we discuss an algorithm by
J. L. Blue for computing the Euclidean norm of a real vector.
The algorithm is portable, accurate, and efficient, avoids
overflow and underflow, and will be included in the next
edition of PORT.

This talk is a survey of recent Bell Labs work on FORTRAN-based software portability. We call a program *portable* if the effort to move it into a new environment is much less than the effort to rewrite it for the new environment. Thus a portable program can be used more widely and for a longer period of time than a program that is developed for a specific computer. Avoiding the necessity of revising a program many times makes it possible and worthwhile to do a better job the first time. When high quality, portable software becomes commonplace, derivative improvements in hardware and languages can be expected.

Our approach to FORTRAN portability is based on our experience with the ALTRAN language and system for symbolic algebra. Although the complete system consists of more than 300 subprograms, and the time required to rewrite it would be at least 2 man years, the typical installation time is only 1 to 2 man weeks. The system has been successfully installed on 9 fundamentally different computers under 13 FORTRAN compilers at approximately 30 sites. Since there is only one implementation, programs written in ALTRAN are readily portable among these sites.

During the past few years we (and others) have collaborated with P. A. Fox in developing the PORT Library of state-of-the-art procedures for numerical mathematics. In making these procedures both portable and efficient, our experience with ALTRAN has proven invaluable.

Both ALTRAN and PORT are written in PFORT, a large, carefully defined subset of ANS FORTRAN. Conformance to the rules of PFORT is mechanically verified by B. G. Ryder's PFORT Verifier, which checks both individual procedures and inter-procedure communication. The Verifier is written in PFORT, and has verified itself.

To make PFORT programs easier to write and easier to read, B. W. Kernighan has developed an extension called RATFOR and a preprocessor to translate it into PFORT. RATFOR adds a carefully chosen set of control statements to PFORT, and also adds DEFINE and INCLUDE statements. The preprocessor is written in RATFOR, and the resulting PFORT has been verified by the PFORT Verifier.

Unfortunately, RATFOR does not extend FORTRAN's meager facilities for data structures, and the preprocessor detects only those syntax errors that occur in the new control statements. For these and other reasons A. D. Hall and S. I. Feldman are currently developing an extended FORTRAN language called EFL.

From a theoretical viewpoint the key to all our work on software portability is a model of the computing environments in which our programs will be expected to operate. If a target computer (or its FORTRAN compiler or operating system) fails to satisfy the assumptions of the model, then the installer will encounter unanticipated (and possibly serious) obstacles. On the other hand, if a target computer has important capabilities that are not included in the model (for example, a high degree of parallelism), then those facilities will almost certainly be underutilized.

We assume a hardware-software environment that fully supports the PFORT language and is characterized by parameters (machine constants) in four basic areas: (1) logical unit numbers, (2) word size, (3) integer arithmetic, and (4) floating-point arithmetic.

By way of illustration, the single-precision floating-point numbers are characterized by parameters β, t, e_{min}, and e_{max} such that zero and all numbers of the form

$$\pm \beta^e (x_1 \beta^{-1} + \cdots + x_t \beta^{-t})$$

$$0 \le x_i < \beta; \ i=1,\ldots,t$$

$$0 < x_1$$

$$e_{min} \le e \le e_{max}$$

are exactly representable. These *model numbers* are a (possibly complete) subset of the *machine numbers*.

For given machine numbers x and y, let

$$z = x \text{ op } y$$

$$\bar{z} = \text{fl}(x \text{ op } y),$$

where $\text{fl}(e)$ denotes the machine number that is obtained by evaluating the expression e in floating-point arithmetic. If x, y, and z are all model numbers, we require that $\bar{z} = z$. If x and y are model numbers, but z is not, we require that $\bar{z} \in [z_1, z_2]$, where z and z_2 are the model numbers that bracket z. Finally, if x and/or y are not model numbers, then one is not entitled to rely on their extra precision; instead, they must be broadened to model intervals, and these rules must be generalized accordingly. If a computer is less accurate than this, we feel that the manufacturer should promptly repair it; if more accurate, our programs will probably benefit, and our error bounds will be somewhat too pessimistic.

To demonstrate the use of this model, we shall discuss the theory of an algorithm by J. L. Blue for computing the Euclidean norm of a real vector. The algorithm makes only one pass over the vector, scaling very small and very large magnitudes to avoid underflow and overflow, respectively. If the vector has too many components or its norm is too large, the procedure sets an error condition; otherwise, it returns an accurate result. The algorithm is highly portable, and will be included in the next edition of PORT.

REFERENCES

A.   Background

1.   W. S. Brown
     Software Portability
     Report of the 1969 NATO Conference on Software Engineering
        Techniques
     Ed. J. N. Buxton and B. Randell
     NATO Science Committee, 1970, pages 80-84.

B.   SNOBOL4

2.   R. E. Griswold, J. F. Poage and I. P. Polonsky
     The SNOBOL4 Programming Language
     Prentice-Hall, Englewood Cliffs, N.J., 1971.

3.   R. E. Griswold
     The Macro Implementation of SNOBOL4 - A Case Study of
        Machine-Independent Software Development
     W. H. Freeman, San Francisco, 1972.

C.   ALTRAN

4.   W. S. Brown
     Altran User's Manual, Third Edition
     Bell Laboratories, Murray Hill, N.J., December 1973.

5.   A. D. Hall
     Altran Installation and Maintenance Manual, Second Edition
     Bell Laboratories, Murray Hill, N.J., September 1972.

6.   S. I. Feldman
     ALTRAN Reference Card
     Bell Laboratories, Murray Hill, N.J., April 1975.

D.   PORT

7.   P. A. Fox, A. D. Hall and N. L. Schryer
     The PORT Mathematical Subroutine Library
     Computing Science Technical Report #47
     Bell Laboratories, Murray Hill, N.J., August 1976.

8.   P. A. Fox
     PORT Mathematical Subroutine Library User's Manual
     Bell Laboratories, Murray Hill, N.J., April 1976.

E.   PFORT

  9.   B. G. Ryder
    The PFORT Verifier
    Software Practice and Experience 4, pp. 359-377 (October 1974).

  10.   A. D. Hall and B. G. Ryder
    The PFORT Verifier
    Computing Science Technical Report #12
    Bell Laboratories, Murray Hill, N.J.
    May 1973, Rev. July 1975, 61 pages.

F.   RATFOR

  11.   B. W. Kernighan
    RATFOR - A Preprocessor for a Rational FORTRAN
    Software Practice and Experience 5, pp. 395-406 (October 1975).

G.   Utilities

  12.   A. D. Hall
    The M6 Macro Processor
    Computing Science Technical Report #2
    Bell Laboratories, Murray Hill, N.J., April 1972, 13 pages.

  13.   A. D. Hall
    SEDIT - A Source Program Editor
    Computing Science Technical Report #16
    Bell Laboratories, Murray Hill, N.J., May 1974, 9 pages.

  14.   A. D. Hall
    FDS: A FORTRAN Debugging System Overview and Installer's Guide
    Computing Science Technical Report #29
    Bell Laboratories, Murray Hill, N.J., April 1975, 38 pages.

PORT — A PORTABLE MATHEMATICAL SUBROUTINE LIBRARY

P. A. Fox

ABSTRACT

The PORT mathematical subroutine library, developed at Bell Laboratories, is described. Library subprograms are written in a mechanically verifiable subset (PFORT) of ANS Fortran. Adaptation to a particular computer/compiler environment is achieved by calls to Fortran function subprograms defining environment-dependent quantities. Scratch storage in PORT is provided by a dynamic storage allocation scheme implemented as a package of simple portable Fortran subprograms. PORT is installed and in use on a wide variety of computers.

Introduction

The first mathematical subroutine library for a computer was written by Maurice V. Wilkes, David J. Wheeler, and Stanley Gill for the EDSAC at the University of Cambridge in England in 1951 {1}. The programs were written in machine language and certainly no thought was given to portability; to have a library at all was remarkable.

In the 25 years since that one, many libraries have been developed, each requiring huge expenditures of time, money and expert programmers. For a brief period it was thought that programming in Fortran would permit mathematical libraries to move from computer to computer, but this was an illusion, and, in fact, each new effort started essentially from scratch. Bell Labs, having many different makes of computers, has been particularly interested in the possibility of moving programs between computers, rather than developing or obtaining essentially redundant libraries to run on the various machines. The increasing use of networking to distribute the computing load between these computers aggravates the problem of using separate incompatible libraries. Early in 1974, a project to develop a portable mathematical subroutine library for use at Bell Labs was started. A library was to be developed which would both insure consistent computation across the computer types on hand, and also extend into the future, adapting to new computer acquisitions.

The first edition of the library, PORT, has now appeared. The library has automatic error handling and dynamic storage allocation, both implemented portably, and is in use on three IBM 360/370 computers, two UNIVAC 1100 Series, two Honeywell 6000 Series, and two PDP 11's. It is being installed on a CDC 6600, and is being considered for a Data General computer.

PORT and Portability

The many excellent papers at the conference have provided definitions of portability ('transportability,' etc.), and have described a multitude of schemes for constructing, adapting and processing programs to enable them to run in disparate computer/compiler environments. In many cases a large subroutine library, operational on one computer class, has had to be the starting point for adaptation. PORT, on the other hand, has had some of the well-known "advantage of coming last." We were able to devote a good deal of time initially to the structure of the library and to building a sound basis for portability.

Our approach is a simple two-pronged one:

(1)   Programs are written in a verifiable subset (PFORT) of ANS Fortran;

(2)   The target environment is specified in terms of machine-dependent quantities.

The PFORT *Language*

PORT subprograms are written to avoid the problem of dialect. The programming language is restricted to the PFORT (Portable Fortran) subset of ANS Fortran specified in {2} and {3}. Programs submitted to PORT are always sent through the PFORT Verifier {2} to guarantee their adherence to this language restriction.

To make the library viable, two non-ANS usages are made in PORT. First, dummy arrays in subprograms often are given a last subscript dimension of 1, under the assumption that there is no runtime subscript range checking. This is valid for the usual production system. Second, for some of the subprograms implementing error handling and dynamic storage allocation (see below), it is assumed that variables, local to a subprogram, that are initialized by DATA statements and then assigned new values, retain their subsequent values from one subprogram invocation to the next. If a program overlay system is being used, care must be taken to keep these subprograms in the main link.

In many minicomputer Fortran systems, the ANS Fortran requirement that LOGICAL, INTEGER, and REAL data are allocated one "word," while DOUBLE PRECISION and COMPLEX data are allocated two "words," is violated. To allow PORT to run on these smaller computer systems we have been careful to make the amounts of storage allocated to different data types independent.

Specification of Machine-Dependent Quantities

In another paper in this volume, Brown and Hall mention the use, in PORT, of machine-dependent quantities to characterize a particular computer/compiler (hardware/software) environment. Specification is provided for logical input-output unit designators used in the local operating system, and quantities are included to describe the number representation for integers and for single and double precision floating-point numbers.

In somewhat more detail, the following quantities are specified:

LOGICAL UNIT NUMBERS
    The standard input unit
    The standard output unit
    The standard punch unit
    The standard error message unit

INTEGER AND CHARACTER STORAGE
    The number of bits per INTEGER storage unit
    The number of characters per INTEGER storage unit

Integer Variables

Let the values for integer variables be written in the s-digit, base-α form;

$$\pm(x_{s-1}\alpha^{s-1}+x_{s-2}\alpha^{s-2}+\cdots+x_1\alpha+x_0)$$

where $0 \le x_i < \alpha$ for $i = 0,\ldots,s-1$.

Then we specify,

> the base, α
> the maximum number of digits, s
> the largest integer, α^s-1.

Although the quantity, α^s-1, can easily be computed from s, and the base, α, it is provided because a naive evaluation of the formula would cause overflow on most machines. (Storage of integers as magnitude and sign or in a complement notation is not specified since PORT subprograms must be independent of the storage mode.)

Floating-Point Variables

If floating-point numbers are written in the t-digit, base-b form:

$$\pm b^e(\frac{x_1}{b} + \frac{x_2}{b^2} + \cdots + \frac{x_t}{b^t})$$

where $0 \le x_i < b$ for $i = 1,\ldots,t$, $0 < x_1$ and $e_{min} \le e \le e_{max}$, then for a particular machine, we choose values for the parameters, t, e_{min}, and e_{max}, such that all numbers expressible in this form are representable by the hardware and usable from Fortran. Note that the formula is symmetrical under negation but not reciprocation. On some machines a small portion of the range of permissible numbers may be excluded.

Then we specify,

> the base, b, for both single and double precision
> the number, t, of base-b digits

In order to accommodate machines (such as the CDC 6000 Series) with the b-point on the right we must concede the possibility that the magnitude of e_{min} may be substantially smaller than e_{max}. Thus, for single-precision floating-point we specify,

the minimum exponent, e_{min}
the maximum exponent, e_{max}

For double precision, b remains the same, but t, e_{min}, and e_{max} are replaced by T, E_{min}, and E_{max}. Normally, we have $E_{min} \leq e_{min}$ and $E_{max} \geq e_{max}$, and $T > t$. However, in machines such as the CDC 6000 Series or the PDP-10 KA Processor, where double precision is implemented by software simulation, small double-precision floating-point numbers carry only t base-b significant digits. In such cases, we take E_{min} to be the exponent of the smallest number with T base-b significant digits, and it may be that, $E_{min} > e_{min}$.

Integer values are used for the above specifications, and of course it is possible to construct the corresponding smallest, largest, etc., floating-point numbers, but because the computations involved can be a little tricky, and just to make things more convenient, we also specify the single and double precision quantities for the following:

the smallest positive magnitude, $b^{e_{min}-1}$
the largest magnitude, $b^{e_{max}}(1-b^{-t})$
the smallest relative spacing between values, b^{-t}
the largest relative spacing between values, b^{1-t}
the logarithm of the base b, $\log_{10} b$

The relative spacing is $|(y-x)/x|$, when x and y are successive floating-point numbers.

The redundancy in these definitions also allows a particular installation of the PORT library to be checked for consistency. After the library has been compiled, a special checking subprogram is called to verify that the integer and floating-point constants satisfy the following conditions:

(1) $t \leq T$

(2) $E_{max} \geq e_{max}$

(3) $E_{min} \leq e_{min}$

(4)   the largest integer agrees with $\alpha^S - 1$

(5)   the floating-point constants agree with those computed from the integer constants

(6)   the largest and smallest floating-point values are closed under negation, i.e.: $-(-x) = x$

If a discrepancy is found, a warning message and the values of the quantities involved are printed. Condition (3) may fail on some machines, even though the specifications are correct; in this case the warning message should be ignored. For 2's complement machines, if e_{min} has been set too small, condition (6) will fail, since the negative of the smallest number will underflow. Note that some of the integer definitions must be used by the checking routine to determine the output unit for error messages and the correct formats for printing.

Installing PORT

Once machine-dependent quantities have been defined for a library, various mechanisms can be used to adapt the library to a given computer/ compiler environment. For example, the quantities can be flagged in

the master version of the library so that a preprocessor can generate
a specialized version with appropriate values inserted at the flagged
locations. In the first edition of PORT we have preferred to avoid
the use of preprocessing or macro processing. We furnish, instead,
three Fortran function subprograms (integer, single precision, double
precision) which can be invoked to determine the basic machine or
operating system dependent constants given above. When the library
is moved to a new environment, only the DATA statements in these three
subprograms need to be changed. Values are provided, in the library,
for the Honeywell 6000 Series, the IBM 360/370 Series, the SEL Systems
85/86, the XEROX SIGMA 5/7/9, the UNIVAC 1100 Series, the DEC PDP 10
(KA and KI processors), the DEC PDP 11, the Burroughs 5700/6700/7700,
and the CDC 6000/7000 Series; others can be added.

Thus, when a PORT tape is sent to a new location, the installation
procedure generally consists only of adjusting the three machine constant
function subprograms by removing the C's (for Comments) from the DATA
statements appropriate to the target computer, and then compiling the
source. In order to keep installation simple, both single and double
precision versions of each program (when appropriate) are included
on the tape.

Dynamic Storage Allocation

A dynamic storage allocator is integrated into the basic PORT
library structure. Other methods of providing scratch storage, such
as compiling workspace directly into individual subprograms, or passing
the names of scratch arrays in the calling sequence are either inefficient
or inconvenient. The allocator in PORT is implemented as a package of

simple portable Fortran subprograms which control a dynamic storage stack. We have found that within the library dynamic storage allocation leads to cleaner calling sequences, improved memory utilization, and better error detection.

The stack resides in a labeled COMMON region and has a default length of 500 single-precision locations. By the nature of a stack, allocations and de-allocations are carried out on a last-in first-out basis. In order to make the stack invisible to most users of library programs, the package is self-initializing, but if one wishes to adhere to a strict interpretation of the Fortran Standard, it is necessary to declare the labeled COMMON region in the main program. The first ten locations of the stack are used for bookkeeping, and then each allocation made in the stack carries an associated space overhead consisting of two control words: a type word to specify the allocation type (INTEGER, REAL, etc.), and a backpointer to point to the end of the previous allocation. Between allocations there may also be padding for type alignment.

The PORT stack-handling routines can be called by the user who may wish to use the scratch facility. The subprograms include, besides stack allocation and de-allocation, a routine to initialize the stack to a length larger than the default value, a routine to find out how much of the stack is still available, and a routine to modify the length of the most recent allocation. It is also possible to find out certain facts about the current state of the stack, such as the number of outstanding allocations, the current active length, the maximum length used so far, etc.

More details on the implementation of the dynamic storage stack can be found in {4}.

Further Aspects of PORT

In addition to dynamic storage allocation, the PORT library provides an automatic centralized error-handling technique. The user is protected, in a sense, from himself, because the calling sequences to the library subprograms do not include error flags as parameters which must be checked on return from the subprogram. In general, the occurrence of an error causes a message to be printed and the run to be terminated.

However the more experienced user is able to recover from errors of a recoverable type. (Blunders, such as specifying -2 for the number of linear equations to be solved, are considered fatal; one can't recover from them.) To recover from errors the user can gain control over the error-handling process by entering the *recovery* mode, and while in this mode can

- determine whether an error has occurred, and, if so, obtain the error number
- print any current error message
- turn off the error state
- leave the recovery mode

Thus it is possible for the user to detect whether a computation is in trouble in a way that can be remedied (for example by easing a too stringent convergence criterion), so that a fix can be put in, the error state turned off, and the computation continued by calling again the subprogram that caused the trouble.

More detail on error handling in PORT is given in {4}.

The fact that the calling sequences in PORT do not need parameters for error handling, or arrays for scratch space simplifies their structure. Every effort has been made to provide clean, simple calls for the average library user, and to describe the use of each subprogram in a brief reference sheet.

These simple calls actually represent, however, a sort of outer or visible layer of the many-layered library. The outer, or top-level, routines often set default values and then call inner, or lower-level, routines containing more parameters. These inner routines are in many cases also documented and available to the user who may wish to exercise more control over the computation at hand.

Status and Summary

The initial emphasis in developing the library has been placed on building the framework for the library, that is, the dynamic storage allocation and automatic error-handling, and in establishing portability through language specification and definition of machine-dependent quantities. Nevertheless the initial edition of the library contains some excellent programs in its twelve areas of computation. Especially notable are Blue's {5} quadrature routine, QUAD, and Schryer's {6} differential equation solver, ODES. The second edition of PORT, due out this year will have several new offerings including programs for the solution of sets of nonlinear equations, a portable uniform random number generator, and a B-spline least squares fitting package. There will also be programs in areas such as Fast Fourier Transform, polynomial root solvers, polynomial evaluation, and linear equations.

The majority of the programs in PORT have been developed within Bell Laboratories; a few have been adapted from published (non-proprietary) sources. In each case the subprogram has been tested for portability, reliability, accuracy, and efficiency, and subjected to a refereeing process before being accepted in the library.

REFERENCES

1. Wilkes, M. V., Wheeler, D. J., and Gill, S., The Preparation of Programs for an Electronic Digital Computer, Addison-Wesley, Reading, Mass., 1951.

2. Ryder, B. G., The PFORT verifier, Software Practice and Experience 4 (1974), 359-377.

3. Hall, A. D., and Ryder, B. G., The PFORT Verifier, Computing Science Technical Report Number 12, (Revised July 1975), Bell Laboratories, Murray Hill, N.J.

4. Fox, P. A., Hall, A. D., and Schryer, N. L., The PORT Mathematical Subroutine Library, Computing Science Technical Report Number 47, (1976), Bell Laboratories, Murray Hill, N.J.

5. Blue, J. L., Automatic Numerical Quadrature - DQUAD, Computing Science Technical Report Number 25, (1975), Bell Laboratories, Murray Hill, N.J.

6. Schryer, N. L., A User's Guide to DODES, a Double-Precision Ordinary Differential Equation Solver, Computing Science Technical Report Number 33, (1975), Bell Laboratories, Murray Hill, N.J.

FORTRAN POISONING AND ANTIDOTES

Brian T. Smith

ABSTRACT

The focus of this paper is a discipline for
writing mathematical software in FORTRAN. This
discipline avoids the more obscure constructs
and aspects of the FORTRAN language and leads
to FORTRAN software that is readily ported
from place to place.

1. Introduction

Two recent papers {19, 27} describe the steps in the development
of numerical software. Taking the approach expressed in {19}, there
are three distinct steps in the development of software: first, the
development of theoretical processes to perform specific calculations;
second, the development of practical computer processes and last, the
implementation of these processes in a particular computer language.
Given that the first two of these three steps have been performed, this
paper describes a discipline for implementing realizations of the desired
computational processes in the FORTRAN language {4, 5}.

The discussion herein focuses upon and emphasizes program realizations
that are easily moved from place to place. Using the terminology for
various realizations defined in {27}, there is at one extreme a portable
program; that is, a program that can be compiled and will execute correctly
without change at each of the computer installations under consideration.
However, such programs are rare since very few computational processes
do not depend upon their environment in some way.

Because the concept of portability is too restrictive, and hence
impractical in general, the concept of a transportable program is formulated;
that is, a program that is sufficiently structured that it can be ported
from place to place by making only automated changes. Thus, transportability
includes portability. Of course, the concept of a transportable program
still restricts the kinds of algorithms that can be used and indeed the
kinds of computations that can be performed, but is satisfactory for
our purposes. To denote such an algorithm, Ford {27} uses the term

adaptable algorithm; that is, an algorithm that can adapt to the particular machine environment under consideration. For example, an algorithm that can be expressed in terms of machine parameters such as those discussed in {2, 15, 21, 25, 28, 39, 40} can readily adjust to varying machine environments, and hence is deemed more adaptable than one that is not expressed in terms of these machine parameters.

The FORTRAN programming discipline described here together with the use of adaptable algorithms leads to transportable software. Such software is less susceptible to restrictions of specific FORTRAN dialects and thus more readily moved from place to place with automated modification.

We choose to call each unclear non-transportable FORTRAN construct a FORTRAN poison and the coding discipline that neutralizes the poison an antidote. There are three sources of these poisons which are emphasized in this paper; first, poorly structured representations of computational processes and tricky coding; second, inconsistent formatting conventions; and third, constructs in FORTRAN that are either ambiguously defined or not defined by the ANSI FORTRAN Standard {4, 5}.

We classify FORTRAN poisons in terms of these three sources and describe a discipline to avoid them. Although our desire is to be complete, that goal is hopeless; there are innumerable compilers and computing environments whose features contribute to the difficulty of creating portable FORTRAN software that would need to be researched. In lieu of completeness, therefore, our goal is a consistent discipline; if unforeseen incompatibilities arise in new environments, they can be identified, and resolved in an automated manner that avoids the offending constructs. Examples of such transformational facilities are described in {10}.

A disciplined approach to writing FORTRAN software has benefits other than automated transformation, such as readability and ease of maintenance. These other benefits improve one's ability to move such software from place to place, not necessarily by automated transformation but by manual analysis and modification. Such a discipline is needed, particularly for the FORTRAN language, for two basic reasons: first, the language does not contain constructs and control structures well suited for numerical algorithms; and second, it does contain constructs too closely tuned to early machine architectures. In the first instance, a disciplined use of and format for the safe FORTRAN constructs replaces the missing constructs and control structures. In the second instance, a disciplined abstention is the antidote.

The disciplined approach to writing FORTRAN software applies to the writer of both individual programs and suites of programs for program libraries. For the individual program, a disciplined coding practice permits effective and reliable manual modification of the software to meet new environments or new applications of previously prepared software. For programs for software libraries, a discipline is essential for the cost effectiveness of the library service, since the library programs are not usually maintained by the author of the programs. Transferring the library to new machine environments, once the language dialect and hardware behavior are fully understood, becomes a relatively straightforward task which can be undertaken in an efficient and reliable fashion.

The presentation of the disagreeable constructs and the discipline that avoids them is organized into three parts. First, various FORTRAN constructs are reviewed, examining them from the point of view of contributing to the clear structure of FORTRAN source text. This material is discussed in Section 2 and its nine subsections. Next, FORTRAN is discussed again, this time reviewing constructs that inhibit the transfer of software from place to place. The emphasis in this review is on constructs that are either ambiguously or poorly defined by the ANSI Standard. This topic is presented in Section 3 and its eleven subsections. Third, the slightly unusual constructs, accepted by the Standard but rejected by certain dialects of FORTRAN are described in Section 4. Section 5 gives two anecdotes illustrating how one can be surprised by not strictly following the rules defined by the Standard. Last, Section 6 concludes with a summary of the disciplined approach to FORTRAN programming.

Throughout this paper, we refer to the reference {4} as the Standard because it is more accessible than the later official (and very slightly different) Standard {5}. Two clarifications {6, 7} interpret several of the ambiguous points within the Standard, and reference {3} gives a brief history of the FORTRAN standardization effort. The new draft proposed Standard {8} is referred to as FORTRAN 77. For emphasis, the Standard {4} is sometimes referred to as FORTRAN 66.

2.   Clearly Structured FORTRAN

There have been many articles describing techniques for writing clear structured programs; cf. {34} for references to much of the literature of this subject and {38} for references to such literature for the FORTRAN language. Concepts such as gotoless programming and top-down programming are familiar ideas. Our purpose here is not to discuss these concepts but to assume that the software involved has been structured in these recommended ways. We do describe, however, those FORTRAN constructs that should be avoided because they encourage poorly structured software.

Before discussing FORTRAN constructs in terms of well-structuring, there is one general concept of good programming structure that needs to be emphasized as it pertains to FORTRAN. Avoid tricky programming, particularly that which relies upon special features of the FORTRAN language. Here is an excellent example taken from {33} that typifies our point. This program's trick relies on the fact that integer division in FORTRAN truncates towards zero for all arguments.

```
      INTEGER  I,J
      REAL  X(10,10)
      DO 1 I = 1,10
      DO 1 J = 1,10
    1 X(I,J) = (I/J)*(J/I)
```

What does this program segment compute? In a non-obvious way, it initializes the contents of the array X to an identity matrix of order 10.

Putting aside any concern for efficiency, this program segment relies upon the truncation of integer division in FORTRAN. It is particularly unfortunate when examined from the point of view of adaptability of the underlying algorithm to non-FORTRAN environments. For languages such as PL/1 and ALGOL, this property of division does not hold, unless special precautions are taken; for instance, in ALGOL, the operator / must be replaced by the operator ÷ and in PL/1, each of the expressions of the form a/b must be replaced by TRUNC(a/b) to avoid fixed point overflow in certain cases.

Kernighan and Plauger {33} give an excellent survey of the general concepts in programming style and discipline. For an excellent survey of the goto issue, read Knuth's recent article {34} on when to use, and when not to use, goto statements.

With the general principle of good programming structure and style in mind, we can now examine specific features of FORTRAN that inhibit good structure and suggest disciplines for avoiding them.

2.1 Type Specification Statements

The ANSI FORTRAN Standard does not require the type specification of variables or arrays in a FORTRAN program. Instead, it permits the type of a variable or array, if not specified, to be inferred from the first letter of its name. This is an unfortunate feature of FORTRAN from the point of view of maintenance. Often when studying a program, one needs to know the type and/or precision of a variable in order to make a correct change. Checking the specification part of a program for the non-occurrence of the type specification of a particular identifier is a time consuming and error prone process; there needs to be a discipline for the explicit type specification of entities.

We recommend the following discipline. Specify the types of all identifiers, declaring the extents of arrays in the type specification statements. Arrange the type specifications in some specific order, say with type INTEGER first, followed by LOGICAL, REAL, DOUBLE PRECISION, COMPLEX, and finally EXTERNAL declarations. The identifiers in each specification statement should be alphabetically ordered. In this discipline, the DIMENSION statement becomes unnecessary; its use is not recommended.

(Also the absence of DIMENSION statements in programs avoids certain statement order restrictions enforced by FORTRAN compilers on some minicomputers such as PDP-15; cf. {35}).

For subprograms, it is further useful to have the type specifications for the arguments of the subprogram appear separately from the local identifiers and the identifiers in common blocks. We recommend that the type specifications of arguments appear just after the FUNCTION or SUBROUTINE statement. Similarly, for each labeled or blank common block, the identifiers present in these blocks should appear in separate type specifications, ordered as above. Finally, the types of local identifiers should all be specified in a similar fashion after the last COMMON statement. Any EQUIVALENCE statements should then follow the specification statements for the local identifiers.

This order for the type specification, COMMON, and EQUIVALENCE statements satisfies many compilers, including the PFORT verifier {44} and the WATFIV compiler {22}; cf. {35} for a list of compilers for minicomputers that limit the order of FORTRAN specification statements. In addition, the general rule that the explicit type specification of an identifier appear before its use is satisfied by this ordering. This rule, in particular, conforms to the IBM FORTRAN G and WATFIV {22} compilers which print a diagnostic for the following sequence of statements.

```
SUBROUTINE  EXAMPL(DIMA,A)
REAL  A(DIMA)
INTEGER  DIMA
    .
    .
    .
END
```

The diagnostic is caused by the entity DIMA being used in the second statement before it is declared as type INTEGER in the third statement. The IBM FORTRAN G and WATFIV compilers even terminate with a diagnostic for the type specification statement

INTEGER B(N),N

Here, N is used before its type is specified, even though the default type is consistent with the specified type. The difficulty can be overcome by separating (and placing ahead) the type specifications for dummy scalar arguments from those for dummy array arguments; cf. the recommendation in Section 2.6.

A frequently used practice for integer and real entities is to select identifier names so that their actual type is the same as the default type (type INTEGER for names beginning with I,J,K,L,M,N and type REAL otherwise). We do not feel this practice to be necessary in general. However, it is worthwhile for dummy argument names of subprograms in libraries, where the user may use the dummy argument names for the actual arguments. For, when he forgets to declare them when necessary, a situation of mismatched types is created, which is seldom diagnosed by FORTRAN compilers. For local variables, adherence to the default type naming convention can make it difficult for identifier names to be mnemonic. We prefer mnemonic names for all identifiers, but for dummy argument names we recommend that the name should further correspond to the FORTRAN default specification.

In addition to explicit type specifications of all entities, we recommend that the type specification of each intrinsic or basic external function appear in every program that uses it. Of course, the type specification here must be consistent with the value returned by the function (cf. 10.1.7 of {4}). Such specification statements should appear in the order given above (INTEGER, LOGICAL, REAL, DOUBLE PRECISION, and COMPLEX) with the function identifiers in alphabetical order in each list. It is recommended that such function type specification statements appear after the type specification statements for the local identifiers, and indeed after the EQUIVALENCE statements if any.

Although type specification statements for functions are neither forbidden nor required by the Standard (cf. 10.1.2 and 10.1.7 of {4}), they were required by WATFOR and early WATFIV compilers for functions whose default type was not the type of the returned value. For instance, the identifier CMPLX, if the intrinsic function CMPLX was used in a program unit, had to appear in a type specification statement COMPLEX since its first letter C, not being among I,J,K,L,M, or N, implied type REAL. On the other hand, for intrinsic functions, there were several compilers that would not permit their declaration. So, for example, when EISPACK was being tested before its first release in 1972, certain CDC 6000-7000 and UNIVAC 1108 compilers miscompiled subprograms with such declarations (assuming intrinsic functions to be externally provided). Our solution to this incompatibility was to convert the offending declaration statement into a COMMENT card for such compilers. This solution was possible because we had used type specification statements for the intrinsic and basic external functions that were distinct from those for other entities.

The above example illustrates that portability over even a wide collection of diverse computing environments is impossible or too restrictive to be practical. The WATFIV compiler required a specification statement that certain other compilers rejected. Such an incompatibility can be resolved in terms of transportable software whereby automated programming aids {1, 11, 13, 24, 36} can recognize that the type of an intrinsic function is being specified and can change the specification statement to a COMMENT statement. The point here is that a disciplined approach to preparing FORTRAN source text facilitates automated or manual modifications to software to overcome the vagaries of certain computing environments.

2.2 DO Loops

In terms of clearly structured programs, there are several aspects of the FORTRAN DO loop that are unfortunate. One such aspect is the multi-purpose role of the labeled statement terminating the loop. Its label can refer to both the end and the beginning of the statement in which it appears (cf. {11}). If several DO ranges are terminated by the same labeled statement, the role of that label is even more complex. In particular, if a transfer occurs to a label which terminates several DO ranges, to which part of the compound statement is the transfer intended? The Standard (7.1.2.8 of {4}) states that any transfer to such a compound statement must come from within the range of the innermost DO loop, and thus such transfers are interpreted to be to the beginning of that statement terminating the innermost DO loop.

Our antidote for this poison is to follow a discipline in which the range of each DO loop is terminated with its own labeled CONTINUE statement. Since the final statement of a DO range will now be in effect a null statement, this discipline has the added advantage, as pointed out in {11}, that transformations to FORTRAN code that require the insertion of statements before the DO range termination statement are now simpler.

Another unfortunate aspect of the DO loop is the extended range feature. Its use clearly hides the structure of a program. Because it is an unusual and restricted feature, it is easily misused by the programmer and subject to misinterpretation by the compiler. Our recommendation is to avoid its use.

A further adverse aspect of the DO loop is in connection with undisciplined exits before the DO loop is satisfied. The only mechanism for exit from a DO range is a GO TO statement or an arithmetic IF statement that transfers to a label outside the range of the loop. However these same statements can be used for transfers within the range of the loop. There needs to be a discipline which distinguishes between these two roles for the transfer statement within DO loops. Several extensions to FORTRAN such as {9, 11, 31, 32, 45} demonstrate the need for an exit statement in FORTRAN by defining an exit-the-loop statement. An alternative that conforms to the Standard is a discipline suggested in {45}. This discipline involves the use of comment and indentation conventions which clearly denote certain transfer statements as exits from ranges of DO loops. For example, in the program segment,

```
                  .
                  .
        DO 20 I = 1,N
                  .
                  .
          DO 10 J = 1,N
                  .
                  .
    C
    C  ....EXIT
    C
              GOTO  30
                  .
                  .
    10    CONTINUE
                  .
                  .
    20 CONTINUE
    30   .
                  .
                  .
```

the transfer to statement label 30 is clearly notated as an exit from
the DO range ending at statement label 20.

In connection with exits from DO loops, the Standard makes a most
important distinction between exits when the loop is satisfied and exits
prematurely terminating the loop. When leaving the range of a DO loop,
the control variable is only defined (cf. 7.1.2.8 of {4}) when the loop
is prematurely terminated via a transfer statement (GO TO or arithmetic
IF statement) out of the loop. Many compilers also define the value
of the control variable when the loop is satisfied, but this action is
not standard. Consequently, we recommend that if the control variable
is desired after leaving the DO loop range, it should be copied into
a local variable within the loop and that local variable be used
outside the loop. For instance, if the control variable is I, the
identifier SAVEI is an obvious mnemonic to denote the local variable.

Such a convention helps to clarify the intent of the programmer. This issue of definition status of control variables is further discussed in Section 3.1 of this paper.

2.3   Arithmetic IF Statement

Avoid its use. For two or three way branches, it introduces an unnecessary label, compared to replacing the arithmetic IF statement with one or two logical IF statements. The flow structure of the program is only obscured by extra labels, for such labels can be transferred to from elsewhere in the program. In our experience, the logical IF statements convey the program flow more clearly.

2.4   FORTRAN Labels

The absence of alphanumeric labels is another unfortunate aspect of FORTRAN. Numeric labels cannot convey the conditions or reasons for a transfer -- numbers, in general, have no mnemonic value. Difficulties associated with the lack of alphanumeric labels are further magnified by the fact that FORTRAN lacks convenient control structures, thus further proliferating labels.

The antidote is a disciplined use of labels. Choose the labels in columns 1 to 5 so that they appear in ascending order in increments larger than 1, say 10, or still larger if many modifications may be expected. This permits the insertion of labels without disturbing the existing statement label order. Ascending order facilitates searches for the transfer points when studying the structure of the program. Avoid the use of dummy labels, that is, labels that are not referenced or are not the destination of some transfer from elsewhere in the code. To distinguish between reference labels (such as in FORMAT and DO statements) and those labels used for the beginning of path segments, we recommend a convention that places COMMENT statements before each of the latter describing the conditions for the transfer to that label; with such COMMENT statements one can approach the feature of a literal label present in many other languages.

Another practice which can make FORTRAN programs easier to modify is the use of labeled CONTINUE statements as the destination of GO TO statements. When additional statements need to be inserted at the transfer point, they can then be placed after the labeled CONTINUE statement without disturbing other lines of source text. However, this practice conflicts with the discipline of Section 2.2 where the labeled CONTINUE statement was reserved for the termination of a DO range. The conflict can be resolved by an indentation scheme that distinguishes between the two uses of the CONTINUE statement, such as the scheme recommended in Section 2.9.

Pertaining to structured programming, haphazard transfers are disquieting, particularly in FORTRAN programs. It is worthwhile to organize the layout of the FORTRAN text so that the majority of explicit transfers, if not all of them, are forward. For backward transfers closing explicit loops, an indentation discipline displaying the range of such loops can improve the readability of the source text.

2.5 Arithmetic Exceptions

Often, programmers use arithmetic exceptions to produce program-detectable diagnostics. Such exceptions may include floating point overflow to indicate that an argument is out of range for a particular computation or a zero divide to indicate that certain values of the input arguments are invalid. This technique of employing diagnostics is not portable, since the nature of an error and its location are not consistently reported over diverse machine environments. For instance, machines with pipeline processing units, array processing units, or parallel processing units often cannot pinpoint an arithmetic exception in the source text. If a second arithmetic exception occurs as a result of the first or otherwise, such processors are sometimes unable to distinguish between the exceptions or even diagnose which occurred first. Indeed, it can happen that the error location given by some systems can be in a different subprogram from the subprogram in which that actual error occurred. Hence in terms of portability and reliability, this technique is not a satisfactory mechanism for giving program diagnostics.

There is another important reason for not using the arithmetic exceptions as diagnostics that is germane to program libraries -- the attitude of the user. Most users when confronted with such diagnostics immediately assume that the subprogram has bugs in it, and may not consult the documentation to determine whether these diagnostics are expected. One of the common causes of such system diagnostics is incorrectly supplied input data to properly written software. Since poorly written software can also behave in this unfortunate manner, it is easy to understand the user's dilemma and his possibly rash decision that the program is faulty.

2.6   Ordering of Statements

This subject was discussed earlier in Section 2.1 in connection with specification statements. To complete the discussion, DATA statements, statement functions, FORMAT statements and RETURN statements need to be considered; the ordering of the remaining statements is largely dictated by the structure of the algorithm being implemented.

The purpose of a disciplined ordering of FORTRAN statements is to improve the readability and clarity of the FORTRAN program, to facilitate reliable maintenance of the source text, and to avoid certain statement order restrictions enforced by some compilers such as the IBM FORTRAN G and WATFIV {22} compilers (cf. Section 2.1). We recommend the following order:

1.   SUBROUTINE or FUNCTION statements (if any).

2.   Type specification statements declaring identifiers used as dummy scalar arguments in the order INTEGER, LOGICAL, REAL, DOUBLE PRECISION, and COMPLEX, followed by those declaring dummy array arguments in the type order.

3.   Comments describing the purpose of the program and defining the role of each dummy argument in the program. We prefer to see a clear specification of the input-output status of each parameter. (Cf. {41} for a discussion of input-output status of arguments.)

4.   Type specification statements for identifiers in a common block followed by the COMMON statement for that block. It is recommended that each common block with its list of associated entities be declared with separate sets of type specification and COMMON statements.

5. Type specification statements corresponding to all local entities, statement functions, and external functions in the order INTEGER, LOGICAL, REAL, DOUBLE PRECISION, COMPLEX, and EXTERNAL followed by any EQUIVALENCE statements. We prefer to see the types of statement functions and external functions specified in separate type specification statements for each type and each class. Comments delineating each class of entities are recommended.

6. Type specification statements corresponding to names of intrinsic or basic external functions used in the program unit.

7. DATA initialization statements for local entities. Note that the Standard (9.1.2 and 9.1.3 of {4}) and many compilers require that DATA statements appear after all specification statements.

8. Statement functions listed in alphabetical order.

9. Other statements (except END) belonging to the program unit. The final statement before the END should be a RETURN statement for a subprogram unit or a STOP or GO TO statement for a main program unit. In addition, we prefer a single RETURN statement for a subprogram; the investigation of the behavior of the subprogram at the point of completion becomes much easier. We recommend that each FORMAT statement accompany the input/output statement from which it is first referenced. Alternatively, another discipline for FORMAT

statements which is often used is to collect them just
before the END statement, ordering them by their numeric
labels. This latter discipline may be preferable if
certain FORMAT statements are referenced from more than
one input/output statement. Our experience has been that
such multi-references are infrequent. For us, the repetition
of the FORMAT statement with a different label is preferable
to searching for the end of the program unit and then for
the appropriate FORMAT statement.

10.   An END statement.

Besides curtailing the poison derived from haphazardly ordered
FORTRAN statements, this ordering is standard conforming, and hence
acceptable to many compilers. For those compilers that enforce an
order for FORTRAN statements that is incompatible with the above order,
the discipline itself will facilitate the implementation of automated
or manual modifications to avoid the limitations of such compilers.

2.7   The COMMON Statement

In the previous section, we dealt with the recommended position
of COMMON statements in the program body. In this section, we discuss
the use of COMMON for interprogram communication.

Historically, the role of the COMMON statement was twofold; for
blank common, to reuse storage (primarily needed by the loader) for
program entities, since blank common, which could not be initially
defined by a DATA statement, could overlay part of the operating
system which was no longer needed to execute the program; for labeled
common, to communicate the values of certain specified entities between

program units. In the former role, the COMMON statement creates a
storage association of entities, possibly of unlike type, for the
purpose of conserving space. In the latter role which nowadays represents
the more important use of the COMMON statement, it creates a mathematical
association of entities in different program units, that is, an association
which causes two or more entities to be equated mathematically.

The COMMON statement's poison is a consequence of its history;
both blank and labeled common can be used for storage association as well
as mathematical association. The result is that the two uses of the
COMMON statement obscure the important use as a method for communicating
values between program units. Thus, we recommend the following discipline
be followed. For communicating global entities, the COMMON statement is
to be used with the same entity list for the blank or labeled common
block in each program unit. For conserving storage space, again the
COMMON statement is to be used with the same entity list for the
blank or labeled common block, with an EQUIVALENCE statement[#] identifying
the storage association.

To illustrate our point, consider the following program segments.
Suppose the segment

```
REAL  A(10), B(10)
COMMON  /EXAMPL/A, B
```

appears in one program unit, and the segment

```
COMPLEX  C(10)
COMMON  /EXAMPL/C
```

[#]It should be pointed out that, according to the Standard,
the EQUIVALENCE statement should not be used to equate
mathematically two or more entities; cf. 7.2.1.4 of {4}.

in another program unit. Is the association of C with A and B a mechanism for communicating values, or is this association for some other purpose such as conserving storage space? In general, we cannot tell what is intended, but the interpretation of the program may depend upon knowing the intended purpose of this association.

Hence, this calls for a discipline in which the EQUIVALENCE statement is used only to overlap storage, while the COMMON statement is used exclusively to communicate global information from program unit to program unit by association of entities with the same names and same type specifications. As an additional benefit this disciplined use of the COMMON statement avoids the accidental misdeclaration of identifiers in the common block since the COMMON statement and the associated declarations for the identifiers can be copied without change into each program unit.

The use of the COMMON statement to cause the same data storage to be used for independent entities points out a major deficiency in the FORTRAN language. There is no clean and convenient mechanism in FORTRAN for such association of entities in different program units. Some of the proposals to the FORTRAN Standard committee X3J3 defining a dynamic equivalence feature for dummy arguments have been presented in the FOR-WARD Newsletter {26, 42}.

More poisons with COMMON statements are given in Sections 3.2 and 3.6.

2.8   FUNCTION Subprograms with Side Effects

Further to the topic of clear program structure, we discuss the
use of FUNCTION subprograms with side effects. The use of such subprograms in FORTRAN can be disastrous, particularly in the presence
of optimizing compilers.

There are basically two ways in which a FUNCTION subprogram in
FORTRAN can have side effects; first, the function subprogram can change
the value of one or more of its dummy arguments; and second, the FUNCTION
subprogram or the subprograms that it calls can modify the values of one
or more entities in common blocks in the calling program. The FORTRAN
Standard permits FUNCTION subprograms to have side effects, but restricts
the use of such functions. Briefly, these restrictions are formulated
so as not to impede optimizing compilers in improving the efficiency of
the compiled code.

The restrictions are somewhat involved but will be stated here
for completeness. First, a FUNCTION subprogram must not have a side
effect that changes the values of any variables in the expression,
assignment statement, or CALL statement in which it appears; cf. 6.4
in {4}. The restriction permits an optimizing compiler to evaluate
the subexpressions in such statements in any order consistent with
mathematical meaning. For instance, consider the expression

$$F(I) + X(I)$$

where F is a FUNCTION subprogram of type REAL with integer argument I,
and X is a vector of type REAL. If F were to modify I, different
values for this expression would be obtained, depending upon whether
X(I) is evaluated before F(I) or after. Indeed, the order of evaluation

may depend on context, that is, dependent on the presence or absence of the expression X(I) in a previous statement.

As another example, consider the expression

$$F(I) + F(I)$$

which may result in object code that doubles the value of a single evaluation of the function F, or sums the values of two evaluations of F.   If the function F modifies I, the two methods for evaluating the expression may produce different results; cf. 6.4, 8.3.2, 8.4.2, and 10.2.9 of {4} for the Standard's treatment of this issue.

The second restriction on FUNCTION subprograms with side effects is that they must not be included in subexpressions in a statement which may not be completely evaluated.   For example, consider the expression

$$(1) \quad J \ .LT. \ 0 \ .OR. \ F(I) \ .EQ. \ 0.0$$

The truth of the condition J less than 0 is sufficient to make the entire expression true.   From the point of view of the optimizing compiler, the function F need not be evaluated.   If F(I) defines (or redefines) I, the above construction is prohibited by the Standard in the sense that, after the evaluation of expression (1), I is undefined for all subsequent uses until it is redefined; cf. 6.4 and 10.2.9 of {4}.

Our discipline here is simple; never create function subprograms with side effects.   If there is a need to modify variables in the argument list or in common blocks, use a SUBROUTINE subprogram instead.

There is no easy way to verify that such FUNCTION subprograms conform to the Standard. Ignoring this discipline may have unexpected or even disastrous results.

2.9   Formatting Conventions -- A Tidy Systematic Style

A tidy, clearly-blocked formatting discipline does wonders for the readability and maintainability of FORTRAN software. Although it is a tedious task to rewrite existing software in such a manner, such a discipline requires little additional effort for new software.

There are many possible formats that provide very readable layouts for FORTRAN source text. Variants in the details of particular formats are mainly a matter of style, personal preference, and in some instances specialized constraints such as the media for publication. Such variants are generally unimportant, but what is important is the level of detail within statements that the formatting discipline addresses. Another important point is that the formatting discipline be applied completely and consistently throughout the source text.

There are several formatters generally available for processing existing software {11, 23}. In what follows, we give an example of a formatting discipline to illustrate the level of detail to which a good discipline should be specified. This discipline was derived from that used by the NATS project {12} in the EISPACK and FUNPACK collections of software {18, 46}.

1.   The SUBROUTINE, FUNCTION, specification (INTEGER, LOGICAL, REAL, DOUBLE PRECISION, COMPLEX, EXTERNAL, COMMON, and EQUIVALENCE) and DATA statements begin in column 7. Two blanks separate the type specifier

and the list, with one blank between the word DOUBLE
and PRECISION.

2. Continuation cards for long FORTRAN statements use X in
column 6 and the continued lines are indented two columns
more than the initial line. A statement requiring continuation
is split after the binary arithmetic (+,-,*,/) or logical
(.OR., .AND.) operators for expressions, after the "/,"
in DATA statements, after the ")," in EQUIVALENCE statements,
after the last entity in a common block, after the comma
in CALL, computed GO TO, and assigned GO TO statements,
and after the comma in long argument lists, long lists in
DATA statements, and long lists of entities in EQUIVALENCE
or COMMON statements.

3. The statements between a DO statement and the CONTINUE
statement terminating the DO loop range are indented an
additional two columns. Indentation of two columns is
also recommended to indicate the 'then' and 'else' parts
of logical IF statements, particularly when the 'then'
or 'else' parts comprise more than two statements.

4. For logical IF statements, we prefer no blanks in the
sequence "IF(", one blank after the opening parenthesis,
one blank before the matching closing parenthesis, two
blanks after the closing parenthesis and the usual spacing
for the statement completing the logical IF statement.
If the statement that completes the logical IF statement
does not fit on the current line, we prefer that statement
begin on the next line indented two spaces.

5.   For the DO statement, we recommend a space before and after the statement number, and before and after the "=" operator, with no blanks in the DO parameter list.

6.   For GO TO statements, we prefer no spaces in the string "GOTO" and one space before the statement label. For the computed GO TO statement, a space after "GOTO" and a space after the closing parenthesis is recommended. For the assigned GO TO statement, a space after "GOTO" and one after the comma following the identifier is recommended. For the ASSIGN statement, we recommend a space before and after the statement label and one before the identifier. Two blanks after the ")" and before the first entity of the entity list of an input/output statement are recommended.

7.   The preferred spacing around operators is:

     a)   a space before and after the binary operators (+,-,*,/) except when they appear in subscript expressions, expressions in CALL statements, FUNCTION subprogram references, and relational expressions where no blanks around the operators are preferred.

     b)   no blanks between the unary operators (+,-,.NOT.) and their operands.

     c)   one blank before and after the logical operators (.AND.,.OR.) and the relational operators (.EQ.,.NE.,.LT.,.LE.,.GT.,.GE.) regardless of context.

    d)   one blank before and after the "=" operator in an assignment statement.

8.   The use of blank comment cards (blank cards except for a C in column one) to display the structure of the algorithm is recommended. In particular, a blank comment card is preferred in the following places:

    a)   after the SUBROUTINE or FUNCTION statement;

    b)   after the type specifications of the dummy arguments of a subprogram;

    c)   after each COMMON statement;

    d)   after the last DATA statement;

    e)   after the last statement function, followed by a demarcation line of (say) asterisks (*);

    f)   before a DO statement;

    g)   after a CONTINUE statement ending a DO loop;

    h)   before a logical IF statement.

This discipline for formatting conventions is only one of many possible disciplines that can result in tidy, clearly structured code. For an example of another formatting discipline, see {17, 46}.

3.   Poisonous FORTRAN Constructs

In this section, we present several common FORTRAN constructs that when used present difficulties in producing portable and transportable software. The resulting discipline that avoids these undesirable constructs defines a subset of FORTRAN that approximates an intersection of the most commonly used FORTRAN dialects. As mentioned earlier, this discipline avoids most of those constructs that are poorly defined by the FORTRAN

Standard or features that are peculiar to the FORTRAN language. Also, the discipline avoids all extensions to the Standard, even those in common usage. Generally speaking, this discipline produces software that is more easily moved from place to place.

There are several verifiers (or filters) and checkers available for FORTRAN {22, 41, 44}. A general discussion of such tools appears in {29}. Some of the poisonous constructs given below are diagnosed by these tools.

3.1 Defined, Redefined, and Undefined Entities

Section 10.2 of the FORTRAN Standard {4} and its subsections describe ways in which entities can become defined, redefined, and undefined. Some of the rules stated there are often violated and are the source of many FORTRAN poisons; cf. {47} for a general treatment of the lifetime of entities and the duration of program states in FORTRAN as well as other languages.

According to the FORTRAN Standard (10.2.2, 10.2.3, 10.2.6 and 10.2.9 of {4}), the lifetime of an entity begins when the entity is defined in any one of six ways and ends when the entity becomes undefined in any one of eight ways. To clarify the Standard, we briefly summarize the ways in which an entity can become defined or undefined. Following this summary are several examples illustrating the more frequent misuses of undefined entities in FORTRAN. However, before discussing the concept of definition of entities in FORTRAN, we need to be aware of the three ways in which entities can become associated in FORTRAN: by use of the COMMON statement, by use of the EQUIVALENCE statement, and by use of subprogram linkage generated for a SUBROUTINE or FUNCTION reference. Association of entities of unlike type is permitted only with COMMON and EQUIVALENCE statements; cf. Section 2.7 of this paper where such associations are discouraged in COMMON statements.

All entities can be considered undefined before execution begins, and become defined in any one of the following six ways. Those entities not defined by one of these methods remain undefined.

D1) An entity which precedes the symbol "=" in an assignment statement becomes defined after the assignment statement is executed.

D2) An entity which is the control variable of a DO loop (for example, I in the statement DO 10 1 = 1,N) becomes defined when the DO statement is executed.

D3) An entity in the list of variables or array elements in a DATA statement becomes defined at the first executable statement in the program unit when the program unit is first executed.

D4) An entity which is the control variable of a DO-implied list in an input/output statement becomes defined when the input/output statement is executed but is only defined within the scope of the particular DO list.

D5) An entity in the list of an input statement becomes defined upon execution of the input statement at the point of its appearance in the list. For example, in the statement

READ(5,10)   (I,A(I),L=1,10)

the entity I is successively defined from values specified
on the input medium and then used to index the array element
A(I).

D6) An entity which is associated with another entity of matching
     type via a COMMON or EQUIVALENCE statement becomes defined
     when that entity of matching type becomes defined, with one
     exception: If the associated entity of matching type is
     defined as per method D5 above, the first entity does not
     become defined until the input statement is complete. An
     entity which is a dummy argument of a subprogram unit
     becomes defined at the first executable statement of the
     subprogram unit, provided the actual argument associated
     with the dummy argument is defined. Similarly, this
     definition applies to array dummy arguments, element
     by element.

An entity is said to be redefined if it is defined and becomes
defined a second time by one of the above methods. For example, at
the beginning of each pass after the first through the DO loop, the
control variable becomes redefined.

An entity remains defined unless it becomes undefined by one
of the following actions:

U1) An entity which is the control variable of a DO loop
     becomes undefined when the DO loop is satisfied. Note
     that it remains defined if a transfer out of the DO
     loop is made.

U2) An entity which is the control variable of a DO-implied list becomes undefined upon the completion of the DO-implied list; note that such a control variable is then undefined within the remainder of the input/output statement and in subsequent statements, until it is defined by one of the six methods D1-D6. For example, in the input statement

    READ(5,10)   (A(I),I=1,10),(B(I,J),J=1,10)

the entity I in the subscripted item B(I,J) is undefined.

U3) An integer entity in an ASSIGN statement becomes undefined for subsequent uses as an integer, until it is defined as an integer by one of the six methods D1-D6.

U4) Consider an entity X that can be defined (or redefined) as a consequence of executing a FUNCTION subprogram; that is, the FUNCTION subprogram has a side effect which modifies X. Suppose that the FUNCTION subprogram is referenced in a subexpression which may not be evaluated: For example, the FUNCTION subprogram F in the expression J .LT. 0 .OR. F(A) .GT. 5.0 may not be evaluated when J is negative; here, the side effect of F is to modify the entity X in a COMMON statement. Then the entity X becomes undefined for all uses after the evaluation of the expression, even for further uses within the same statement (cf. 10.2.9 of {4}). For example, in the statement

    IF(J .LT. 0 .OR. F(A) .GT. 5.0)   Y = X + 1.0

the entity X in the assignment statement completing the IF statement becomes undefined, since the FUNCTION subprogram F modifies X, and F may not be evaluated when J is negative (cf. Section 2.8 of this paper).

U5) An entity associated with an entity of different type becomes undefined when the associated entity is defined.

U6) An entity associated with an entity of matching type becomes undefined when the associated entity becomes undefined.

U7) All local entities (not in common blocks and not dummy arguments) become undefined upon execution of a RETURN or END statement. This rule also extends to entities in a common block when the program unit to which the return is made does not contain the common block and is not currently invoked from some sequence of program units that contains the common block.

U8) All initially defined entities (that is, entities appearing in a DATA statement) which are redefined in the program unit become undefined upon execution of a RETURN or END statement.

Implicit in the above discussion of the definition states of entities is the important rule (cf. 10.3 of {4}) that the use of an entity before it is defined or after it becomes undefined is prohibited. There are two frequent violations of this rule. Often, programmers (because their compiler and/or operating system initializes all storage locations to zero) assume that before the first statement of a program unit is executed all arithmetic entities (items of type INTEGER, REAL, DOUBLE PRECISION,

and COMPLEX) have the value zero, and all logical entities have the
value .FALSE. The dangerous aspect of this assumption is that the
actual values such undefined items have may behave like zero for
some operations but not for other operations on the same machine,
or behave like zero on some machines but not on other machines.
For example, if X is a small unnormalized number, it can cause the
expression X .NE. 0.0 to be true, and yet provoke an exception if used
as a divisor; that is, the statement

$$IF(X .NE. 0.0) \quad Y = 1.0/X$$

can result in a zero divide; cf. {20}. Clearly, assumptions for the
initial values of entities are dangerous and non-portable.

Of course, the antidote is to initialize explicitly all variables
in a program before they are used. We are helped in this goal by
several verifiers and checkers such as DAVE {41} and WATFIV {22}
that diagnose uninitialized entities. Another mechanism for diagnosing
uninitialized entities is present in many CDC operating systems. There,
all of core can be initialized to a quantity treated as undefined by
the operating system. Any operation with this quantity is diagnosed
as an error by the hardware and operating system. However, most of these
checking techniques are not foolproof as, for instance, the particular
test data may not cause the algorithm to follow a path that fetches
the undefined variable; DAVE is an exception in that it does diagnose
all uninitialized entities.

The second violation of the above rule involves uses of previously
defined but currently undefined entities, uses that in most cases are
diagnosed by DAVE but by few other FORTRAN checkers. The more frequent
FORTRAN constructs that violate this rule are:

C1) Use of the control variable of a DO loop or a DO-implied list after the DO is satisfied. This construct is diagnosed by DAVE.

C2) Re-use without redefinition in a subprogram of a local variable which has become undefined upon the previous execution of the RETURN or END statement, on subsequent invocations of the subprogram. The implicit assumption that local variables retain their values between consecutive executions of a subprogram is not supported by the Standard. DAVE diagnoses this construct, including the more subtle case when initially defined entities are redefined.

C3) Use of local entities which have become undefined for reasons of association. The difficulty here is that most programmers are unaware of what causes entities to become undefined by association and the effect that using such undefined quantities may have on the proper execution of the program.

The difficulty with construct C1 is that it makes the program non-portable in that the actual value of the control variable in such circumstances is system dependent, sometimes being the terminal loop value plus the increment loop value, sometimes being the terminal loop value itself, and sometimes being plain unpredictable.

The antidote for this poison is the same as discussed earlier in Section 2.2. It involves the disciplined use of an integer identifier, say SAVEI, to retain the value of the control variable I outside the DO loop range. For example, SAVEI in the program segment

```
        INTEGER   I,SAVEI
           .
           .
           .
        DO 10 I = N1,N2
           SAVEI = I
             .
             .
             .
 10 CONTINUE
           .
           .
           .
```

retains the most recently defined value of I in the loop. Using SAVEI
rather than I at the completion of the DO loop avoids the difficulty.
Also, it clearly illuminates this exceptional usage of the DO control
variable.

The construct C2 is often used for state indicators of an algorithm
or to save rather costly or difficult-to-recompute data from program
call to program call. For instance, a SUBROUTINE subprogram may be
estimating an integral in two phases; one computes nodes and weights
based upon some property of the integrand and the second computes the
integral using these weights and the values of the integrand at the
nodes. Several integrands with the same behavior may be desired;
much computational effort can be saved if the nodes and weights are
calculated just once. To save this information, local variables are
used into which the nodes and weights are computed and stored on the
first call of the subprogram and then retrieved on subsequent calls
of the subprogram. A local variable which has been initialized in a
DATA statement is used to indicate the status of the program; that is,
whether the initial call has been made.

In such a situation, the assumption is that the storage used for
the computed nodes and weights and for the status variable is left
unaltered between successive invocations of the subprogram. The Standard,
however, indicates that this assumption is not valid and that values in
local storage remain undefined when referenced upon subsequent invocation
of the subprogram. In practice, the error from this assumption can occur
in an overlay environment. If, after the initial call is made, the space
occupied by the subprogram and its data areas is used by another program,
then the operating system is required to bring in a fresh copy of the
subprogram. However, this fresh copy will not have the computed nodes
and weights in its storage areas, nor will the initially defined entities
have their redefined values.

The antidote, of course, is to avoid the use of such entities.
Instead, either add the corresponding identifier names as arguments
or specify them in blank or labeled common blocks. However, if
labeled common blocks are used, be aware of another restriction
which if ignored creates undefined quantities; that labeled common
blocks must be specified (in a COMMON statement) in every program unit
that directly or secondarily invokes the subprogram in question.

For the future, the new FORTRAN 77 {8} recognizes the need to
save the values of certain entities between successive references to
subprograms, and for this purpose provides the SAVE statement. This
is a welcome change.

To illustrate construct C3, consider the following program
segment:

```
                                                        cols.
                                                        79-80
C                                                         1
C        THIS IS AN EXAMPLE OF ASSOCIATION OF ENTITIES    2
C        OF UNLIKE TYPE IN WHICH THE DEFINITION OF ONE     3
C        OF THE ASSOCIATED ENTITIES CAUSES                4
C        THE OTHER TO BECOME UNDEFINED.                    5
C                                                          6
         REAL  AZ(2),SIZEZ,ZI,ZR                          7
         COMPLEX  Z                                       8
         EQUIVALENCE  (ZR,Z,AZ(1)),(ZI,AZ(2))             9
C                                                         10
           .                                              11
           .                                              12
           .                                              13
         ZR = -3.0                                        14
         ZI = 1.0                                         15
         Z = CMPLX(2.0*ZR,-2.0*ZI)                        16
         SIZEZ = ABS(ZR) + ABS(ZI)                        17
           .                                              18
           .                                              19
           .                                              20
         END                                              21
```

The definition of Z at line 16 undefines the entities ZR and ZI
(cf. clarification number 13 of {7}); consequently, the resulting
value for SIZEZ in line 17 is unpredictable. Specific compilers
may reasonably produce either 4 or 8, the 4 by using the values
of ZR and ZI saved in registers from the previous statements, and
the 8 by accessing the values of ZR and ZI from storage. The
clarification of this point in {7} reveals that the Standard does
not even specify that the real and imaginary parts of Z are stored
with the real part first followed by the imaginary part in consecutive storage locations.

A similar poison lurks when associating the first storage unit
of a double precision entity with an entity of type REAL. The Standard
nowhere specifies that the first storage item is a valid entity of type
REAL, nor that the first item is a less precise approximation to the
double precision entity.

The antidote for such FORTRAN constructs is to avoid their use.
These constructs are definitely not supported by FORTRAN 66, although
a partial association of entities of type REAL and COMPLEX is supported
by FORTRAN 77.

3.2 Unsafe Associations

Recall that entities can be associated in any one of three ways
in FORTRAN: by use of the COMMON statement, by use of the EQUIVALENCE
statement, and by use of subprogram linkage generated for a SUBROUTINE
or FUNCTION reference. Association of entities of unlike type is permitted
only with COMMON and EQUIVALENCE statements.

In certain circumstances, the definition or redefinition of
associated entities is illegal. First, if two entities with distinct
identifiers in the same subprogram unit become associated by (two
instances of) argument association, or by argument and common association
simultaneously, definition or redefinition of either is prohibited (cf.
8.3.2 and 8.4.2 of {4}). Second, for argument association, if the
actual argument is an expression or constant, the redefinition of the
dummy argument is prohibited. In practice, the result of such definitions
is unpredictable in the sense that what happens depends upon which of
the several methods of implementing argument association is used. We
give three examples of this illegal use, illustrating the possible outcomes
corresponding to the usual implementations of argument association. The
PFORT verifier diagnoses these illegal constructs; cf. {44} for an
alternative discussion of these points.

Consider first the following example program.

```
C
C      THIS IS AN EXAMPLE OF ILLEGAL ARGUMENT
C      ASSOCIATION IN WHICH TWO DIFFERENT IDENTIFIERS IN
C      A SUBPROGRAM ARE ASSOCIATED WITH EACH OTHER BY
C      ARGUMENT ASSOCIATION.
C
       INTEGER  NIN,NOUT
       REAL   INPUT,COMPTD,EXPECT
       DATA   NIN,NOUT/5,6/
C
C
C
       READ(NIN,10)   INPUT
    10 FORMAT(E16.8)
       COMPTD = INPUT
       EXPECT = SQRT(INPUT)
C
       CALL   EXAMPL(COMPTD,COMPTD,COMPTD)
C
       WRITE(NOUT,20)   INPUT,EXPECT,COMPTD
    20 FORMAT(21H0THE INPUT VALUE IS   ,E16.8/
      X        24H0THE EXPECTED VALUE IS  ,E16.8/
      X        24H0THE COMPUTED VALUE IS  ,E16.8)
       STOP
       END
       SUBROUTINE  EXAMPL(X,Y,Z)
C
       REAL  X,Y,Z
C
C
C
       Y = X**2
       Z = SQRT(X)
C
       RETURN
       END
```

The CALL statement in this main program unit is strictly forbidden,
since it causes the association of three different dummy arguments
of the subprogram EXAMPL, two of which are redefined. In practice,
the value of the result COMPTD, after the call to EXAMPL, is undefined,

since it will depend on the particular method for implementing argument association. For instance, suppose the value of INPUT read from the data file is 4.0. Then, if the address of COMPTD is used for the locations of X, Y, and Z, 4.0 would be the result for COMPTD (whether or not the execution order for the statements defining Y and Z is preserved). On the other hand, the method for implementing the argument association may be to copy the input values for the actual arguments corresponding to X, Y, and Z into temporary storage locations internal to the subprogram EXAMPL and then at the RETURN statement, copy the final values of X, Y, and Z obtained from the temporary storage locations into the locations occupied by the actual arguments. Depending on the order in which the values of X, Y, and Z are restored to the actual arguments, the possible values for COMPTD are 2.0, 4.0, or 16.0.

Now consider another example, which illustrates a misuse of simultaneous argument and common association.

```
C
C      THIS IS AN EXAMPLE OF ILLEGAL ARGUMENT AND COMMON
C      ASSOCIATION IN WHICH TWO DIFFERENT IDENTIFIERS
C      IN A SUBPROGRAM ARE ASSOCIATED SIMULTANEOUSLY
C      BY ARGUMENT AND COMMON ASSOCIATION.
C
       INTEGER  NIN,NOUT
       REAL  Y
       COMMON  /EXAM/Y
C
       DATA  NIN,NOUT/5,6/
C
C
C
       READ(NIN,10)  Y
    10 FORMAT(E16.8)
C
       CALL  EXAMPL(Y)
C
       WRITE(NOUT,20)  Y
    20 FORMAT(20H0THE VALUE OF Y IS  ,E16.8)
       STOP
       END
       SUBROUTINE EXAMPL(B)
C
       REAL  B
C
       REAL  Y
       COMMON  /EXAM/Y
C
       B = Y**2
       Y = SQRT(Y)
C
       RETURN
       END
```

If the value read into Y from unit 5 is 2.0, then the possible values of
Y printed by the main program unit are 4.0 and 2.0, again depending upon
the method of implementing argument association. If the value of the
actual argument is stored temporarily in the subprogram, then the result
is 4.0, as the assignment to B in EXAMPL does not change the first
storage unit in the labeled common block EXAM until the RETURN statement
is executed. If the address of the actual argument Y is used, then the
value is 2.0.

The third example involves simply argument association, but this time the actual argument is a constant and corresponds to a dummy argument that is redefined within the subprogram. Consider the program:

```
C
C      THIS EXAMPLE ILLUSTRATES THE ILLEGAL USE OF
C      ARGUMENT ASSOCIATION WHEN THE ACTUAL ARGUMENT
C      IS AN EXPRESSION THAT IS NOT A VARIABLE, ARRAY
C      NAME, OR AN ARRAY ELEMENT NAME.
C
       INTEGER  NOUT
       REAL  ONE,TWO
       DATA  NOUT/6/
C
C
C
       ONE = 1.0E0
C
       CALL  EXAMPL(1.0E0)
C
       TWO = ONE + 1.0E0
       WRITE(NOUT,10)  TWO
   10  FORMAT(19H0TWO HAS THE VALUE ,E16.8)
C
       STOP
       END
       SUBROUTINE  EXAMPL(X)
C
       REAL  X
C
C
       X = 2.0E0
C
       RETURN
       END
```

This program is illegal because the dummy argument is redefined, and the actual argument is a constant. In practice, the resulting combination of program and subprogram produces unpredictable results. This time, the results do not depend upon how the subroutine handles argument association but upon how the calling program treats expressions (or in this case, constants) when communicated as arguments. Some compilers store the value of the expression (or constant) into a temporary location and communicate to the subroutine the address of that temporary location. In that case, the result of 2.0 would be obtained for TWO. If, on the other hand, the address of the constant 1.0E0 itself is passed to the subroutine, then the unexpected result of 3.0 for TWO will be obtained, since the redefinition of the dummy variable X in the subroutine EXAMPL changes the "constant" 1.0E0 to 2.0E0.

The antidote for these constructs is simply to avoid such unsafe associations. Although, in most cases, problems only occur when the appropriate entities are defined or redefined, the unsafe construct should be avoided; there is no guarantee that a later modification will not cause crucial entities to be inadvertently defined.

The last three examples are diagnosed by the interprogram analysis of the PFORT {44} and DAVE {41} verifiers. Waite {47} gives an excellent survey of the possible techniques for implementing FORTRAN argument association. Most of the unsafe constructs involving argument association implicitly assume that a particular one of these techniques is used. Consequently such constructs are not portable. Indeed, besides being explicitly prohibited by the Standard, they violate an implicit rule that states that no FORTRAN construct is permitted that is capable of distinguishing between the various techniques for implementing argument association, both within the 'CALLing' program and the 'CALLed' subprogram unit.

3.3 Type Conversions

Type conversions occur in program units in two ways: implicitly in mixed mode expressions and in assignment statements $v = e$ where the types of v and e are different; and explicitly with the intrinsic functions INT, FLOAT, IDINT, SNGL, DBLE, REAL, and CMPLX. It is our opinion that implicit conversions obscure the meaning of the statements, since they require a reference to the appropriate type specification statement (if provided) for comprehension. In addition, such implicit conversions make it difficult to move software from place to place when precision conversion is necessary.

Our antidote is to make all conversions explicit by use of the intrinsic functions INT, FLOAT, IDINT, SNGL, DBLE, REAL, and CMPLX, and by the introduction of the arithmetic statement function[#] DFLOAT. We recommend the use of these functions as indicated by the following table.

Conversion Across the Assignment Operator

Type of v	Type of e	Function to be Used
INTEGER	REAL	v = INT(e)
REAL	INTEGER	v = FLOAT(e)
INTEGER	DOUBLE PRECISION	v = IDINT(e)
DOUBLE PRECISION	INTEGER	[#]v = DFLOAT(e)
REAL	DOUBLE PRECISION	v = SNGL(e)
DOUBLE PRECISION	REAL	v = DBLE(e)
REAL	COMPLEX	v = REAL(e)
COMPLEX	REAL	v = CMPLX(e,0.0E0)

[#]DFLOAT is not a Standard basic intrinsic function, but we recommend that, when it is needed, it be defined as an arithmetic statement function in the following way:

```
DOUBLE PRECISION DFLOAT
INTEGER I
DFLOAT(I) = I
```

Alternatively, the implicit conversion could be avoided by using the statement function

```
DFLOAT(I) = DBLE(FLOAT(I))
```

but this may restrict the domain of integers that can be exactly converted to double precision data.

We do not recommend the use of the intrinsic function IFIX; it is the same as the basic intrinsic function INT.

Mixed mode expressions are limited by the FORTRAN Standard to contexts where REAL and DOUBLE PRECISION operands are combined by arithmetic and relational operators, and where REAL and COMPLEX operands are combined by arithmetic operators. We recommend that the intrinsic functions displayed in the previous table be used to make even these data conversions explicit, as well as avoiding more general mixed mode expressions permitted by compilers which extend the FORTRAN 66 language.

The intrinsic functions INT and IDINT should be used with care in portable or transportable software. The range of values for these functions is severely limited on some machines, particularly minicomputers. A range of values of less than 32768 in magnitude is not uncommon.

A discipline that avoids the use of the intrinsic functions MAX1, MIN1, AMAX0, and AMIN0 is recommended, partly to avoid the implicit conversion between the types of the arguments and the types of the results. But more importantly, these functions do not all have double precision analogues, and the missing analogues cannot be supplied as statement functions because of the requirement that they function correctly with a variable number of arguments. When such functions are used, automated conversion of software between single and double precision is difficult. (The NAG precision converter {24} does handle even these functions by converting the arguments appropriately using the basic intrinsic functions.) In place of MAX1(e), MIN1(e), AMAX0(e), and AMIN0(e), we recommend the use of the composite functions FLOAT(MAX0(e)), FLOAT(MIN0(e)), INT(AMAX1(e)), and INT(AMIN1(e)) respectively where e is a list of expressions of the appropriate type.

When using constants in expressions, we recommend that the constant be of the appropriate type. For example, when a double precision quantity is compared with zero, we prefer to represent zero as 0.0D0 and not 0.0 (or 0 which many compilers accept). In addition, a real number, say c, in a complex expression should be represented by the complex constant (c,0.0). This practice should also be extended to the constants in DATA statements.

3.4 Constants and the DATA Statement

Constants are often involved when one is changing the precision of source text from single to double precision or vice versa. In order to localize the extent of such changes, we recommend a discipline in which a constant used in a program unit be represented by an identifier whose value is specified in a DATA statement. In the DATA statement, precision should be indicated by use of the exponent E0 or D0 for constants near one and by the appropriate integer exponent for constants much different from one in magnitude. No more digits should be specified than are implied by the type (and so precision) of the identifier. Some compilers treat excess digits as an error, while others truncate (or round) the decimal expansion or machine representation of the constant. Thus, there is no consistent treatment of the excess digits and hence we recommend that they not be specified. However, for ease of constant modification when porting software from place to place, we do recommend that the most precise representation of the constants ever needed be given in COMMENT lines that precede their appearance in DATA statements.

With this approach, constants can more easily be adjusted to accommodate
the word size of the particular processor on which the software is to
run.

In most computer environments, there are two distinct programs
for conversion of constants from decimal format to internal representation.
One program is used to convert source text constants (such as those that
appear in DATA statements or arithmetic expressions) and the other program
is used to convert decimal constants read as data. Since these two
programs can use slightly different algorithms for rounding, avoid
constructions that depend on comparisons of the two conversion processes.
For example, consider a program that stops after a data value, say 0.99,
is read into the identifier STOPFL. The statement

                    IF(STOPFL .EQ. 0.99)   STOP

may fail to terminate the program; the real datum obtained from the
conversion of 0.99 by the input routine may not equal the real datum
obtained from the conversion of 0.99 by the compiler conversion routine.

A discipline that avoids these difficulties is the use of the
integer form of constants where possible. For example, rather than
multiplying by 0.1E0, dividing by 10.0E0 may avoid many difficulties.
When reading data from input files, represent the data in integer format
instead of floating point format if it can be scaled conveniently,
particularly if the software being used is sensitive to small changes
in the data. This avoids many difficulties when validating the installation
of software at a new site.

A final poison is the use of the DATA initialization statement for
purposes other than defining that certain identifiers have specific
values. Despite its name, the DATA statement should not be used to

initialize variables. In certain cases (as described in Section 3.1), it leads to variables becoming undefined in terms of the Standard, at which point compilers are free to do what they wish with such variables. See the anecdote in Section 5 for a graphic example.

3.5 DO Loops Revisited

In Section 2.2, we discussed the use of the DO loop in terms of clear structured programs. There are two other aspects of the DO loop that should be emphasized here.

First, the FORTRAN Standard (Section 7.1.2.8 of {4}) does not define the action of a DO loop when the initial parameter is less than the terminal parameter. Although most compilers execute the loop once, this cannot be assumed in portable software. (Indeed, FORTRAN 77 {8} specifies that the DO loop range will not be executed in this case). Similarly, programs that modify the initial, terminal, or increment parameter of the DO loop range are not Standard conforming. Indeed, what actually happens is compiler dependent.

Our second poison concerns the use of the DO-implied list. Sometimes, programmers overlook or are unaware of the fact that after the execution of a DO-implied list with control variable, say I, I becomes undefined. Hence, subsequent references to I without first defining it are illegal. Most compilers do not diagnose this error. Consequently, we recommend a discipline that tends to avoid the problem (and also the problem of I becoming redefined unintentionally by the DO-implied list); that is, use a distinct name for the control variable of a DO-implied list, say IMPI, that reduces the likelihood of reusing this identifier within the program unit.

3.6   The COMMON Statement Revisited

There are three poisons concerning the COMMON statement that we present here; Section 3.2 describes another in connection with argument association.

First, according to the Standard (Section 7.2.1.3 of {4}), the size (length in number of storage units) of a labeled common block must be the same in each program unit in which the common block is specified. If this rule is violated, the result in terms of portability may be disastrous, as the actual space for the common block may be equal to the size of the common block in the program unit first loaded that contains it. As there is no portable way to control the order in which program units are loaded, the space may not be sufficient to hold the largest block of that name.

Second, the order of the entities in common blocks affects the portability of the program unit. Certain manufacturers (particularly IBM) require for efficient utilization of the machine that entities which occupy two storage units (COMPLEX and DOUBLE PRECISION types) begin on even word boundaries (that is, their word addresses are even numbers). If this rule is not adhered to, diagnostics for alignment errors are produced at execution time, or the efficiency of the program is severely degraded in order to avoid this limitation on data fetches. Consequently, we recommend a discipline that orders the entities in common blocks essentially in reverse order to that specified in Section 2.1 for the order of type specification statements; that is, entities in the order COMPLEX, DOUBLE PRECISION, REAL, INTEGER, and LOGICAL. Another solution is to have a separate common block for each

data type, but this is less desirable as it separates the natural grouping
of global entities communicated between program units.

In addition, either of these practices for COMMON statements aids
transportability in another context. Some compilers have a shortened
format for integer and logical entities (fewer than the Standard's one
storage unit) in order to conserve data storage. For example, IBM
compilers permit one half of a storage unit for integers and one quarter
of a storage unit for logical entities. Such savings in storage space
for large integer or logical arrays may be significant. When moving
software to environments that can use such space saving features, or from
environments that do use these features, usually just the type specifications
need be modified when either of these disciplines is followed.

Third, entities in labeled common blocks may become undefined
upon the execution of a RETURN (or END) statement. In order that
this not happen, there must exist a COMMON statement for that block
in at least one of the higher level program units in the chain of
active programs at the time the RETURN statement is executed.

3.7   The EQUIVALENCE Statement

The legal uses of the EQUIVALENCE statement are not clearly
defined by the Standard {4}, and consequently the EQUIVALENCE
statement is often abused. Rather than describe the misuses, we
concentrate first upon a discipline that avoids all difficulties
from the point of view of portability.

The FORTRAN Standard states in Section 7.2.1.4 of {4}:

"The EQUIVALENCE statement is used to permit the sharing of storage by two or more entities."

"The EQUIVALENCE statement should not be used to equate mathematically two or more entities."

In other words, the EQUIVALENCE statement should be used to conserve storage only; it should not be used to enable a storage unit defined with a value using one identifier to have that value referenced using a different identifier in the same program unit. In particular, the EQUIVALENCE statement should not be used to access the real and imaginary parts of entities of type COMPLEX, or the more significant parts of double precision variables; cf. topic 13 in the clarification {5}. The rationale behind these restrictions is that the Standard nowhere specifies which storage unit for a complex entity contains the real part, or for double precision entities which storage unit contains the most significant part (or even how complex and double precision numbers are represented).

We can further appreciate the Standard's restrictive approach towards (or more properly, interpretation of) the EQUIVALENCE statement by considering this statement from the point of view of the compiler designer. In producing object code according to the Standard's interpretation of the EQUIVALENCE statement, extensive optimization strategies can be used, essentially treating the equivalenced entities as independent, since their values cannot be equated mathematically. To produce object code while allowing the equating of entities via their addresses, the

optimization strategies must be curtailed or more extensive source code
analysis is required to preserve the identification of entities with
the same addresses.

No compiler or verifier that we know of is this pedantic about
the EQUIVALENCE statement. To exemplify our point, consider the
following program:

```
C
C      THIS IS AN EXAMPLE OF ILLEGAL USE OF EQUIVALENCE
C      THAT IS NOT DIAGNOSED BY MOST FORTRAN COMPILERS.
C
       INTEGER DIMS,DIMT,I,NIN,S(10)
       EQUIVALENCE   (DIMS,DIMT)
C
       DATA   NIN/5/
C
C
C
       DIMS = 7
       READ(NIN,10)   DIMT,(S(I),I=1,DIMS)
    10 FORMAT(15I4)
C
          .
          .
          .
       END
```

According to 7.2.1.4 of the Standard {4}, this program is illegal,
since the entities DIMS and DIMT are being equated mathematically in
the READ statement. On the other hand, according to 10.2.3 of the
Standard {4}, this program is to be interpreted so that DIMS in the
READ statement takes its value before the READ statement is executed,
namely 7, and does not take the value of the first integer field of
the data. The current versions of the IBM FORTRAN G, G1, H, and WATFIV
compilers and PFORT FORTRAN analyzer at Argonne National Laboratory
do not diagnose this construct; indeed, contrary to the Standard, the

first four compilers use the value of DIMT read from the input file
for DIMS. How does your compiler interpret this program segment?
Does it diagnose the construct as illegal?

Another difficulty with the EQUIVALENCE statement is the same
alignment problem previously described for the COMMON statement.
However, the difficulty only occurs when identifiers of differing
type are equivalenced. It arises because entities of type COMPLEX
and DOUBLE PRECISION need to be aligned to even word addresses on
certain machines for efficient utilization of the hardware and the
EQUIVALENCE statement can cause a conflict. For example, consider
the specification statements:

```
REAL  A(10), B(7)
DOUBLE PRECISION  D(5), E(3)
EQUIVALENCE  (A,D), (A(4),B), (B,E)
```

The above statements would require that the double precision entities
D and E both could not have even word addresses. Such uses of the
EQUIVALENCE statement must be avoided in portable software to avoid
diagnostics or extreme program inefficiencies.

Because of the ambiguous description of the EQUIVALENCE statement
in the Standard, we recommend that it not be used except to conserve
storage, as suggested by the Standard. A test for adherence to this
rule has been suggested by J. M. Boyle {14}; the program unit with
EQUIVALENCE statements should give results which are the same as
those obtained from the same program without EQUIVALENCE statements
for all possible values of the data. If the FORTRAN program passes
this test, it is using the EQUIVALENCE statement properly. However,
as stated, this test is not practical; but as a conceptual model for
analyzing the role of the EQUIVALENCE statement in a program, it is
very illuminating.

3.8   Certain Intrinsic Functions

The intrinsic functions AMOD, MOD, SIGN, ISIGN, and DSIGN, and
the basic external function DMOD are not defined when their second
argument is zero. Avoid such usage of these functions, as problems
in portability will be encountered; cf. 8.2 of {4}.

3.9   Relational Expressions

Certain digital computers (notably CDC) implement the relational
operators by subtracting the values of the two expressions to be
compared and testing the difference against zero. This process can
yield arithmetic exceptions or possibly produce incorrect conclusions
from relational tests. For example, the conclusion of equal operands
may be incorrectly made when subtraction of the two comparands results
in underflow or a number near underflow with a zero exponent. On the
other hand, the comparison of large floating point numbers may result
in overflow exceptions.

An example of this phenomenon is reported in {20}. There, much
effort was expended to reformulate the usual algorithm for determining
the Euclidean norm of a vector. One purpose of the reformulation was to
avoid spurious overflows and underflows when forming the squares of the
components of the vector. Scaling by the largest element in magnitude
was done to avoid overflow. However, to avoid a possible divide check
when scaling all components by the largest one in magnitude, a test for
a zero divisor was needed. But on machines which raise the divide check
error when the exponent of the divisor, say X, is zero, but treat the
expression X .NE. 0.0 as true when the fraction of X is nonzero, no
safe test exists.

Consequently, for software that is to be ported to machines with this property, care should be taken to avoid algorithmic processes that require comparisons between floating point numbers that are small in magnitude. Reworking the algorithm using scaling or the other techniques exemplified in {20} is recommended.

3.10 Characters Strings

It is precisely to avoid the difficulties described in the previous section with relational expressions and floating point numbers that we recommend that software which manipulates characters be implemented with entities of type INTEGER. There still exist some difficulties such as fixed point overflow exceptions that can occur when comparing non-numerical INTEGER entities. Also, manipulations that attempt to extract characters from or pack characters into INTEGER words must accommodate to the representation of integers in 1's or 2's complement notation as well as sign-magnitude notation. Both of these difficulties can be overcome by representing one fewer character than is maximal in an integer word, choosing for the first character position a value that makes the integer word positive and small in magnitude.

Concerning the treatment of character strings, we follow the recommendation of the PFORT verifier {44} in specifying that 1H Hollerith strings in DATA statements and A1 edit descriptors in FORMAT statements be used. The resulting software can probably then be made portable with no assembly language modules. Admittedly, such a discipline is very wasteful of storage compared with using assembly language modules for packing, unpacking, and comparing of character strings. If the assembly language approach is taken, it relinquishes complete portability but only to the extent of writing these rather easily specified modules. The DAVE analyzer is written in this fashion. The documentation for the PFORT analyzer recommends that certain of its modules be rewritten in assembly language to improve the time to perform its analysis.

The character data type has been defined for FORTRAN 77 {8}. It will be a welcome addition to the FORTRAN language.

3.11 Input/Output Statements

A discussion of the portability of input/output statements must be limited to certain special classes of software, mainly because of the basic inconsistencies present in the interface between the customary specification of algorithmic processes and the special cases of peripheral input/output devices. If input/output operations are the major functions or at least time consuming components of the software, then special non-portable constructs will often be essential. Indeed, for such applications, the use of other languages, better suited to the efficient treatment of files, is probably in order.

Consequently, for this topic, we limit discussion to the situation where the FORTRAN input/output statements are used to communicate a minor amount of data and results to and from peripheral devices. Numerical software is such a category of software with FORTRAN input/output statements principally for validation. In contrast, in the PSTAT package of statistical software {16}, input/output is a major concern; in {16}, specialized FORTRAN constructs are used to solve its input/output problems in a transportable manner.

First, concerning the interface between FORTRAN and input/output peripherals, there are several simple practices to follow which avoid many difficulties with portable and transportable software. We recommend that logical units for input/output devices be specified as values of entities of type INTEGER such as NIN, NOUT, NERR, NPUNCH, NSCRAT, NWORK1, etc.; for instance, the first three might be used to denote the standard input, output, and diagnostic message units which tend to be environment dependent. The values for such entities should be given in DATA statements. For example, the standard IBM FORTRAN systems would use

```
INTEGER  NERR,NIN,NOUT,NPUNCH
DATA  NERR,NIN,NOUT,NPUNCH/6,5,6,7/
```

Such a discipline readily permits modification of the unit numbers when moving the software from place to place, for instance when logical units 5 and 6 may be reserved for something else or unit 7 may not be the system punch peripheral.

Another discipline is to avoid the use of non-standard input/output constructs. The PRINT statement is not standard and at some installations is reserved for messages to operators. The * or ' characters are not standard delimiters for Hollerith strings in FORMAT statements. Use of the E edit descriptor to read or write double precision entities is not permitted by the Standard and indeed may give incorrect results. For example, on machines that use a format for single precision entities that is different from that for double precision entities (say different exponent ranges), the printed value of a double precision entity using an E edit descriptor may bear no apparent relation to its correct value.

A major weakness in FORTRAN with regard to the input/output area is the lack of a convenient mechanism whereby the varying number of significant digits in arithmetic entities between machines can be expressed in the format edit descriptors. For example, consider printing the value of a real entity X to full significance, say n digits. We would like to use an edit descriptor, say En+8.n where n is a machine dependent variable, but the FORTRAN syntax rules prohibit variables in edit descriptors. An alternative, permitted by the Standard, would be an edit descriptor of the form E32.24 if 24 were the maximum number of significant digits over all machines to be considered. However, such an edit descriptor is often diagnosed speciously, insisting that one should not request more significant digits than are warranted by the precision of the entity.

We do not have a completely satisfactory antidote for this poison. There are at least three approaches, however, to address the difficulty. For transportable software, one approach is to have a simple macro substitution process, implemented by a preprocessor, that substitutes into symbolic expressions for the machine dependent parameters. For instance, the edit descriptor E8+%NDIGIT.%NDIGIT in a format statement would become, for a machine with a value of 7 for NDIGIT, E15.7. This approach is taken by IMSL {2} and Krogh {36}.

A second approach is to write the software (the validation tests for numerical software) so that all the significance in the printed entities appears in the first few digits. For example, in FUNPACK {18} and to some extent EISPACK {46}, the validation tests print entities that measure the accumulated errors in the respective computations. Consequently, only the first few digits in the printed results are significant, and it is sufficient to use edit descriptors such as E12.4 universally.

A third approach, which is potentially portable, requires the construction of an array of characters (of type INTEGER) to be used as a format field descriptor; cf. {30} for the same technique applied to printing error messages. Proceeding from a template of the format, the characters corresponding to the variable part of the format are inserted. However, this approach requires the determination at run time of the correct value of NDIGIT in the above example. Such a value may be derived from environmental inquiry subroutines in the run time library or the subroutine library, or may be derived from the parameter MACHEP, a measure of the machine precision, which can be computed at run time {37}. NAG (cf. Chapters X01 and X02 of {40}) and PORT (cf. Chapter Utility of {43}) are examples of subroutine libraries that have such environmental inquiry procedures. To illustrate this technique, consider the following sample program:

```
C
C       THIS EXAMPLE ILLUSTRATES A TECHNIQUE FOR GENERATING
C       VARIABLE FORMAT DESCRIPTORS AT RUN TIME IN A PORTABLE
C       WAY.   THIS EXAMPLE RELIES ON THE EXISTENCE OF CERTAIN
C       ENVIRONMENTAL INQUIRY TECHNIQUES PRESENT IN SUCH LIBRARIES
C       AS NAG {40} AND PORT {43}.
C
        INTEGER  DIMFMT,FMT(8),NDIGIT,NOUT,WIDTH
        REAL  PI
C
C       FORM THE TEMPLATE   (EWW.DD)   FOR AN OUTPUT FORMAT.
C
        DATA  FMT(1),FMT(2),FMT(5),FMT(8)/1H(,1HE,1H.,1H)/
        DATA  DIMFMT,NOUT/8,6/
C
C
C
C       CALL THE ENVIRONMENTAL INQUIRY ROUTINE ENVIRN
C       TO DETERMINE THE NUMBER OF SIGNIFICANT DIGITS IN
C       REAL DATUM.  A FIRST ARGUMENT VALUE OF 22 EXTRACTS THIS PARAMETER.
C
        CALL ENVIRN(22,NDIGIT)
C
C       INSERT THE WIDTH AND DECIMAL DIGIT FIELDS INTO THE
C       FORMAT  FMT.
C
        WIDTH = 8 + NDIGIT
        CALL   INSERT(WIDTH,3,2,DIMFMT,FMT)
        CALL   INSERT(NDIGIT,6,2,DIMFMT,FMT)
C
C       COMPUTE AN ESTIMATE OF PI DERIVED FROM THE ARCTANGENT
C       FUNCTION.
C
        PI = 4.0E0*ATAN(1.0E0)
C
C       PRINT THE VALUE OF PI TO FULL SIGNIFICANCE PERMITTED
C       BY THE REAL DATA TYPE.
C
        WRITE(NOUT,10)
     10 FORMAT(38H0VALUE OF PI DERIVED FROM ARCTANGENT   )
        WRITE(NOUT,FMT)  PI
C
        STOP
        END
```

```
      SUBROUTINE  INSERT(VALUE,FIRST,WIDTH,DIMFMT,FMT)
C
      INTEGER  WIDTH,VALUE,FIRST,DIMFMT
      INTEGER  FMT(DIMFMT)
C
C     THIS SUBROUTINE INSERTS DECIMAL (OR BLANK) CHARACTERS
C     INTO A VALID FORMAT ARRAY, FORMING A FORMAT DESCRIPTOR.
C
C     ON INPUT-
C
C        VALUE    IS THE NUMERIC VALUE TO BE INSERTED INTO THE
C                 FORMAT ARRAY  FMT.
C
C        FIRST    IS THE FIRST POSITION OF  FMT  IN WHICH THE DECIMAL
C                 CHARACTERS OF THE FIELD DESCRIPTOR ARE TO BE PLACED.
C
C        WIDTH    IS THE WIDTH OF THE FIELD DESCRIPTOR.
C
C        DIMFMT   IS THE EXTENT OF THE FORMAT ARRAY  FMT.
C
C     ON OUTPUT-
C
C        VALUE,FIRST,WIDTH,DIMFMT
C                    ARE UNCHANGED.
C
C        FMT         IS THE FORMAT ARRAY INTO WHICH CERTAIN DECIMAL
C                    (AND BLANK) CHARACTERS ARE INSERTED TO FORM A
C                    VALID FIELD DESCRIPTOR.  THE LEADING POSITIONS
C                    OF THE DESCRIPTOR FIELD CONTAIN BLANKS UP TO THE
C                    FIRST NONZERO CHARACTER, OR UP TO A SINGLE ZERO
C                    IN THE RIGHTMOST POSITION IF VALUE IS ZERO.
C
      INTEGER  BACKI,DIGIT,DIGITS(11),I,LAST,VAL
      DATA  DIGITS(1),DIGITS(2),DIGITS(3)/1H0,1H1,1H2/,
     X      DIGITS(4),DIGITS(5),DIGITS(6)/1H3,1H4,1H5/,
     X      DIGITS(7),DIGITS(8),DIGITS(9)/1H6,1H7,1H8/,
     X      DIGITS(10),DIGITS(11)/1H9,1H /
```

```
C
C
C
      LAST = FIRST + WIDTH - 1
      VAL = VALUE
C
      DO 30 I = FIRST,LAST
        BACKI = LAST + FIRST - I
C
C       CHECK WHETHER THE ZERO OR BLANK CHARACTER IS TO BE GENERATED.
C
        IF( VAL .EQ. 0 .AND. BACKI .LT. LAST )  GOTO 10
C
C         GENERATE THE APPROPRIATE DECIMAL CHARACTER.
C
          DIGIT = MOD(VAL,10) + 1
          VAL = VAL/10
          GOTO 20
C
C         GENERATE BLANKS FOR THE LEADING ZEROS IN THE NUMERIC
C         CHARACTER FIELD.
C
   10     DIGIT = 11
C
C       INSERT A CHARACTER INTO THE FORMAT ARRAY.
C
   20   FMT(BACKI) = DIGITS(DIGIT)
   30 CONTINUE
C
      RETURN
      END
```

Although this last approach is portable to many machine environments, it is cumbersome and obscure. Since the character strings are stored with 1 character per word, it is wasteful of space if many variable formats are needed. It still requires machine parameters such as NDIGIT or MACHEP, and relies fundamentally on format scanners ignoring blanks between characters in format edit descriptors. However, it is more versatile than the other approaches.

4.   Non-Standard Dialects of FORTRAN

A major concern in developing portable FORTRAN software is to
know what constructs in the Standard are subject to restrictions or
tend to be extended in various dialects. For, a goal of portable
software is to avoid these kinds of constructs. A very interesting
study {35} graphically illustrates the difficulties of producing
portable source code. The study consists of a survey of 15 dialects
of FORTRAN, including IBM's FORTRAN G, H, and H extended, the ANSI
full and basic languages, and 10 dialects for minicomputers. The
study discovered that of 90 FORTRAN statements and items which were
classified and analyzed, only four are not extended or restricted
in some manner in this set of 15 dialects; the four items are: the
CONTINUE statement, the GO TO statement, the INTEGER type specification
statement, and the integer format edit descriptor I. For portability,
it is the intersection of such dialects that has been discussed in the
previous sections of this paper.

There remains one major inconsistency among the various dialects
which is documented in {35}. It involves the ordering of the specification
statements, such as type specification, DIMENSION, COMMON, and EQUIVALENCE
statements within a program unit. Suffice it to say, that if all these
statements are present in a program unit, there is no fixed order that
satisfies all 15 FORTRAN dialects. Not using DIMENSION statements avoids
some of the conflicts. However, the only approach that eliminates the
problem entirely is to avoid COMMON statements which, of course, may
be too heavy a price to pay for portability. Another alternative is

to pretend that these exceptional dialects do not exist, or hope that they will fade away. For transportability, an alternative is an automated reordering process tuned for the exceptional dialects.

For completeness, we list some of the more frequent extensions to and restrictions in the various dialects of FORTRAN. Most of these items were encountered in the development of software for the NATS {12} and NAG {27} projects. We refer the reader to {35} for a complementary list of such restrictions and extensions in FORTRAN dialects.

Most of the restrictions are related to limitations in table sizes within the compilers. We list these first followed by two restrictions that are more properly attributed to spurious compilers:

L1) A limitation on the number of comment cards. For instance, the IBM 360-370 FORTRAN G compiler will not include in the source listing more than 30 consecutive COMMENT cards between the SUBROUTINE or FUNCTION statement and the first executable statement.

L2) A limitation on the number of parentheses in expressions and in FORMAT statements. For instance, the ICL 4100[#] FORTRAN compiler limits the number of paired parentheses in arithmetic or logical expressions to 12. Certain FORTRAN compilers for CDC equipment limit the depth of nested parentheses in FORMAT statements to 2.

[#]The ICL 4100 is an early computer of International Computers Ltd. of the United Kingdom. It is in use at a few British Universities.

L3)   A limitation on the number of DO loops or nested DO
loops in a program unit. Although many compilers
impose limitations, the limits are rather generous.
For example, the IBM 360-370 FORTRAN G and H extended
compilers limit the number of nested DO loops to 25,
and the WATFIV compiler limits the number of DO loops
to 255 within one compilation. The report {35} lists
compilers for which the limit is 9.

L4)   A limitation on the number of dummy arguments in a
subprogram. For instance, the ICL 4100 FORTRAN compiler
restricts the maximum number of arguments to 31.

L5)   A limitation on the number of DATA initialization
statements, again notably for the ICL 4100 compiler.

L6)   A limitation on the number of continuation cards for a
statement. The Standard limits it to 19, but the
report {35} lists compilers for which the limit is 5.

L7)   A limitation on the largest statement label. The
report {35} identifies two PDP-8 compilers for which the
limit is 2047 but the more common limitation, if any, is
32767.

L8)   A limitation on the ways to complete a logical IF statement.
Some compilers require that it be completed on the same line
(i.e., no continuation), while others require that if completed
by an assignment statement, at least the assignment operator
'=' must be on the initial line. Still other compilers
restrict the type of statement that may complete a logical
IF statement; for instance, not permitting an arithmetic IF
statement.

L9)   A limitation on the juxtaposition of operators. For example, the Burroughs 3500 FORTRAN compiler diagnosed the expression IP .EQ. -1 as having a syntax error, requiring IP .EQ. (-1) instead.

L10) A limitation on the values of subscripts. Some program checkers (early versions of WATFIV) require that subscript values be within the overall extent specified for the dummy argument. For instance, if

REAL A(1)

is specified for the dummy argument A, a reference to A(I), where I has the value 2, would be flagged as a subscript error despite the fact that the actual argument associated with A has a larger extent, say 10. Indeed, such compilers may diagnose zero or negative extents of dummy arguments at execution time regardless of whether in the particular execution sequence reference is made to those arguments. The Standard does imply that positive extents are required for dummy arrays which are referenced at execution time. However, if a dummy array is not referenced, the issue is not clear.

L11) A limitation on the form of the actual arguments corresponding to dummy array arguments. Actual arguments of the form W(2*N+1) are treated by some compilers (notably some early PDP-10 FORTRAN compilers) as expressions, thereby communicating to the subprogram the address of a temporary location into which the value of W(2*N+1) has been placed. Despite the fact that such action is a violation of the Standard, it does impinge upon the portability of software.

Some of the first nine limitations, (and many others like them), affect the size of a program unit, which indeed reinforces one of the usual arguments for clearly structured programs. The latter two restrictions are avoided by adjusting interfaces between program units; for L10, one can add arguments to the subprogram unit, or modify existing arguments so that nonpositive values of extents are impossible; for L11, one defines the identifier TWONP1 with the value 2*N+1, and uses W(TWONP1) in place of W(2*N+1).

With regard to extensions, many of which are in common use, we should be aware that they are not defined by the Standard, and are thus not suitable for portable software. We give a list of the more common extensions that are not part of FORTRAN 66, which we therefore recommend be avoided in portable software.

E1) Complex double precision, half word integer, part word logical, and character data types. Among the various extensions, neither the syntax of the corresponding type specification statements nor the names used to identify the corresponding intrinsic and basic external functions are standard.

E2) The ENTRY and multiple RETURN features. These statements are not standard, particularly with respect to dummy arguments defined by one ENTRY statement and used when entering the subprogram elsewhere.

E3) Extensions permitting general expressions in subscripts. Such extensions include: arbitrary expressions that yield noncomplex, nonpositive values for subscript expressions; expressions in dummy array declarators; nonunit lower extents for dummy array declarators.

E4) Mixed mode expressions which are still not accepted by many compilers, particularly compilers on small machines.

E5) A different number of arguments in the CALL statement from that specified in the corresponding SUBROUTINE or FUNCTION statement; this feature is not possible on many compilers unless the well-established linkage conventions between subprograms is changed. This feature is thus not available on many compilers.

E6) DO-implied lists and array names in DATA initialization statements.

E7) Expressions for the DO parameters in DO statements.

E8) Multiple statements per source record.

E9) The dollar-sign character $.

E10) Non-use of parentheses to define the association for the exponentation operator ** in expressions such as A**B**C. Some compilers scan such expressions left-to-right while others scan them right-to-left, giving the two distinct parses

$$A^{\left(B^C\right)} \quad \text{and} \quad A^{(B*C)}$$

respectively; cf. {35} for examples of such compilers.

The current FORTRAN 77 defines most of these extensions, with the notable exceptions of half word integers, part word logicals, and double precision complex entities. Other extensions not included are multiple statements per source record and variable number of arguments. In the long term, the definition by the Standard of such popular and useful features will improve portability of FORTRAN software, but in the short term it will do little for portability until the old compilers are retired.

5. Two Interesting Anecdotes

Our approach to preparing portable or transportable software has been the safe conservative discipline of restricting the source text to a subset of the FORTRAN Standard 66 and avoiding FORTRAN constructs that rely on features in the Standard that are vague or ambiguous. Our approach is admittedly restrictive, but in extending it beyond the current Standard, great care must be taken. We give two anecdotes to illustrate the nature of the problems that can be encountered in trying to extend the Standard to non-conforming constructs. The first example is taken from experience with EISPACK {46}.

In the final stages of testing the second release of EISPACK, we were notified of a problem in the subroutine BANDR, which reduces a real symmetric band matrix to a symmetric tridiagonal matrix. The problem only occurred with certain CDC optimizing compilers when the eigenvalue calculator TQL1 followed BANDR. The coding in BANDR had been as follows:

```
SUBROUTINE  BANDR(NM,N,MB,A,...,E,E2,....)
    •

    •
REAL   ...,A(NM,MB),E(N),E2(N)
    •

    •
M1 = MB - 1
    •

    •
DO 850 J = 2,N
    •

    E2(J) = A(J,M1)**2
    E(J)  = A(J,M1)
    •

    •
850 CONTINUE
    •

    •
END
```

When BANDR precedes TQL1, E2 is not needed and in order to save space,

we stated in our documentation that E2 could be identified with E

thereby returning just the vector E. Since the dummy arguments E

and E2 are defined within BANDR, this recommendation is a violation

of the Standard (cf. 8.4.2 of {4}). However, we believed we were

safe because in our experience the references to E and E2 would be

by address to the actual arguments corresponding to these arrays, and

hence no problem of the kind discussed in Section 3.1 would occur with

respect to the timing of the references to these arguments. But upon

execution of our test software that identified these arguments, the

combination of BANDR and TQL1 failed to give the correct results. It

failed, not because our assumption about references by address was

incorrect, but because the CDC compiler had interchanged the order of

execution of the two assignment statements to E2 and E. Such a change

in the execution order could have been prompted by a desire to avoid refetching A(J,M1), that is, storing it in E before squaring it and storing the result in E2. Of course, the compiler was not in error; we had violated a rule.

From our point of view, the problem was annoying, requiring us to rewrite our documentation to include a warning. We also modified the source text to BANDR by inserting several other statements between the two assignment statements. The compiler (and all other compilers covered under the certification) then did assign to E before assigning to E2, thereby protecting users who may identify the actual arguments.

The second anecdote involves an attempted use of an undefined variable. This difficulty was revealed in a subtle way on certain optimizing compilers for CDC equipment, and does not show up in most other environments. Consider the following program segment:

```
      SUBROUTINE  SUBR(...)
        .
        .
      LOGICAL   FIRST
      DATA   FIRST/.TRUE./
C
C
   10 IF( .NOT. FIRST )   GOTO 20
        FIRST = .FALSE.
          .
          .
        GOTO 30
C
C         TRANSFER HERE WHEN NOT THE FIRST EXECUTION OF SUBR.
C
   20    CONTINUE
          .
          .
   30 CONTINUE
        .
        .
      END
```

The action of the CDC compiler in this case was to effectively eliminate the assignment statement after statement 10 and the statements between 20 and 30. The rationale is that FIRST is a local variable with the initial value .TRUE., and cannot have any other legal value at the first executable statement of the subprogram unit. The Standard states (cf. 10.2.5 of {4}) that the redefined values of entities appearing in DATA statements become undefined for subsequent executions of the program unit. The compiler implementors here assumed that the initial value was retained (or undefined) and so optimized the compiled code by removing the statements between 20 and 30 entirely. The effect of this action is that the computation for the first call is repeated for every call, and it may be very difficult to tell that the program is wrong.

Hence, be aware of the Standard. One can hardly expect to be aware of the subtle assumptions made by the implementors of FORTRAN compilers in areas where they have no Standard to follow.

6.   Conclusions

By surveying the pitfalls in writing numerical software in FORTRAN, we have examined many deficiencies in the FORTRAN language. But our purpose is not to belittle FORTRAN, for despite its inadequacies, FORTRAN has one major advantage over other languages -- its general availability throughout the computing community. Instead, our purpose is to suggest that at least some of FORTRAN's deficiencies can be overcome by a disciplined approach to FORTRAN programming.

Such an approach to programming FORTRAN has two aspects. First, there is the programmer's attitude. The disciplined approach to FORTRAN encourages orderly and well structured programs. Avoidance of certain unfortunate constructs within FORTRAN (such as those discussed in Sections 2.1 through 2.8) as well as a tidy formatting scheme for source text (discussed in Section 2.9) can make the intent of the software clear from just reading it. The attitude that is thus created is one of adaptability; that is, the software is written so that needed modifications to adjust to new environments are easily recognized, reliably implemented, and thus generally straightforward.

Second, this disciplined approach is a practice; that is, a way of writing programs that is designed to avoid many problems with portability without requiring a detailed knowledge of many computing environments. In Sections 3.1 to 3.11, FORTRAN concepts and constructs have been surveyed, pointing out the dangers in using various FORTRAN features, and recommending a practice that avoids the difficulties. This practice is restrictive, further limiting the facilities in the FORTRAN language, but at the same time making the language easier to use safely in the sense of portability. Sometimes, the restrictions make certain features in the language cleaner and more understandable rather than more exceptional; cf. the discussion on the EQUIVALENCE statement in Section 3.7.

The word discipline suggests obedience, even blind obedience, but the latter is not intended. A practice, to be useful in changing computing environments, must be flexible, still retaining the properties of readability, ease of maintenance and transportability. The two anecdotes in Section 5, rather than suggesting that extensions of the discipline are ill-advised, are chosen to illustrate the great care that must be taken in extending the discipline, particularly beyond the Standard. All the subtleties and assumptions inherent in the original design of the discipline must be understood before extensions are considered. For that reason, the rationale behind the discipline has been described. Also much of the experience and knowledge of existing computing environments that has gone into the formulation of this discipline has been stated in order that the consequences of any extensions can be analyzed.

In summary, the disciplined approach represents an attitude towards programming and a practice in writing programs; together, they make the tiresome prospect of following a fixed programming style worthwhile.

Acknowledgments

I would like to thank James Boyle, Jim Cody, Wayne Cowell, Brian Ford, Burton Garbow, and Steve Hague for their constructive criticisms of this paper; the paper is much improved by their helpful suggestions and encouragement.

This paper represents a synthesis of experience gained by the developers and collaborators of the NATS and NAG projects. I wish to thank all the unnamed individuals who, through their efforts to produce transportable numerical software, have contributed to much of the experience reported in this paper.

REFERENCES

1. Aird, T. J., The FORTRAN Converter's User Guide, International Mathematical and Statistical Libraries, Houston, (Jan. 1976).

2. Aird, T. J., Battiste, E. L., Bosten, N. E., Darilek, H. L., and Gregory, W. C., Name Standardization and Value Specification for Machine Dependent Constants, Signum Newsletter 9, 4 (1974).

3. American National Standards Institute, History and Summary of FORTRAN Standardization Development for the ASA, Comm. ACM 7, 10 (Oct. 1964), 590.

4. American National Standards Institute, FORTRAN vs. Basic FORTRAN, Comm. ACM 7, 10 (Oct. 1964), 591-625.

5. American National Standards Institute, X3.9-1966 FORTRAN, (Mar. 1966).

6. American National Standards Institute, Clarification of FORTRAN Standards -- Initial Progress, Comm. ACM 12, 5 (May 1969), 289-294.

7. American National Standards Institute, Clarification of FORTRAN Standards -- Second Report, Comm. ACM 14, 10 (Oct. 1971), 628-642.

8. American National Standards Institute -- X3J3 Subcommittee, The Draft Proposed ANSI FORTRAN, SIGPLAN Notices 11, 3 (Mar. 1976).

9. Baker, B. S., Algorithm for Structuring Flowgraphs, J. ACM 24, 1 (Jan. 1977), 98-120.

10. Boyle, J. M., Mathematical Software Transportability Systems -- Have the Variations a Theme?, this volume, Proceedings of Argonne Workshop on Automated Aids, W. R. Cowell, Ed., Springer-Verlag, Berlin, (1977).

11. Boyle, J. M., and Matz, M., Automating Multiple Program Realizations, Proceedings of the Symposium on Computer Software Engineering XXIV, Polytechnic Press, Brooklyn, (Apr. 1976), 421-456.

12. Boyle, J. M., Cody, W. J., Cowell, W. R., Garbow, B. S., Ikebe, Y., Moler, C. B., and Smith, B. T., NATS -- A Collaborative Effort to Certify and Disseminate Mathematical Software, Proceedings 1972 National ACM Conference, II, ACM, (1972), 630-635.

13. Boyle, J. M., Portability Problems and Solutions in NATS, Proceedings of the Software Certification Workshop, W. R. Cowell, Ed., Argonne National Laboratory, (1973), 80-83.

14. Boyle, J. M., A Test for the Proper Use of EQUIVALENCE, Private Communication, (June 1976).

15. Brown, W. S., A Realistic Model of Floating-Point Computation, to appear in Mathematical Software III, J. R. Rice, Ed., Academic Press, New York, (1977).

16. Buhler, R., P-STAT Portability, Computer Science and Statistics: 8th Annual Symposium on the Interface, University of California, Los Angeles, (Feb. 1975), 165-172.

17. Cline, A. K., Algorithm 476, Comm. ACM 17, 4 (April 1974), 220-223.

18. Cody, W. J., The FUNPACK Package of Special Function Subroutines, TOMS 1, 1 (Mar. 1975), 13-25.

19. Cody, W. J., An Overview of Software Development for Special Functions, Lecture Notes in Mathematics 506, Numerical Analysis Dundee 1975, G. A. Watson, Ed., Springer-Verlag, Berlin, (1976), 38-48.

20. Cody, W. J., Mathematical Software -- Why the Fuss?, Presentation at the Spring SIAM Meeting, Chicago, (June 1976).

21. Cody, W. J., Machine Parameters for Numerical Analysis, this volume, Proceedings of Argonne Workshop on Automated Aids, W. R. Cowell, Ed., Springer-Verlag, Berlin, (1977).

22. Cress, P. H., Dirksen, P. H., McPhee, K. I., Ward, S. J., Wiseman, M. A., WATFIV Implementation and User's Guide, University of Waterloo Report, Waterloo, Ontario, Canada.

23. Dorrenbacher, J., Paddock, D., Wisneski, D., and Fosdick, L. D., POLISH, A FORTRAN Program to Edit FORTRAN Programs, University of Colorado Computer Science Technical Report, #CU-CS-050-74, (1974).

24. DuCroz, J. J., Hague, S. J., and Siemieniuch, J. L., Automated Aids in the NAG Project, this volume, Proceedings of Argonne Workshop on Automated Aids, W. R. Cowell, Ed., Springer-Verlag, Berlin, (1977).

25. Fox, P. A., Hall, A. D., and Schryer, N. L., Machine Constants for Portable FORTRAN Libraries, Computer Science Technical Report No. 37, Bell Laboratories, Murray Hill, N. J., (1975).

26. Ford, B., Reid, J. K., and Smith, B. T., The MAP Statement, FOR-WARD, FORTRAN Development Newsletter 2, 4 (October 1976), 29.

27. Ford, B., The Evolving NAG Approach to Software Portability, Software Portability, An Advanced Course, Cambridge University Press, London, (1977).

28. Ford, B., Preparing Conventions for Parameters for Transportable Software, this volume, Proccedings of Argonne Workshop on Automated Aids, W. R. Cowell, Ed., Springer-Verlag, Berlin, (1977).

29. Griffith, M., Verifiers and Filters, Chapter IIIA, Software Portability, Cambridge University Press, (1977), 33-51.

30. Hall, A. D., and Schryer, N. L., A Centralized Error Handling Facility for Portable FORTRAN Libraries, Computer Science Technical Report No. 37, Part 2, Bell Laboratories, Murray Hill, N. J., (1975).

31. Kernighan, B. W., RATFOR -- A Preprocessor for a Rational FORTRAN, Software Practice and Experience 5, 4 (1975), 395-406.

32. Kernighan, B. W. and Plauger, P. J., Software Tools, Addison-Wesley, Reading, (1976).

33. Kernighan, B. W. and Plauger, P. J., Elements of Programming Style, McGraw-Hill, (1974).

34. Knuth, D. E., Structured Programming with GO TO Statements, Computing Surveys 6, (1974), 261-302.

35. Konberg, E., and Widegren, I., FORTRAN Dialects -- A Selection, Swedish Institute of National Defense Report (FOA 1), C1500-M4, (Feb. 1973).

36. Krogh, F. T. and Singletary, S. A., Specializer User's Guide, Section 914, Memorandum 404, Jet Propulsion Laboratory, Pasadena, (Feb. 1976).

37. Malcolm, M. A., Algorithms to Reveal Properties of Floating-point Arithmetic, Comm. ACM 15, 11 (Nov. 1972), 949-951.

38. Muxworthy, D. T., A Review of Program Portability and FORTRAN Conventions, Technical Paper Series of EUROCOPI Report No. 1, Ispra, Italy, (Sept. 1976).

39. Naur, P., Machine Dependent Programming in Common Languages, BIT 7, (1967), 123-131.

40. Numerical Algorithms Group, Mark V Manual, Oxford University, Oxford, UK, (Aug. 1976).

41. Osterweil, L. J. and Fosdick, L. D., DAVE -- A Validation Error Detection and Documentation System for FORTRAN Programs, Software Practice and Experience 6, 4 (1976), 473-486.

42. Perrine, J., Correspondence from J. Perrine, FOR-WARD, FORTRAN Development Newsletter 3, 1 (Jan. 1977), 7.

43. PORT Mathematical Subroutine Library Manual, Bell Telephone Laboratories, Inc., Murray Hill, N. J., (Jan. 1976).

44. Ryder, B. G., The PFORT Verifier, Software Practice and Experience 4, 4 (1974), 359-377.

45. Sedgwick, A. E. and Steele, C., DEFT, University of Toronto Computer Science Technical Report No. 62, (1974).

46. Smith, B. T., Boyle, J. M., Garbow, B. S., Ikebe, Y., Klema, V. C., and Moler, C. B., Matrix Eigensystem Routines, EISPACK Guide, Lecture Notes in Computer Science 6, Springer-Verlag, Berlin, (1974). (Cf. 2nd edition 1976, and the guide for 2nd release of EISPACK.)

47. Waite, W. M., Theory, Chapter II.A, Software Portability, Cambridge University Press, (1977), 7-19.

TWO NUMERICAL ANALYSTS' VIEWS ON THE DRAFT
PROPOSED ANS FORTRAN

C. L. Lawson*

and

J. K. Reid

1.   INTRODUCTION

The American National Standards Committee X3J3 has developed and
approved a draft proposed American National Standard language, and this
draft has been published by the Association for Computing Machinery.

Our aim here is to examine this draft from the point of view of
a numerical analyst wishing to supply portable numerical software.
Such software is often provided to the user in the form of a library
of subprograms all of which have already been compiled under an
optimizing compiler and carefully tested in this compiled form.
We therefore pay particular attention to this library situation.

2.   NEW FEATURES

In this section we present a list of most of the changes that
have been made with respect to the 1966 language and comment on
their importance from our point of view. We return to discuss
particular aspects in later sections. The list is not exhaustive
but we have attempted to include all features that particularly

*This note presents the results of one phase of research carried
out at the Jet Propulsion Laboratory, California Institute of
Technology, under Contract NAS 7-100, sponsored by the National
Aeronautics and Space Administration.

affect us. We have not attempted to place the items in order of importance because it is so much a matter of opinion as to whether an obviously overdue facility (such as mixed-mode arithmetic) is more important than a facility (such as 'END=') that gives a new capability.

In our brief description of new or changed facilities we use the letter F when referring to the full language explicitly and the letter S for the subset language. Alongside we give a brief remark on the usefulness of the facility from our point of view.

Facilities Present	Remarks
Mixed mode expressions	Obviously overdue
Subscript expressions	Obviously overdue
F: May be real or double precision	Unimportant
S: Must be INTEGER	O.K.
Must not contain array elements and function references.	Annoying restriction. Affects regularity of language.
DO loops:	
May be executed zero times and increment may be negative.	Very welcome. Mathematical software frequently needs this.
F: Controlled by expressions	Useful
F: These expressions may be real or double precision.	A frill
Control parameter defined on normal termination	Welcome
Generic functions (F only). E.g. MAXO, AMAX1, DMAX1 may all be referenced as MAX.	A real aid to portability since precision changing is often needed.

Dimension statements

    F: Lower bound other than 1 — Useful for zero

    F: Upper bound may be
       expression that does not
       include function or array
       element reference.

Really useful in library situation
since it can reduce the number of
subroutine arguments. The restriction is a nuisance.

DATA statements

    Array names permitted — Very useful

    F: Implied DO loops — Very useful

Direct-access input-output

Very welcome: will permit portable
programs to use modern hardware
more effectively.

'END=' to recognise ends of files

Important: No such facility in
old language and needed for
preprocessors.

Constants may have more digits
    than the processor will actually
    use.

Most welcome since it allows
such constants to be written
in a processor-independent way.

Character data type

Useful for preprocessors and
error messages. We regret
absence of collating sequence.

IMPLICIT statement

Will help portability when change
of length is needed.

SAVE permits variables and arrays
    to retain their values between
    subprogram calls.

Very welcome for situation
where data has to be calculated
just on first entry to subprogram.

PARAMETER statement allowing
    constants to have symbolic
    names (F only)

Useful in driver test decks and
for environmental parameters.

ENTRY statement (F only) — Mildly useful for sharing of code.

Alternate RETURN (F only)

Seems unnecessary since a flag
can have same effect.

In FORMAT statements the edit
    descriptors E,D,F, and G apply
    to both single-precision and
    double-precision data. (F only)

Simplifies conversion of a program
between single- and double-precision.

The number of digits to be
input or output in an exponent
field can be specified. (F only)
Examples: E20.6E2,E30.15D3.

Provides uniform performance on
different computer systems.

3.   MISSING FACILITIES

Here we list a number of features that are not in dpANS FORTRAN
but would be very useful for the production of mathematical library
subroutines.

Missing Facilities	Remarks
DOUBLE PRECISION COMPLEX	Effectively bars the use of COMPLEX since precision change is often needed when a program is moved.
A method of subdividing dynamic array.	Library subprograms need this for workspaces (see section 4)
Dimensions of arrays cannot be taken implicitly from calling subprograms although the analogous feature of passing the length of a character variable by writing "*" is provided.	Would enable argument lists to be shortened.
IF ( ) THEN-ELSE-ENDIF	Permits structured coding of two-way branches without the need for statement numbers. (Since publication of dpANS FORTRAN in March 1976, X3J3 has voted to include this feature.)
Access to environmental parameters such as the relative precision of REAL and D.P. arithmetic, overflow and underflow limits, etc.	Presently these numbers must be written into codes where needed and changed when the code is converted to a different machine. See section 6.
More flexible use of named COMMON.	See Section 7.

4. WORKSPACE ARRAYS

In this section we discuss the provision of workspace arrays for library subprograms, a real problem under the old standard which has not been relieved significantly by the new one. It happens frequently that a library subprogram requires one or more work arrays whose lengths vary dynamically from call to call. The only way to give the user full control over such space (so that he can, for example, make several subroutines share available storage) is to pass the storage down through the argument list. A common practice is to require a single workspace. This may be divided up by using explicit displacements. For example, if we would really like to refer to the Ith element of array B we might refer instead to W(IB+I), where IB is an offset computed by the subroutine from problem dimensions. Alternatively we may obtain the effect of a dynamic EQUIVALENCE by embedding our subroutine inside a dummy, as in the following example.

```
SUBROUTINE    QQ01A(N,A,W)
DIMENSION     A(N), W(N,2)
CALL     QQ01B(N,A,W,W(1,2))
RETURN
END

SUBROUTINE    QQ01B(N,A,B,INDEX)
DIMENSION     A(N), B(N), INDEX(N)
(Code which uses B and INDEX is workspace)
END
```

Both lead to inefficient code; the first solution may result in extra overhead on every subscript reference and the second has an extra subprogram call. Furthermore the change of type between W and INDEX violates the old and proposed new standards although no actual information is passed during the subprogram call. It is disappointing that no dynamic form of EQUIVALENCE is available.

A proposal made by the members of the IFIP Working Group on Numerical Software to the standards committee X3J3 June 1975 involves a MAP statement which permits a dummy array to be dynamically subdivided. If this were available the above code would be written as:

```
SUBROUTINE   QQO1A(N,A,W)
DIMENSION  A(N),W(N,2)
MAP/W/ B(N), INDEX(N)
        (Code which uses B and INDEX as workspace)
END
```

Acceptance of this proposal would enormously ease the library workspace problem.

5.   CALLING USER'S CODE

Another problem that we would like to discuss is that of interfacing a user's code with a library subroutine. We will do this through the example of requiring the solution of a single non-linear equation $f(x)=0$ in the range $x\varepsilon(a,b)$ with given accuracy and with given limit on the number of iterations. The usual approach would be through the library subprogram calling a user-written subroutine

```
SUBROUTINE  CALFUN(F,X)
COMMON/CALFN/ (quantities needed to calculate f(x))
    (User's Code)
END
```

which is given x in X and places $f(x)$ in F. The subprogram that calls the library subroutine would have the form

```
EXTERNAL  CALFUN
COMMON/CALFN/
_____
_____

_____
    (Code to find A,B,ACC,MAXIT,initial x,quantities in
    CALFN)
CALL  NQO1A(A,B,ACC,MAXIT,X,IFLAG,CALFUN)
    (test value of IFLAG)
_____
_____

END
```

This form is convenient for the author of the library subroutine, who simply calls CALFUN whenever he wants it. However, it is not convenient for the user. First he must remember to declare CALFUN in an EXTERNAL statement and then he must construct the common block CALFN in order to communicate to CALFUN that data needed to find f from x. An impossible situation occurs if the user wishes to pass a dummy array to CALFUN, particularly likely where the "user" is in fact another library subroutine.

A commonly used method of overcoming this difficulty is to return to the user's code wherever a function value is wanted. In our example, the code might take the form

```
        _____
        _____
        _____
        (Code to find A,B,ACC,initial x,other quantities
         required to find f(x) from x)
    DO  10   ITER=1,MAXIT
        (Code to find f(x) from x)
    CALL   NQO1B(A,B,ACC,X,F,IFLAG)
        IF(IFLAG.NE.0) GO TO 20
 10 CONTINUE
        (Code to deal with case that needs more iterations)
 20     (Code to follow termination)

        _____
        _____
        _____
```

This form may at first seem clumsy (and indeed is rather awkward for the author of the library subroutine) but is in fact very convenient for the user. It does not need an EXTERNAL statement and all the quantities needed to calculate f(x) are at hand. He can use a "MAXIT" termination criterion as above, or could replace it by some other test such as central processor time. And of course, there is no problem in dummy arrays.

The proposed new standard language does not assist the solution of this problem.

If we attempt to abstract the situation being discussed we may say that we wish to have a segment of user code A call a library subroutine B which in turn transfers control to a segment of user code C. It is frequently the case that A and C need to have common access to some data, D, that is not needed by B. The amount and structure of the data constituting D depends on the application for which A and C are written and is of no concern to B.

Generally one call from A to B will generate a large number, possibly hundreds, of transfers of control between B and C. Thus the control linkage between B and C should be as efficient as possible. It should not cause repeated initialization of argument associations that will in fact be the same for the whole sequence of transfers between B and C.

One sometimes needs to cascade this structure. For instance the user code C may call another library subroutine E which may transfer control to another segment of user code F.

One mechanism that has some desirable features for supporting this coding situation is the "Internal Subroutine" provided by Univac FORTRAN V. This construct however introduces some concepts of scope and block structure that might well be regarded as major philosophical changes to FORTRAN. Thus at this point we can only describe the problem and hope that FORTRAN language specialists will give some attention to it.

6.   ACCESS TO ENVIRONMENTAL PARAMETERS

The purpose of having a FORTRAN standard is "to promote portability of FORTRAN programs for use on a variety of data processing systems" (quoted from dpANS FORTRAN). In the EISPACK package of eigenvalue-eigenvector subroutines, where portability was a prime consideration, the only line of code that was left open to be completed differently for different machines was

MACHEP = ?

The FORTRAN standard could close this portability gap by providing some mechanism whereby a program could access environmental parameters such as the relative precisions of single-precision and double-precision arithmetic, overflow and underflow limits, etc.

A straightforward way to accomplish this would be to introduce additional named constants following the example of .TRUE. and .FALSE.. Thus for example the relative precision of single-precision arithmetic could be denoted by .SRELPR. and that of double-precision arithmetic by .DRELPR.. Then the EISPACK example above could be coded as

MACHEP = .SRELPR.

Using the PARAMETER statement of dp ANS FORTRAN this assignment could be made at compile time:

PARAMETER MACHEP = .SRELPR.

The use of special identifiers such as .SRELPR. could be restricted to PARAMETER statements if this would provide any advantage to the compiler.

Other mechanisms for accessing environmental parameters, such as special intrinsic functions, would also be possible. Ideally the mechanism should be executable at compile time so that it does not require execution time every time a library subroutine using the mechanism is called.

7. MORE FLEXIBLE USE OF NAMED COMMON

Generally a named common block having a particular name, say NAME1, will be declared in more than one program unit of an executable program. dpANS FORTRAN requires that all of these declarations must designate the same amount of storage. Named common would be much more useful in library subroutines if this requirement were not so rigid.

As an alternative approach consider the following notion of primary and secondary declarations of common. Each unique named common block would have a primary declaration (as a possible syntax consider COMMON/NAME1*/...) in one program unit and secondary declarations in zero or more other program units.

The primary declaration would allocate the storage for the block. In the secondary declarations indefinite dimensioning would be permitted. For example one could write

COMMON/NAME1/A(10),B(5),C(*)

The asterisk could only be used with the last item in the list. It indicates that C is an array of unknown length. The maximum legal length for C would be determined by the total length of the common block NAME1 as specified by its primary declaration.

8. DISSOCIATE STORAGE LENGTHS OF VARIABLES OF DIFFERENT TYPES

In 1966 ANS FORTRAN it is specified that INTEGER, REAL, and LOGICAL variables each occupy the same amount of storage and that DOUBLE PRECISION and COMPLEX variables each occupy twice that amount of storage.

Conforming to these requirements leads to very inefficient use of computer storage in some cases. For example a LOGICAL variable

actually carries only one binary bit of information but will occupy a
60-bit word of storage in some machines.

This potential inefficiency has caused some computer manufacturers,
particularly in the minicomputer class, to violate this aspect of the
standard in the FORTRAN compilers they provide. Thus it is not uncommon
to find FORTRAN compilers in which INTEGER variables occupy less storage
than REAL variables or in which DOUBLE PRECISION variables occupy less
than twice the storage of a REAL variable.

The dp ANS FORTRAN introduces the CHARACTER data type and specifically
permits the ratio of its storage requirement to that of the non-character
types to be different on different machines. This is reasonable in view
of the diversity of computers on which one may hope to support FORTRAN.

I propose that it would similarly be more realistic to permit the
relative storage requirements of INTEGER, REAL, LOGICAL, and DOUBLE
PRECISION variables to be different on different computers. The storage
for a COMPLEX variable should remain twice that of a REAL, since this
is consistent with the mathematical meaning of a complex number.

This proposal has implications for the association of variables of
different types through EQUIVALENCE, COMMON, and subprogram argument lists.
I believe rules covering these situations could be developed that would
be less restrictive than those that have been introduced in dp ANS FORTRAN
to handle the fact that character storage units and non-character storage
units are related differently on different machines.

9.   CONCLUDING REMARKS

The new language is much more complicated than the old one, as is
illustrated by the relative size of the reports. It remains to be seen

whether it achieves the wide acceptability of its predecessor. It is to be regretted that the increased power has been obtained by so many ad hoc extensions with varying restrictions.

The subset language suffers much less from these defects. It shares with the full language some very useful improvements (mixed mode expressions, improved DO, array names in DATA statements, direct-access input-output, END=, character data type, IMPLICIT, SAVE, PARAMETER). Unfortunately, it is not an extension of the old standard because it does not include DOUBLE PRECISION and COMPLEX data types and BLOCK DATA subprograms. If these were included, we would feel more hopeful for its universal acceptance.

Despite these reservations and the gaps that we mentioned in Section 3, we believe that the new language will be a much better language for writing mathematical software. Several facilities (e.g., direct-access input-output, recognizing end-of-file) were totally unavailable under the 1966 standard and many more are now provided in a much more convenient form. We sincerely hope that it achieves the success of its predecessor.

INTERMEDIATE LANGUAGES: CURRENT STATUS

W. M. Waite

ABSTRACT

Intermediate languages are often used or proposed to
increase the portability of a system. This paper examines
the general characteristics and current status of the technique, and indicates areas in which further work is needed.

The use of an intermediate language to achieve portability was first suggested by a committee of SHARE {27}, but this effort never reached fruition. There were many reasons, both political and technical, for its failure. Since that time there have been a number of other attempts, some of which have been highly successful. The purposes of this paper are to indicate how intermediate languages are used, review the state of the art, and make modest projections for this technique in the milieu of mathematical software. The treatment must necessarily be broad, and the reader is referred to the bibliography for details of the various systems discussed.

1. Basic Concepts

The variety of programming languages is bewildering to many users. It is difficult to compare and contrast languages because there is no universal frame of reference, no "taxonomy" by which a language may be characterized. I shall therefore begin by outlining a semantic characterization of programming languages which is useful for our purposes. I shall then discuss the process of translating from one language to another in the light of this characterization. Finally, I shall focus upon the roles played in this process by intermediate languages. The purposes of this section are to isolate the properties of intermediate languages which are important in the context of portability, and to provide a terminology for discussing them.

1.1. Semantic Characterization

Much of the diversity of languages arises from what Christopher Strachey termed syntactic sugar - the form taken by the code which is written by a programmer. In this section I shall ignore questions of syntax and concentrate on the properties of a language which determine the <u>meaning</u> of a program written in that language.

One can engage in endless philosophical debate about the meaning of "meaning." For the purposes of this paper let us agree that the "meaning" of a program is the transformation which it carries out; the mapping which it provides from data into results. The program is thus a realization of a function, and its meaning is the function which it realizes.

The realization of a function by a program is expressed in terms of three interlocking components, the <u>primitives</u>, <u>composition rules</u> and <u>state</u>. Together, these three components provide a semantic characterization of a programming language. In this section I shall briefly summarize the role of each component and give examples from current languages. (See {35} for an extensive bibliography of semantic description techniques.)

1.1.1. <u>Primitives</u>. The primitives of a language are the elements which cannot be explained using the concepts of that language. Every language has such elements: the "point" in plane geometry, "to be" in English, the "real datum" in Fortran. Primitives are defined by an appeal to intuition, experience, faith or some formal system other than the one containing the primitive. For example, the "real datum" of Fortran is defined as "a processor approximation to the value of a real number" {1}. This definition appeals to our experience ("processor approximation") and to formal mathematics ("value of a real number").

Ultimately, the meaning of a sentence in a language is determined by the meaning of the primitives which it contains. When we implement a program on a new machine, we provide definitions of the primitives which it contains in terms of concepts provided by that machine. If the meaning of the entire program is to remain invariant, the meaning of the primitives must remain invariant. This may or may not be easy to arrange; it depends upon both the original meaning of the primitives and the facilities provided by the target computer.

We begin with an unrestricted set of values, abstract entities upon which operations can be performed. Modes classify values according to the operations possible upon them. An object is a concrete instance of a value; it is an elementary object if that value is a primitive of the language. The FORTRAN data types logical, integer and real are examples of primitive modes which have obvious counterparts on most computers and in most languages.

Since the meaning of a program depends upon the meaning of the primitives, the particular properties ascribed to such modes as logical, integer and real are crucial. Considerable care must be taken to avoid both over- and under-specification: If the modes are under-specified, then it will be impossible to guarantee properties of the source language on all computers, whereas if they are over-specified it will be impossible to guarantee efficient implementations on all computers.

The simplest way to define a mode is to enumerate the values which belong to it. Unfortunately, this only works for finite modes such as "logical" in FORTRAN. Most languages rely almost completely upon intuition and experience to define infinite modes such as integer and real. ALGOL 60 {29} is typical: "Decimal numbers have their conventional meaning."

Usually we must rely upon our experience to remind us of the finite nature of machine arithmetic. In some cases, however, this point appears specifically in the language definition. For example, the PASCAL User Manual {21} states:

"There exists an implementation-dependent standard

identifier <u>maxint</u>. If a and b are integer expressions,

the operation:

   a <u>op</u> b

is guaranteed to be correctly implemented when:

   abs(a <u>op</u> b) $\leq$ maxint,

   abs(a) $\leq$ maxint, and

   abs(b) $\leq$ maxint"

Note that this avoids the problem of asymmetric integer range which occurs
in 2's-complement machines.

An operand of one primitive mode may be equivalent to an operand
of another under some appropriate mapping. Some of these mappings are
defined by the language; others are described, but their definition is
left up to the implementor. For example, the ALGOL 68 Revised Report
{36} asserts that for every integer of a given length there is an
equivalent real of that length; the FORTRAN Standard implies a relationship
between integer and real objects by its definition of assignment, but does
not define it precisely.

An elementary action is defined in terms of a relation which
exists among values: Given some value(s) as operand(s), the elementary
action returns the related value(s) as its result(s). Relations among
elements of a finite set, like the sets themselves, can be defined by
enumeration.

Arithmetic operators depend upon a small set of relations which
hold among integer or real values. Although the operators can be defined
in terms of these (see the ALGOL 68 Standard Environment, for example),

such a definition does not usually add to the understanding which we already have from our experience in mathematics. There is, however, one important exception: integer division. Number theorists recognize two kinds of integer division, one which truncates toward zero (-3 divided by 2 yields -1) and the other which truncates toward negative infinity (-3 divided by 2 yields -2). Both definitions give the same result for positive operands.

A survey of 20 machines {12} revealed that only one, the ICL KDF9, truncates the result of an integer division toward negative infinity. Language definitions are considerably less forthright. Some (e.g., BCPL {31} avoid the question by defining integer division only for positive operands; some (e.g., ALGOL 68) give algorithms which define integer division in terms of other operators and hence specify the results completely. Many, however, simply rely upon a (non-existent) intuition. LISP {25}, for example, defines the result as "the number-theoretic quotient," and PASCAL defines the action of the operator to be "divide and truncate (i.e., value is not rounded)."

The equivalence relations defined between items of different modes form the basis for the class of elementary actions known as transfer functions. Most transfer functions are straightforward, but again there is an exception: Truncation of a negative number could be interpreted as toward zero or toward negative infinity.

1.1.2. Composition. Taken by themselves, the primitives of a language have little expressive power. In order to extend the universe of discourse, a language must provide composition rules - methods by which primitives can be combined to express a wide variety of ideas.

These composition rules define the allowable structures, and permit them to be re-analyzed and interpreted in terms of the primitives from which they were built.

The two common composition rules for objects are the array and the record (or structure). Each allows the user to construct a composite object from a collection of immediate components. The immediate components of the composite object are denoted by unique simple selectors. (See {24} for a formal definition of these composition rules.) Arrays and records differ in their methods of specifying the simple selectors, and usually also in the restrictions placed upon the immediate components.

An array is a sequence of N elements (N>0). A particular element is selected by an integer index in some range [1,u], whose value may be computed during execution. Multi-dimensional arrays may be thought of as arrays whose immediate components are themselves arrays. Unfortunately, there is no consensus on which dimension is "outermost." In other words, is A[1:3,1:4] an array with three arrays as immediate components or four arrays as immediate components? The answer may be important for efficiency reasons, since the entire object may be paged {26, 28}. ALGOL 60 is silent on this point, PASCAL defines A[I,J] as equivalent to A[I] [J], and FORTRAN specifies a mapping function which effectively defines A[I,J] as equivalent to A[J] [I]. (Actually, none of these definitions precludes any hardware mapping, since they merely specify the logical order of the elements. FORTRAN virtually forces a hardware mapping by allowing the user to reference a multidimensional array as a one-dimensional one.)

A record is a non-empty set of fields. The field selector cannot be computed during execution; it must be known at the time the

program is translated. In general, records may have <u>variants</u>: Not
every instance of the record will display the same structure. Some
set (possibly empty) of fields is common to all instances of the
record while the remainder (possibly empty) will differ from one
variant to another. Each variant may itself have variants.

The simplest composition rule for actions is the <u>sequence</u>.
A source-language expression is another example of a composite action.
Its structure may be regarded as being very similar to that of a composite
object: An elementary action with a number of <u>immediate operands</u>, each
of which may be an elementary object or an expression. This structure
is established by the language definition, according to some set of
rules (which may include the normal mathematical ones of operator
precedence and parenthesization, or may be based upon a strict
left-to-right or right-to-left sequence).

Other composite actions are the <u>conditional</u> and <u>iteration</u>. As
in the case of expressions, the structure and semantics of these
composite actions vary widely from one source language to another.
Many machines provide special instructions for iteration, but these
also show little uniformity.

A procedure is a composite action which has been "packaged"
so that it can be invoked without restating its definition. Information
must be exchanged between the procedure and the invoker; this exchange
may take place through parameters and (for functions) a result, or by
access to a common environment. We can distinguish five major parameter
mechanisms:

1. Value: The value of the argument is stored in the address
   space of the procedure. This object may be altered by the
   procedure without affecting the caller in any way.

2. Result: The value of the parameter is undefined upon entry to the procedure. Upon return to the caller, the parameter value is stored in the address space of the caller.

3. Value/Result: The value of the argument is stored in the address space of the procedure. This object may be altered by the procedure without affecting the caller. Upon return, the current value is stored in the address space of the caller.

4. Reference: The address of the argument is made available to the procedure. Any operations performed within the procedure use the addressed object directly.

5. Name: A procedure for computing the address of the argument is made available to the procedure. Any operations performed within the procedure must use this procedure to compute the argument address at the time reference is made to the argument.

Method (1) is sometimes constrained so that the procedure is not permitted to alter the argument, and the context in which the address computation of method (5) is carried out may vary. (ALGOL 60, for example, uses the caller's context while LISP uses that of the called procedure.)

All of these mechanisms can be synthesized from (1), provided that the proper data types are available. To obtain the effect of (2) - (4), we pass the address of the argument; this assumes that the concept "address as a value" can be expressed in the language {17}. (ALGOL 68 ref modes are an example of this concept, which is absent in FORTRAN

and ALGOL 60.) To obtain the effect of (5), we pass the procedure
for computing the argument address; this assumes that the concept
"procedure as value" is available in addition to that of "address as
value." (Remember that the value returned by the procedure must be
an address.)

1.1.3. States. The so-called "operational" or "interpretive" formalisms
for semantic description consider a computation as a sequence of state
transitions, each implementing a single elementary action. The state
itself may be described as a single composite object. It is assumed to
contain all of the information necessary to determine the future course
of the computation. For example, the state of a program might consist
of the entire memory contents of the computer upon which it is running,
plus the contents of all files and their current status. Loosely
speaking, the concept of a state embodies the environment within which
the computation takes place.

A computation may modify the state in three ways as it progresses:

1. New components may be created and added to the state.
    (Example: Addition of local storage on block entry in
    ALGOL 60.)

2. Components may be deleted from the state. (Example:
    Deletion of local storage on block exit in ALGOL 60.)

3. Components of the state may be altered. (Example:
    Assignment of 2 to an integer variable whose value
    was previously 3.)

The third possibility could be regarded as a deletion followed by
addition, but this is usually not done.

Each state component has a definite lifetime in the history of a computation, called its _extent_. The extent begins when the component is added and terminates when it is deleted. Extent is a property of a state component rather than a component of the program. For example, consider the local storage for an ALGOL 60 block. When the block is entered, the extent of a state component corresponding to that storage begins. This extent terminates when the program exits the block. If the program now re-enters the block, a _new_ component of the state is created. This new component is not a "reincarnation" of the previous one, and hence it has a completely distinct extent.

Note that extent is based upon the _execution history_ of the program rather than its textual representation. In most languages the extent of an object is related to the textual representation of the program (as in the case of ALGOL 60 local storage, which is related to the textual block), but they are really quite distinct properties. Most languages follow a contour model {22} for most state components, with some languages providing more unstructured extents as well.

If the language permits the user to specify initial values for state components, then care must be taken when the extent of the component is less than the entire execution history of the program: The initial values are associated with the _textual description_ of the state component, rather than the component itself. But we have already seen that the textual description may apply to many components, as in the case of the local variables of an ALGOL 60 procedure. Are all such components to have the specified initial value? If not, to which components does the initial value apply?

1.2.   Translation

Translation is the process of creating a program in a target language which has the same meaning as an existing source language program. Whether performed by a human or by a machine, this process can be decomposed into three distinct subprocesses which I shall call analysis, code generation and assembly. (My terminology is drawn from the jargon of compiler construction; see {3}.) In this section I shall describe each of these subprocesses, relating them to the task of the human programmer. The purpose of this discussion is to establish parameters for the translation process which will aid us in evaluating the scope and effectiveness of particular intermediate languages.

1.2.1. Analysis. The first step in any translation is to determine the meaning of the source program, and this is the task of the analysis phase. Conceptually, the analyzer uses the source language composition rules to obtain both the structure of the state in terms of elementary objects and the structure of the program in terms of elementary actions. This structure, in conjunction with the meaning of the source language primitives, embodies the meaning of the program.

Analysis depends only upon the source language. The analyzer accepts as input the representation of the source program (usually in the form of a sequence of characters) and produces an abstraction of the meaning in terms of source language concepts. This abstraction may take any of a number of forms: If the translator is a human, then it is his "understanding" of the algorithm expressed by the program; if the translator is a computer, then it is a data structure (normally a tree or directed acyclic graph). Whatever its form, it is consistent with the rules of the source language.

The lexical, syntactic and semantic analysis modules of a compiler comprise the analysis phase. Lexical and syntactic analyzers may be constructed automatically, and represent about 15% of the normal compiler. The size of the semantic analyzer depends critically on the complexity of the language being translated, and may be as much as 10% of the final product. Thus about 15-25% of the compiler is concerned with analysis. For typical languages, construction of the analyzer may require 2-3 man-months.

1.2.2. Code Generation. This subprocess is the heart of the translation. Given the abstraction which embodies the meaning of the source program, it creates a new abstraction which embodies the "same" meaning but is stated in terms of target concepts. We must be careful about our interpretation of "same" in this context: Recall that the meaning of a program rests upon the meanings of the primitives of the language. Since the new abstraction must be stated in terms of target primitives, a mapping from source primitives to target primitives is implied. "Same" must be interpreted in the light of this mapping. For example, the meaning of the machine code translation of an addition of two reals must be interpreted in the light of the finite representation of a real and the peculiarities of the hardware add instruction.

No techniques currently exist for the automatic construction of code generators, and the nature of the process makes it unlikely that any satisfactory ones will be developed. The code generator represents 60-70% of a typical compiler, and may account for 90% if highly optimized code is being generated for a difficult machine. Construction of the code generator (including all of the preliminary design) may require 12-15 man-months for typical language pairs.

Unlike the analyzer and assembler, the code generator depends upon both the source and target languages. The relationship between the two languages, in conjunction with some measure of the quality of target code desired, determines the complexity and cost of the code generator.

1.2.3.   Assembly. The final step in the translation is the encoding of the target abstraction created by the code generator. This subprocess is the reverse of analysis, resulting in a concrete representation of the program which is acceptable to some further process. Typical further processes are interpretation (perhaps by a computer or microprogram) and translation.

Assembly depends only upon the target language. It accepts a target abstraction and produces an encoded form such as a sequence of bits or characters. In the process, it creates objects to represent the abstract values manipulated by the program. For example, it may assign machine addresses to labels and create bit patterns for constants.

Although there are no techniques for generating assemblers automatically, the amount of effort required is usually quite small. The assembly module represents 5-10% of a typical compiler and can be built for normal target computers in about 1 man-month. (Note that the assembly module is not the same as a symbolic assembly program. The symbolic assembler is a complete translator made up of analysis, code generation and assembly modules. Code generation in a symbolic assembler may be quite sophisticated, encompassing macro expansion and a wide range of pseudo operations.)

1.3. Use of Intermediate Languages

An intermediate language is used to split a translation into two steps: Source programs are first translated into intermediate language programs; these are, in turn, translated into programs in the target language. In this section I shall first discuss two specific examples of intermediate languages which are quite common. They lie at opposite ends of a broad spectrum, and illustrate two major reasons for employing this technique. Finally, I shall present the general criteria under which an intermediate language can be used to enhance portability.

1.3.1. FORTRAN Preprocessors. New problem- and procedure-oriented languages are being created almost daily in response to a wide variety of stimuli. In most cases, the inventor has neither the time nor the knowledge to create a compiler which will produce object code for his computer. He therefore factors the translation, using FORTRAN as an intermediate language.

This strategy results in a preprocessor - a translator which converts the source program into a FORTRAN program. The FORTRAN program is then submitted to the standard FORTRAN compiler for further translation. In many cases the preprocessor is very simple. All difficult syntactic and semantic analysis is left to the FORTRAN compiler, with the preprocessor performing only the straightforward lexical analysis. Of course, all questions of machine architecture and careful optimization are avoided. RATFOR {23} is one example of this approach; PDELANG {15} is another, more complex processor.

FORTRAN as an intermediate language has the advantage of ubiquity. It is available on most computers, and by using a minimal subset of the standard language {34} one can virtually guarantee that the preprocessor output will run on any machine. The preprocessor itself can be written in this subset, thus ensuring the portability of the entire system.

1.3.2. <u>Relocatable Object Code</u>. A computer system should be capable of combining code and data created by different people using different languages into a single running program {11}. This requires uniform conventions for communication and linkage, which constitute a definition of a language. The system designer therefore factors the translation, using relocatable object code as an intermediate language.

This strategy does not really reduce the effort required to create a translator, since the relocatable object code is very close to the executable code of the target computer. (Some "loaders" or "link editors" may, however, provide facilities which ease translation {19}. Although relocatable object code can lay no claim to universality, it certainly permits access to all facilities of the target computer.

1.3.3. <u>Portability Enhancement</u>. Let T denote the amount of effort required to get a "portable" program running in a given environment; let E denote the amount of effort required to implement the program "from scratch" in the same environment. The <u>portability factor</u> of the program is then E/T. A program's portability might be enhanced by using an intermediate language to reduce T in one of two ways.

- Translate the source program into a language which is ubiquitous.
- Translate the source program into a language which is easy to implement.

In practice, the first approach restricts the user to FORTRAN or COBOL; I do not see this situation changing in the foreseeable future.

The second approach rests upon the premise that it is possible to design a simple intermediate language which contains sufficient information to permit implementation of a program on a variety of computers with acceptable run-time efficiency. The translation from source language to intermediate language would thus reduce the complexity of the code, yet not require any representational decisions which would lower the efficiency of the final product. Section 2 discusses several current languages in the light of this premise.

Quantitative verification of such a vaguely-worded premise is impossible. I have come to believe, however, that it does not hold under any reasonable interpretation. The problem is that each decision which is not made by the source-to-intermediate translator increases the amount of effort required to code the intermediate-to-target translator. I shall return to this question in Section 3.

2. Current Intermediate Languages

In this section we shall examine several representative intermediate languages, classifying them according to the taxonomy of Section 1.1 and evaluating them from the standpoint of portability enhancement. My purpose in these examples is not to provide exhaustive descriptions of any of these languages, but to indicate the state of the art. I shall use these examples as a basis for prediction in Section 3.

2.1.   FORTRAN

The FORTRAN Standard is currently in the final stages of revision. A new version has not yet been adopted as of this writing, and is certainly not in widespread use. I shall therefore confine my remarks to the existing standard, X3.9-1966.

FORTRAN was originally designed to produce optimum code for the IBM 704 {2}. Any constructs which were difficult to optimize (e.g., arbitrary subscript expressions) or inappropriate to the 704 hardware (e.g., DO loops with negative increments) were omitted from the language. Subsequent extensions have corrected some of these omissions, but FORTRAN remains semantically impoverished.

Because it is widely available, FORTRAN is often used as a source language for portable programs {34}. A number of preprocessors (Section 1.3.1) also use FORTRAN as an intermediate language.

2.1.1. _Primitives_. Five primitive modes (integer, real, double precision, logical and Hollerith) are used to classify FORTRAN values. The usual arithmetic and Boolean operators are provided, along with a number of intrinsic and standard functions. No operators are available for Hollerith-mode objects; careful reading of the standard indicates that such objects may _only_ be read and written (even assignment and comparison for equality are explicitly forbidden).

Logical values are defined by enumeration, and Hollerith values by an implied enumeration ("any characters capable of representation"); the remaining mode definitions are based upon intuition and experience. Restrictions caused by the finite machine representation are mentioned but not quantified.

Integer division results in the "the nearest integer whose magnitude does not exceed the magnitude of the mathematical value"; the effect of truncation is never defined explicitly, but the description of real assignment leads me to believe that truncation toward zero is intended.

2.1.2.   <u>Composition</u>. FORTRAN provides one composite mode, complex, upon which elementary actions are defined. Arrays are available, but no operators may be applied to them. An array may have up to three dimensions but there is an explicit mapping which converts every array to a one-dimensional equivalent.

Actions may be composed as sequences, expressions and procedures; simple iterations and conditionals are also available. A procedure may not invoke itself, either directly or indirectly. The parameter mechanism is not stated explicitly. By careful reading of the standard it is possible to conclude that the implementor has the choice of value/result or reference. The name mechanism cannot be synthesized because FORTRAN does not have the concept of "address as value," but the effects of value and result can be obtained by suitable combinations of local variables and assignments.

2.1.3.   <u>State</u>. The standard implies that the state of a FORTRAN computation conforms to a restricted contour model. Local variables and labelled common blocks have extents determined by subprogram entry and exit. Initialization of these state components is possible, and is effected by assignments to the initialized variables. By careful wording, the standard permits the implementor to either follow the contour model exactly or make the extents of all state components the entire execution history of the program.

Direct access to a particular state component is either restricted to a single program unit or (potentially) allowed to all program units. Indirect access to any component may be provided by appropriate parameter passing. The minimum extent of a component is determined by the execution of those program units which access it directly.

2.1.4. Evaluation. FORTRAN is ubiquitous. It is available at many
installations, and it is likely to be the major language at those installations.
This means that extensive software libraries will be available and programming
advisors will be familiar with FORTRAN problems. Because of its descent,
FORTRAN bears a close resemblance to the architecture of many computers
(Burroughs machines are a notable exception). Existing compilers therefore
often produce excellent object code. Program units are usually compiled
separately, and it is easy to introduce subprograms written in assembly
language which create new elementary operations {41}.

Several crucial concepts are missing from the semantic model
upon which FORTRAN is based. The most important are probably those of
recursion and "address as value." Lack of a record composition rule
hampers introduction of new operands {40}.

With care, FORTRAN can be used successfully as an intermediate
language for a wide class of problems. It is not suitable if the source
language permits recursion or general procedure parameters. Most pointer
manipulation can be handled by creating a virtual store for all operands
and managing it with central routines, but this may result in unacceptable
performance. (General recursion and procedure parameters can theoretically
be implemented within a similar framework but the cost is astronomical.)

2.2.   OCODE and INTCODE

BCPL {31} was an outgrowth of a major language design project
undertaken by the Universities of Cambridge and London in the early
1960's. It is a typeless, block-structured language used primarily for
system software. The BCPL compiler uses OCODE {32} as an intermediate
language; the BCPL implementor may bring up the compiler by writing
a translator which accepts OCODE programs and produces programs for

his target computer. In the final product, the two stages of the translation are merged into a single compiler and the OCODE text never appears
explicitly.

Although OCODE is a simple language, creation of a translator
is sufficiently difficult to compromise Richards' portability goals.
He therefore developed a second intermediate language, INTCODE {33},
which is to be implemented via an interpreter.

2.2.1. Primitives. There is no classification of values in
OCODE and INTCODE - all belong to a single mode, the Rvalue. Some
elementary actions (add, subtract, multiply, divide, remainder, negate
and the relational operators) are defined in terms of relations among
Rvalues identical to those among integers; others (shift, and, or,
exclusive or, equivalence and not) are defined in terms of relations
identical to those among bit strings. No explicit limitation on the
number of Rvalues is given by the language definition, although the
BCPL Manual states that the length of an Rvalue is usually between
24 and 36 bits. (To my knowledge, BCPL has been implemented with
Rvalues ranging from 16 to 60 bits in length.)

2.2.2. Composition. There are no composition rules for objects
in either OCODE or INTCODE, and the sequence is the only composition
rule for actions.

2.2.3. State. The state of an OCODE computation is defined by a
linear array of storage cells and two additional values, the stack
frame pointer (which is an index to the array of storage cells) and
the stack frame size.

All state components have the entire execution history of the program as their extent, and all may be accessed directly at any point in the program.

INTCODE computations have a state similar to that of an OCODE computation, with the addition of two accumulators. An auxiliary address register replaces the stack frame size.

2.2.4. Evaluation. INTCODE is trivial to implement on a new computer (efforts of the order of several days are typical). Richards estimates the time penalty due to interpretation as a factor of 10, but explains that INTCODE "is still a useful tool for the initial implementation of BCPL on a new machine since it allows the first production code generator to be written directly in BCPL." OCODE is more difficult to implement, but provides all of the information which it is possible to derive from a BCPL program. Thus no optimizations are precluded by the intermediate representation. New elementary actions can easily be added to both OCODE and INTCODE, provided that they are defined in terms of relationships among single Rvalues. (Several implementations, including that at Bolt, Beranek and Newman {4}, have added real arithmetic.)

The states of OCODE and INTCODE are very rigid in their structure, which causes serious problems if the machine architecture is not closely matched to that structure. For example, an efficient implementation on the Burroughs 6700 is virtually impossible {42}. A somewhat less important deficiency is the lack of distinction among elementary operands {16}. For example, on many computers an integer does not require as much storage as a real and a character needs even less. Since all are represented in OCODE by Rvalues, however, all must be allocated the maximum storage required by any Rvalue.

These languages are well-suited to their original purpose - to ease the implementation of BCPL on "normal" computers. With modest extension they could be used for mathematical software on the same class of machines. Even on more exotic computers they can be implemented efficiently enough to be used during a bootstrap operation.

2.3.   ZCODE

ALGOL68C is a variant of ALGOL68 developed by S. R. Bourne at the University of Cambridge. The compiler produces ZCODE, which is then translated to the language of the target machine. ZCODE was designed to represent machines with general purpose registers, and thus its semantic model differs markedly from those discussed so far. The first version of ZCODE appeared in 1973, and it has evolved as experience has been acquired from implementations on several computers {6}. A new version has been proposed, and will be introduced when the current ALGOL68C implementation is completed at Cambridge {7}.

The design of ZCODE reflects the needs of ALGOL68C and the desire of its authors to perform register optimization in the translation from source to intermediate form. It is a more "general purpose" language than OCODE, and relies upon a more complex abstraction.

2.3.1.   Primitives.   The ZCODE manual {6} does not list the primitive modes of the language explicitly. By inference from the operators, and reference to the ALGOL68C Implementors' Guide {5}, I deduce the existence of integer, real, character, address and bit string modes. Boolean values are represented by 0 (false) and 1 (true). The usual arithmetic and Boolean operators are provided, along with transfer functions between integer and real. Shift, and, or, exclusive or and complement operators are available for bit strings.

An implementor must rely upon his intuition and experience for the properties of all of the primitives. Since ZCODE is designed for an ALGOL68 implementation, he can use the ALGOL68 report as a basis. There is no mention of constraints upon the sizes of integers, reals or addresses in the ZCODE manual, but the Implementors' Guide notes that the compiler requires at least 16 bits per integer.

2.3.2. <u>Composition</u>. There are no composition rules for objects in ZCODE, and the sequence is the only composition rule for actions.

2.3.3. <u>State</u>. The state of the ZCODE computation is defined by a linear array of storage cells and a number of registers (fixed for each implementation). Two of the registers have fixed values 0 and 1 respectively, one each is used to address the global data storage, current stack frame base and stack top, and one is used during procedure calls. The remaining registers are used for working storage, and their properties may vary from one implementation to another.

All state components have the entire execution history of the program as their extent, and all may be accessed directly at any point in the program.

2.3.4. <u>Evaluation</u>. ZCODE is reasonably well matched to ALGOL68C. Because of its explicit register structure, certain kinds of optimization can be done at a level above the intermediate language. An initial implementation of ZCODE would use the minimum register set, with standard properties. This implementation would run ALGOL68C programs, although not as efficiently as one tailored to the register structure of the target machine. In particular, it would run the ALGOL68C compiler. By resetting the parameters in the compiler to reflect

the actual register configuration and then using the running version
to recompile the compiler we obtain a tailored version. The register
optimization algorithms are built into the compiler itself, and thus
need not be rewritten for each machine.

New primitive modes could easily be added to ZCODE by defining
operators upon the values which belong to them and, if desired, denotations
for their values. The elementary objects would have to be represented
in terms of the existing ZCODE state.

ZCODE suffers from the same basic deficiency as OCODE and INTCODE:
Its state has a rigid structure which may not match that of the target
computer. The variable number of registers provides a good deal of
flexibility, but only within the general class of register-oriented
machines. Storage is a single array, as in OCODE, but there is no
restriction that an object must occupy a single cell. The compiler
can take variable sizes and alignments (as required by the target
machine) into account. Unfortunately, every increase in flexibility
results in an increase in implementation difficulty, and ZCODE is much
more difficult to map onto a computer than OCODE.

ZCODE could be used as an intermediate language for mathematical
software destined to be run on register-oriented computers with linear
memories. In order to make it an economically-viable process, however,
implementation tools and procedures must be much more carefully developed.

2.4.   Janus

The concept of a universal intermediate language {27} is still
appealing, and Janus is an attempt to realize this goal. Design work
began in 1970, and major changes in both syntax and semantics were made

in 1972 {9} and 1973 {10}. Since then the changes have been evolutionary, concentrating upon weaknesses or inconsistencies revealed by implementations {39}.

Unlike OCODE and ZCODE, Janus was not designed in conjunction with a specific translator. The basic architecture stems from several theoretical semantic models, and the details are derived from an extensive survey of existing languages and computers. No attempt has been made to anticipate changes in hardware or language concepts - we felt that the problem of adequately reflecting existing technology was difficult enough!

2.4.1. <u>Primitives</u>. Six primitive modes (Boolean, character, integer, real, address, and closure) are used to classify Janus values. The usual arithmetic and Boolean operators are provided, along with appropriate transfer functions. (Addresses embody the concept "address as value" and closures are used to represent "procedure as value.")

Boolean values are defined by enumeration, and character values by a partial enumeration ("set of characters is finite ... at least one alphabet of the letters and the digits is representable"); the definitions of integer and real modes are based upon intuition and experience. Restrictions on integers due to the finite machine representation are partially quantified ("maximum positive value ... not less than the maximum number of elements permitted in any array"), restrictions on reals are mentioned. Janus provides range and precision declarations which can be used to guarantee particular capacities.

Integer division truncates toward zero, and two truncation operators are provided.

2.4.2.   Composition.   Janus provides both array and record composition rules for operands.   Array indices are in the range [0,N-1] for an array of size N.   No explicit facilities are available for multi-dimensional arrays.

Actions may be composed as sequences or procedures.   Only value parameters are permitted, with all other mechanisms being synthesized by the source language compiler using addresses and closures.

2.4.3.   State.   The Janus state is a general composite object {24}. Any object accessible to the Janus program is a component of the state. A particular object is accessed by applying a (composite) selector to the state.   Such selectors are represented by address-mode objects.

The state of a Janus computation conforms to the contour model. Local variables have extents determined by procedure entry and exit. Initialization is permitted only for static storage, which has the entire execution history as its extent.   Explicit category specifications in the Janus code are used to distinguish the state components being referenced.

Direct access to a particular state component is controlled by scope rules implied by the contour model.   This access environment is determined from the current state or is given explicitly by a closure (which is a combination of an address and an environment).

2.4.4.   Evaluation.   Janus seems to provide an adequate model for the essential behavior of a wide class of languages.   Because it does not require many representational decisions, it does not preclude efficient implementation on most computers.   It is easily extended by the addition of new primitives. The category mechanism also makes possible extension by addition of new access mechanisms to the state. Janus does not imply any particular register or memory organization for the representation of its state, and hence it is easily matched to machines such as the Burroughs 6700 as well as to more conventional hardware.

As noted in Section 2.3.4, flexibility increases the cost of implementation. More decisions must be made, more optimization undertaken, and there is more chance for error. With current tools, bringing up Janus requires about 6 man-months by a very knowledgeable and dedicated implementor. It is essentially impossible for a novice.

Janus could be used as an intermediate language for mathematical software targeted for any non-vector machine. (I have not had sufficient experience with vector hardware to judge the possibilities for Janus there.) The goal of current Janus development work is to improve the tools and techniques for program transport. At the moment it is too difficult to implement Janus on a new machine.

3. Prognosis

I see an uncertain future for portability techniques in general and intermediate languages in particular. They are certainly here to stay, but economic considerations may significantly slow their development and completely block their acceptance in many applications. In this section I shall first outline the technical problems which remain to be solved, and then discuss the psychology and economics of portability. Perhaps my view will seem overly pessimistic; I leave you to make the final prediction.

3.1.   Technical Problems

        Given today's computers and languages, I believe that we have
sufficient theoretical understanding of the portability problem and the
use of intermediate languages to solve it. All that remains is the
reduction of theory to practice. This involves selection of the proper
intermediate language(s), construction of step-by-step implementation
procedures, production of the supporting documentation, and distribution
of the system.

   3.1.1.   Language Selection. In Section 1.3 I asserted the falsity of
the premise upon which portability via a single intermediate language was
based. I hope that the examples of Section 2 have reinforced this assertion.
The solution is a hierarchy of languages {30} in which each level represents
a major representational decision. Each such decision should be parameterized
to include as wide a class of machines as possible, consistent with the
decision itself. For example, in the framework of the examples given
in Section 2 we might have the following hierarchy:

   — Janus.

   — J1 {40}. State mapped onto a linear memory. Parameters are
       the sizes and alignments of the various data objects, plus
       packing information for bit and character arrays.

   — ZCODE. Stack and processor mapped onto a particular register
       configuration. Parameters are the properties of the available
       registers.

   — Machine language. Elementary operators mapped onto instruction
       sequences. Parameters are the definitions of the operators in
       terms of machine operators.

For portability, the "machine language" might be INTCODE.

This hierarchy would only be useful for a register-oriented machine. For a machine like the Burroughs 6700 one might map directly from J1 to machine code, or introduce another intermediate step. The best selection of languages for each case is an open question, one which is by no means trivial to answer. It will almost certainly require the design of several new intermediate languages, and a good deal of experimentation on many machines.

3.1.2. <u>Implementation Procedures</u>. It is not enough to simply define hierarchies of languages. Some of the transformations are nontrivial, and will require careful thought and sophisticated algorithms. The designer must supply tools which are reasonably efficient and easy to use, tools which accept the necessary parameters and then perform the transformation without further action on the user's part.

The obvious question is how to implement these tools. Clearly we must invoke the portability process recursively! One possible approach is to write FORTRAN programs which perform the desired transformations, and rely upon the ubiquity of FORTRAN. Another possibility would be to use BCPL, or to include a FORTRAN-INTCODE translator and distribute INTCODE versions of the tools.

I strongly advocate the full bootstrap {37} method of implementation, in which all work is carried out on the target computer. The half bootstrap, in which some work is done on the designer's computer, involves insuperable communication problems in most cases. (An exception is where both computers are under the control of the implementor and are either in close physical proximity or linked by a communication channel.)

3.1.3. <u>Documentation and Distribution</u>. The documentation of the implementation process must be analogous to that supplied by the manufacturers of electronic kits for the home hobbyist {18}: Every step detailed, with exhaustive tests at crucial points. Feedback is important here, since the designer is too close to the project to see the problems. "Portable software is like good wine - it improves with age." {14} Problems with the physical distribution are covered elsewhere {38}.

3.2. <u>Psychological and Economic Problems</u>

The psychological barriers to acceptance of portable software are characterized by the "not invented here" (NIH) syndrome and the belief that code efficiency is king. I shall not discuss these in detail because I hope that a programmer outgrows them as he matures. Even if he does not, they are irrational in most cases and hence not subject to reasoned discourse.

There are, of course, situations in which portability is irrelevant: Routines such as interrupt handlers and device drivers are concerned with details of machine operation and apply only to one computer. In his arguments for the use of assembly language to write system software, Fletcher {13} gives a good characterization of such routines. There is considerable evidence, however, that even a large part of an operating system can be made portable {8}.

Economic problems are more serious. They can be summarized by the statement that nobody wants to pay for portable software development. A computer manufacturer does not want to develop portable software which his customers could use on foreign hardware. Forward portability, the ability to bring existing software up on a new range of machines, is important to a manufacturer in his capacity as a software vendor. Software vendors do not want to develop techniques for portable software which their competitors could use. Thus any progress made by these vendors will be held as a trade secret. The involvement by university research groups may be nearing an end due to the routine developmental nature of the technical work: Computer Science Departments will not award Ph.D.'s for solving the problems outlined in Section 3.1!

We are thus left with the user who faces a change of machine every five years or so and who wants to distribute his programs to colleagues. Unfortunately, the development of a complete system of any degree of sophistication is costly and time-consuming. The result is that each user with a portability problem will develop a partial solution which is adequate for his personal use, but probably not for anyone else.

REFERENCES

1. ANSI (1966). FORTRAN. ANS X3.9-1966, American National Standards Institute, New York.

2. Backus, J. W. and Heising, W. P. (1964). "FORTRAN," IEEE. Trans. on Electronic Computers, Vol. EC-13, p. 382.

3. Bauer, F. L. (Ed.) (1974). Compiler Construction. Heidelberg: Springer-Verlag. (Lecture Notes in Computer Science, Vol. 21.)

4. BBN. (1974). BCPL Manual. Cambridge: Bolt, Beranek and Newman, Inc.

5. Birrell, A. D. (1976). ALGOL68C Implementors' Guide. Second Edition, Computing Laboratory, U. of Cambridge.

6. Bourne, S. R. (1975). ZCODE - A Simple Machine. Edition 7, Computing Laboratory, U. of Cambridge.

7. Cheney, C. J. Private Communication.

8. Cox, G. W. (1975). Portability and Adaptability in Operating Systems Design. Ph.D. Thesis, Purdue University.

9. Coleman, S. S., Poole, P. C., and Waite, W. M. (1974a). The Mobile Programming System: Janus, Software - Practice and Experience, Vol. 4, p. 5.

10. Coleman, S. S. (1974b). Janus: A Universal Intermediate Language. Ph.D. Thesis, University of Colorado.

11. Dennis, J. B. (1975). "Modularity," in Software Engineering, Bauer, F. L., (Ed.) Lecture Notes in Computer Science, Vol. 30, p. 128.

12. Dunn, R. C. and Waite, W. M. (1976). "Truncation and Integer Division." In preparation.

13. Fletcher, J. G. (1975). "No! High Level Languages Should Not be Used to Write Systems Software," Proc. ACM Natl. Conf., p. 209.

14. Gabriel, J. (1976). Remarks at the Los Alamos Workshop on Software Portability.

15. Gary, J., and Helgason, R. (1972). "An Extension of FORTRAN Containing Finite Difference Operators," Software - Practice and Experience, Vol. 2, p. 321.

16. Gentleman, M. (1976). Private Communication.

17. Goos, Gerhard (1974). Some Thoughts on Variables. Bericht nr. 19/74, Fakultaet fuer Informatik, Universitaet Karlsruhe.

18. Heath. (1968). Assembly and Operation of the Laboratory 5 Oscilloscope Model 10-12. Benton Harbor: Heath Company.

19. Hedberg, R. (1963). "Design of an Integrated Programming and Operating System Part III: The Expanded Function of the Loader," IBM Systems J. Vol. 2, p. 298.

20. Hoare, C. A. R., and Lauer, P. E., (1974). "Consistent and Complementary Formal Theories of the Semantics of Programming Languages," Acta Informatica, Vol. 3, p. 335.

21. Jensen, K., and Wirth, N., (1974). PASCAL User Manual and Report. Heidelberg: Springer-Verlag.

22. Johnston, J. B., (1971). "The Contour Model of Block Structured Processes," SIGPLAN Notices, Vol. 6, No. 2, p. 55.

23. Kernighan, B. W. (1975). "RATFOR - A Preprocessor for Rational Fortran," Software - Practice and Experience, Vol. 5, p. 395.

24. Lucas, P. and Walk, K. (1970). "On the Formal Definition of PL/1," Ann. Rev. in Automatic Programming, p. 105.

25. McCarthy, J. (1960). "Recursive Functions of Symbolic Expressions and their Computation by Machine, Part 1," CACM, Vol. 3, p. 184.

26. McKellar, A. C. and Coffman, E. G. Jr. (1969). "Organizing Matrices and Matrix Operations for Paged Memory Systems," CACM, Vol. 12, p. 153.

27. Mock, O., Olsztyn, T., Strong, J., Steel, T., Tritter, A., and Wegstein, J., (1958). "The Problem of Programming Communication with Changing Machines: A Proposed Solution," CACM, Vol. 1, p. 12.

28. Moler, C. B. (1972). "Matrix Computations with FORTRAN and Paging," CACM, Vol. 15, p. 268.

29. Naur, P. (Ed.) (1963). "Revised Report on the Algorithmic Language ALGOL 60," Computer J., Vol. 5, p. 349.

30. Poole, P. C. (1971). "Hierarchical Abstract Machines," Proc. Culham Symp. on Software Engineering, p. 1.

31. Richards, M. (1969). "BCPL: A Tool for Compiler Writing and Systems Programming," Proc. AFIPS SJCC, Vol. 34, p. 89.

32. Richards, M. (1971). "The Portability of the BCPL Compiler," Software - Practice and Experience, Vol. 1, p. 135.

33.   Richards, M. (1972). INTCODE - An Interpretive Machine Code for BCPL. Computer Laboratory, U. of Cambridge.

34.   Ryder, B. G. (1974). "The PFORT Verifier," Software - Practice and Experience, Vol. 4, p. 359.

35.   Tennent, R. D. (1976). "The Denotational Semantics of Programming Languages," CACM, Vol. 19, p. 437.

36.   van Wijngaarden, A., Mailloux, B. J., Peck, J. E. L., Koster, C. H. A., Sintzoff, M., Lindsey, C. H., Meertens, L. G. L. T., and Fisker, R. G., (1975). "Revised Report on the Algorithmic Language ALGOL 68," Acta Informatica, Vol. 5, p. 1.

37.   Waite, W. M. (1973). Implementing Software for Non-Numeric Applications. Englewood-Cliffs: Prentice-Hall.

38.   Waite, W. M. (1975a). "Hints on Distributing Portable Software," Software - Practice and Experience, Vol. 5, p. 295.

39.   Waite, W. M. and Haddon, B. K. (1975b). A Preliminary Definition of Janus. Tech. Rept. SEG-75-1, Dept. of Electrical Engineering, U. of Colorado.

40.   Waite, W. M. (1976). Janus Memory Mapping! The J1 Abstraction. Tech. Rept. SEG-76-1, Dept. of Electrical Engineering, U. of Colorado.

41.   Weizenbaum. J. (1963). Symmetric List Processor. CACM, Vol. 6, p. 524.

42.   Whitby-Strevens, C. (1976). Private Communication.

COMPUTER-ASSISTED PORTABILITY

It is a brave man who suggests or indeed believes that the portability of numerical algorithms can be achieved with complete automation.

— J. Bentley and B. Ford (1976)

The brave men who write in this section make bold claims about the use of computers to assist in the construction of portable numerical codes. They are testing the limits of automatic portability within the present state of computing but they stop short of achieving portability with *complete automation*.

James M. Boyle introduces the subject with a survey of current work on mathematical software transportability systems. He takes a broad view of automated aids and, in his search for commonalities among existing systems, he identifies a theme that characterizes current work and points to future directions. The remainder of the section is devoted to descriptions of several systems by individuals who have participated in their design, development, and use.

It is significant that all of the work described here arises from attempts to satisfy practical needs within mathematical software development projects: software development at the Jet Propulsion Laboratory (Krogh), IMSL (Aird), NAG (DuCroz, Hague, and Siemieniuch), and NATS (Dritz). At the same time, the work has a strong research flavor since the practical problems being addressed originate in fundamental questions about the semantics of programming languages.

MATHEMATICAL SOFTWARE TRANSPORTABILITY SYSTEMS
-- HAVE THE VARIATIONS A THEME?

James M. Boyle

ABSTRACT

Since about 1970 a number of individuals and
groups developing mathematical software have begun
employing at least moderately elaborate software
tools to assist in making their software available
on a variety of computers. In this paper four of
these systems are compared and contrasted by considering for each the setting in which its developers work, the objectives set for the system, the
approach taken and features included to meet the
objectives, and an assessment of the performance
and contributions of the system.

When the systems are viewed in this way, it
becomes apparent that many of the differences
between them result from differences between the
settings in which they were developed, and that
in fact there is a common theme of which they are
variations. In the latter part of this paper this
common theme is generalized and its implications
for the future evolution of tools for mathematical
software development are discussed.

In particular, the concepts of *program realization* and *realization function* are introduced as
abstractions of the common processes carried out by
these systems. The generality inherent in these
concepts suggests how transportability systems can
evolve into software tools which will make it possible
to achieve substantial economies of scale in the
mathematical software development process.

1.   Introduction

The year 1970 marked a qualitative change in the development of mathematical software libraries. Prior to that time, most library development efforts concentrated on preparing a library for a single machine. Substantial use of such a library was assured either if no large library existed for that machine, or if the new library provided programs for solving a number of problems not addressed by its predecessors. Around 1970, several groups interested in developing mathematical software (libraries, chapters of libraries, or individual programs) began to recognize that a kind of "saturation" had occurred: large libraries existed for the computers then in most widespread use. They reasoned that the mathematical software they were developing, which solved problems in areas already covered by existing libraries, would only achieve widespread acceptance if it offered substantial improvements over similar programs in those libraries.

One obvious improvement was to make the new mathematical software available on a number of different manufacturers' computers. The developers of application programs, many of whom had already encountered the need to convert such programs from one machine to another, might thus be induced to employ the new software because using it would simplify such conversion. Another improvement was to make the programs easier to use by providing user-oriented documentation and greater structural uniformity within a package. Users thus might find it easy to begin using programs from a package and certainly would find it easy to use other programs once they had used one. Finally, the most important improvement was to make the software considerably more reliable than that in existing libraries, primarily through careful evaluation of the algorithms used and extensive testing of the programs implementing them. The intent here was to avoid the bad reputation acquired by some of the earlier libraries and to convince users that they could obtain reliable answers via the new library with much higher probability (and much more quickly) than by writing and debugging their own programs. (A further discussion of some of these improvements can be found in {25}.)

Nearly all of the mathematical software developers who set themselves the above goals also observed that the effort to make the software widely available could, under the right conditions, contribute to its reliability. In contrast to developing mathematical software for a single machine, making it available on a number of machines compels a deeper analysis of the arithmetic properties of the software and subjects it to testing under a much broader range of conditions. In order for these steps to contribute to reliability, it is, of course, necessary that the versions of a program for different machines not *diverge*, i.e., that they remain closely (ideally, systematically) related to one another {18}.

A basic step toward avoiding the divergence of several versions of a program is to maintain a single, master version of the program, which is augmented as necessary to permit all needed versions to be mechanically derived therefrom. A limiting case of this approach is, of course, to write completely portable programs, so that different versions are not required for different machines; this approach has recently been applied in the development of the PORT library at Bell Telephone Laboratories {16, 9}. Most of the other mathematical software developers have felt that achieving complete portability blocks attainment of their other design goals (notably efficiency {23}, and even operation on certain systems {26}). Moreover, they have felt that the benefits of only having to maintain one master version of a program could be extended beyond machine versions to single and double precision versions by employing suitable software tools to derive them. One should note that maintaining a master version of a program (by whatever means) also offers substantial economic advantages, as well as increased reliability, to the mathematical software developer; these economies of scale are discussed further in Section 6 (see also {7}).

This, then, is the rationale for employing mathematical software transportability systems: they facilitate maintaining but a single master version for several closely related programs, thereby making possible both increased reliability in the program and reduced cost in its development.

In the following sections I present an overview of four mathematical software transportability systems currently in active use: the systems of Krogh {20, 21}, IMSL {3, 1, 2}, NAG {18, 14} and NATS {4, 5, 13}. For each, I attempt to characterize the setting in which its developer finds himself, for I believe this setting has exerted a substantial influence on the evolution of each system. I then review the objectives for and design of each system. Finally, I attempt to assess each system, not only with respect to how well it meets its design objectives, but also with respect to three criteria which help one discern the evolutionary progress of transportability systems. (These criteria -- which involve the possibility of expressing programs abstractly, the knowledge of target environments required, and the contributions of the systems to the reliability of the transformation process -- are defined in detail in Section 2.4, where examples are available to motivate the definitions.)

In the assessment I also attempt to show how many of the differences between the systems arise naturally from differences between the settings in which they were developed. In particular, there has been considerable debate about whether the (augmented) master version of a program should be maintained in compilable form or in a form which must be processed before it can be compiled; this has often been cited as a fundamental difference between systems. From the point of view taken in this discussion, this difference arises naturally from historical differences in setting and use of one or the other approach and does not reflect a fundamental difference in the capabilities of the systems.

2.   The "Specializer" System of Krogh

2.1   Setting

The Specializer is unique among the four systems discussed here in having been developed as a tool for use primarily by an individual numerical analyst (working on software for the solution of ordinary differential equations) rather than for use by a group developing a mathematical software library. Thus the Specializer evolved in a setting in which a tool was needed which would make it easy for an individual developing a program to make it available to other individuals using computers different from his own. This use of the Specializer by individuals rather than by library developers also meant that there was no emphasis in its design on using it as a tool to control distribution of programs (see Sections 3 and 4); indeed the emphasis was quite the opposite.

The characteristics of the programs with which the Specializer was to be used also influenced its design. It was felt that these programs (many of which were ordinary differential equation solvers) would be more useful if they contained various options permitting selection of alternate versions of the source program (alternate versions more complicated than those for single or double precision) at certain points. These options could be set to obtain a version of the program tailored, or optimized, somewhat for a particular application. (In this respect the requirements for the Specializer were somewhat similar to those which motivated the development of the Program Generator at Moscow State University {6}.)

Finally, I relate an anecdote which I think had a profound effect on the design of the Specializer: A computer user at Jet Propulsion Laboratory sought the advice of Fred Krogh on converting an ordinary differential equation solver program which he had received from a GE-635 installation for use on JPL's Univac 1108. Upon further investigation, it was discovered that the GE-635 version had been converted from an IBM-360 version, which, it turned out, had been prepared from a Univac 1108 program obtained from JPL! This near-circular conversion would have resulted in a substantial degradation in the quality and reliability of the program in question. The Specializer was developed in the hope of avoiding a similar occurrence in the future.

2.2　Objectives

In light of the setting described above, the objectives set for the Specializer may be summarized as follows:　(1)　It should provide a formalism for augmenting a Fortran program so that it can be automatically converted for use on different machines, including conversion to an appropriate arithmetic precision.　(2)　It should provide a means to augment a program to specify different program options.　(3)　The process of converting, or specializing, a program should be relatively inexpensive, and the processor for it should be transportable to various machines. (4)　Objectives (1) to (3) should be realized in a way which minimizes the possibility of "bootleg" conversions, i.e., it should be obvious to anyone possessing a copy of a program produced by the Specializer what other machine versions were planned by the author and how they differ from the one he has.

2.3　Approach

The general approach followed to meet the above objectives is to maintain the master version of a program in executable form, encoding the information describing different versions and options in specially marked comments.　A preprocessor, written in Fortran and called the Specializer, is provided.　It is capable of interpreting the specially marked comments in the master version and using them to produce a new version.　Marked comments are of two kinds:　(1)　those representing conditional Fortran statement selection, which cause a block of the following Fortran statements to be marked as comments or not, depending on the values of selection parameters; and (2)　those representing statement replacement, which cause all or part of the following Fortran statement to be replaced by text generated by the Specializer from the comments. When the Specializer interprets specially marked comments it makes use of the values of *environmental parameters* characterizing the machine for which the program is being specialized (for common machines these values are stored in a table in the Specializer) as well as *Specializer variables* which may be given values in specially marked comments.

The operation of the Specializer may be clarified by some examples, drawn from {20}. Suppose the master version of a program contains the statements:

```
        DO 100 I = 1, N
C$$     IF ((.IMP. .EQ. 'S') .AND. (.EPS. .GE. 1.E-10))
C       SUM = SUM + DBLE(X(I))*DBLE(Y(I))
C$$     ELSE
        SUM = SUM + X(I)*Y(I)
C$$     ENDIF
   100 CONTINUE
```

Conditional selection is based on the values of the environmental parameters .IMP. (specifying the basic precision is single or double) and .EPS. (specifying the relative accuracy of .IMP. precision arithmetic). In the above example, evidently .IMP. $\neq$ 'S' or .EPS. < 1.E-10, so the selection condition is false and the single precision accumulation of the inner product is selected; the machine might be a CDC 7600 for example. If .IMP. = 'S' and .EPS. $\geq$ 1.E-10, say 1.E-8, then the Specializer places a C in column 1 of the statement:

```
        SUM = SUM + X(I)*Y(I)
```

and removes the C in column 1 of the statement:

```
C      SUM = SUM + DBLE(X(I))*DBLE(Y(I))
```

thus selecting double precision accumulation of the inner product; the machine might be a Univac 1108 for example.

As an example of statement replacement, suppose the master version of a program contains the statements:

```
C$     $.TYPE.$ A,B($IDIM$,8), WORK($IDIM**2+5*IDIM$)

       0REAL A,B(10,8), WORK(150)
```

(Here $ delimits expressions to be evaluated by the Specializer.) Evidently then .TYPE. was REAL and IDIM had the value 10. If this program is processed by the Specializer and .TYPE. is DOUBLE PRECISION and IDIM is 20, the second line above is replaced by:

```
       0DOUBLE PRECISION A,B(20,8), WORK(500)
```

Statement replacement is thus useful for specifying those optional versions of a program which can be parameterized.

Finally, implicit replacement is available to simplify tailoring of formats and Hollerith data. An example is:

```
C$I    DATA MESS/'ERROR IN INPUT'/

       0DATA MESS(1)/6HERROR /,MESS(2)/6HIN INP/,

       $      MESS(3)/2HUT/
```

which is for a machine capable of storing six characters per word. The DATA statement becomes:

```
       0DATA MESS(1)/4HERRO/,MESS(2)/4HR IN/,

       $      MESS(3)/4H INP/,MESS(4)/2HUT/
```

on a machine capable of storing four characters per word.

2.4   Assessment

An important consequence of the approach taken in the Specializer
is that every executable version of a program produced by the Specializer
contains the augments which specify the other versions which are available.
Thus someone receiving a copy of the program can immediately see what other
versions were contemplated by the author, and the possibility of bootleg
conversions is greatly reduced. (A contemplated modification to the
Specializer would permit deleting the specially marked comments from
the output program to make it more readable; presumably such a version
would not be allowed to circulate.) Thus it is apparent that the idea
of having the master version of the program be at the same time an
executable one is a natural and necessary result of the objectives
set for the Specializer.

The Specializer has undergone considerable evolution since its
first version was prepared {19}. This evolution has increased the
variety of Specializer statements available and has made available
a higher degree of parameterization and automation of the alternate
version specifications. In my view, this evolution has moved it away
from its original goal of providing a formalism for specifying versions
of a program for different machines, which formalism could be interpreted
either by the Specializer *or by hand* to convert a program from one machine
to another. The formalism specified in {20} is probably too complicated
to be reliably interpreted by hand, making the provision of a transportable
version of the Specializer a necessity.

The evolution mentioned above has undeniably increased the utility
of the Specializer with respect to the first of the criteria of evaluation
mentioned in Section 1. This is the *degree of abstraction* permitted, which

is the degree to which differences between versions of a program can be parameterized, e.g., in terms of the precision of machine arithmetic, etc. Such parameterization implicitly makes use of the idea of programming for a "model machine" as discussed by Brown {9}. In the Specializer it is provided by the environmental parameters (and also by the Specializer variables), which permit selections to be made on the basis of properties of the target arithmetic rather than on the specific machine type, as illustrated by the examples in Section 2.3. Moreover, features of the Specializer not discussed here permit many of the differences between single and double precision versions of a program, and even some of those between complex and real arithmetic versions, to be parameterized.

On the other hand, the Specializer does not permit very extensive user-defined abstractions beyond those which can be specified in terms of Specializer variables, e.g., dimensions. That is, it is not possible, for example, to define an abbreviated form of some machine- or option-dependent construction which occurs repeatedly in a program. Thus the facilities available for abstraction are those built into the Specializer, and these are not extensible. (A macro facility for the Specializer is proposed in {20}, but it has not been implemented; provision of such a facility would provide user-defined abstraction and increase the similarity to the Program Generator described in {6}.)

The second evaluation criterion mentioned in Section 1 is the *extent to which the writer of a program must understand the target environments.* A programmer using the Specializer must provide explicitly for the various target environments in the sense that he must plant the specially marked

comments in the program where needed. (The Specializer leaves a program
without specially marked comments completely unchanged.) However, to the
extent that the machine dependencies can be parameterized in terms of an
abstract model as discussed above, it is not necessary for the programmer
to understand or even to anticipate all potential target environments.
Where such parameterization is not possible, he must understand each
target environment in detail and write appropriate Specializer statements
to cater to it.

The third criterion useful in evaluating transportability systems
is how they contribute to *reliability*. Of course, the systems themselves
are assumed to operate reliably; what is being assessed here is how they
contribute to the reliability of the process of transporting programs.
In my view, reliability is primarily related to the degree of abstraction
a system permits in its specification language and the extent to which
it permits global specifications. Perhaps this relationship can be
clarified by considering the example of implicit replacement discussed
in Section 2.3. It is evident that the Specializer statement:

      C$I   DATA MESS/'ERROR IN INPUT'/

does not completely capture the program modifications necessary to
handle storing this message with differing numbers of characters per
word, since the appropriate DIMENSION statement for MESS must also
be generated. As the Specializer does not provide a method of
specifying two related but physically separated program modifications
in the same Specializer statement, two independent acts are required
of the programmer, thereby increasing the possibility of error. In a

similar fashion, as more and more options are added to a program it becomes
increasingly difficult to verify that certain combinations of them do not
interact in unanticipated ways to produce an incorrect program. Thus the
Specializer docs not provide any particular help in insuring that the
program versions it constructs are reliable. (In its defense, however,
it must be mentioned that none of the other systems provides a facility
comparable to implicit replacement.)

In summary, the Specializer seems to meet its design objectives
well. By providing fairly extensive means for expressing program options
and alternate forms, it greatly decreases the amount of menial work a
numerical analyst must do to prepare a complicated program for use.
And since these options and machine versions are expressed in the source
program it should provide an effective means of avoiding bootlegged conversions.

3. The "Converter" System of IMSL

3.1 Setting

The IMSL Converter was produced in the context of a large commercial
library development effort, a setting substantially different from that
which produced Krogh's Specializer. It is thus perhaps surprising
that the two are remarkably similar.

Originally, IMSL produced a large mathematical and statistical
library for the IBM 360 computer; this was followed by a sequence of
adaptations of the library to other computers. These adaptations
were carried out manually without the benefit of specialized software
tools. As copies of the library for different computers began to
accumulate, however, it became evident that some sort of transportability
system would be useful, at least for the development of new programs,
and development of the Converter was begun.

Some important aspects of the IMSL setting include: Machine-specific programs are somewhat of a tradition; this results primarily from efficiency considerations and from the need to convince a customer that performance of the library on his machine has not been sacrificed to achieve transportability to a number of other machines. Nonetheless, in a substantial fraction of the library such machine-dependent code, if present at all, is well localized and therefore amenable to automatic derivation. Another facet of the IMSL setting is that its programs are developed in a central location by paid personnel; the final polishing and adaptation to other machines are handled in the same way. In Section 4 this setting will be contrasted with that of NAG, the other large library development effort represented here. Finally, to IMSL it is important that users receive copies of the library from IMSL itself; thus, in contrast to Krogh, IMSL desires protection against secondary conversions by users from one machine to another.

3.2 Objectives

Against the background of this setting the objectives set for the Converter may be summarized as follows: (1) It should provide a means to simplify the preparation of transportable Fortran programs by in-house personnel with a minimum of additional training. (2) It should cater for explicit machine dependencies in a program. (3) It should simplify the conversion of a program from one precision to another. (4) Its use in the preparation of programs for the library should not be apparent to library subscribers. (5) Insofar as consistent with the above objectives, it should be inexpensive to run and be transportable. (6) The approach chosen for the converter should permit rapid design and implementation (since it was needed immediately).

Perhaps as a consequence of (6), it was decided that no attempt would be made to include tools in the Converter system to facilitate automatic modification of existing library programs for use with it. Instead, they would be modified for use with the Converter by hand as each required maintenance or enhancement. In Section 4 this situation will be contrasted with that of NAG.

3.3 Approach

Although the design and initial development of the Converter proceeded entirely independently of that of Krogh's Specializer, the general approach is quite similar. Again, the master version of a program is an executable version augmented with specially marked comments which specify the changes to be made to produce other versions. In order to cater to objective (4), the Converter can operate in one of two modes: It can produce a *basis deck*, which is a version of a program that executes in a specific computer environment but still contains Converter statements. Or it can produce a *distribution deck*, which is a version of a program that executes in a specific computer environment and does not contain Converter statements; it is thus suitable for inclusion in a copy of the IMSL library.

To facilitate use of the Converter by a number of people with a minimum of training, as stated in objective (1), its control language features a small number of easily understood facilities. These facilities permit the user of the Converter to specify the target machine for which it is to produce output, to define or redefine Converter constants and parameters, to use the values of Converter

constants in DATA and assignment statements, and to use Converter
parameters to select one or another group of statements to be
included in the program.

As an example of the use of Converter constants in DATA statements
(see {1}), consider a basis deck containing:

C$ DATA EPS/SEPS/,C1/1.23456789012345678901234567/

       DATA EPS/Z3C100000/,C1/1.234568/

In this case the basis deck is for IBM single precision. The appropriate
value for the Converter constant SEPS is contained in the Converter Master
Table and is selected by a C$ OPTIONS statement:

C$ OPTIONS,COMPUTER=IBM,PRECISION=SINGLE

placed ahead of the basis deck. The Converter uses the value of another
constant in the Master Table, ISIGS, to round the representation of
the algorithm-dependent constant value for C1 to seven significant
digits. If the basis deck had been preceded by the statement:

C$ OPTIONS,COMPUTER=CDC,PRECISION=SINGLE

the DATA statement would become:

       DATA EPS/16414000000000000000B/,C1/1.2345678901235/

As an example of conditional statement selection based on Converter
parameters, consider:

       DO 100 I = 1, N

C$ IF (SSIG10 .LE. 10) 1 LINE, 1 LINE

C   SUM = SUM + DBLE(X(I))*DBLE(Y(I))

       SUM = SUM + X(I)*Y(I)

   100 CONTINUE

which is simply the Converter analogue of the first Specializer example in Section 2.3. Here the parameter SSIG10 represents the number of significant decimal digits in single precision (6.3 for IBM, 14.4 for CDC). If SS1G10 is less than or equal to 10, the C is removed from the first line following the C\$IF statement and a C is placed on the second line.

Finally, the design of the Converter includes a feature which has no counterpart in the Specializer: the Scan option for precision conversion. The purpose of this option is to convert a single precision basis deck to a double precision one and *vice versa*. Unlike the other operations performed by the Converter, precision conversion requires no specially marked comments to trigger its operation; rather the Fortran program is scanned character by character and type declarations, constants, and Fortran function names are changed as required. The Scan option also changes Converter statements, constants, and parameters in the basis deck in an appropriate way, see {1}.

3.4 Assessment

Some of the assessments of the Specializer given in Section 2.4 apply equally well to the Converter. As with the Specializer, the Converter basis deck is always in executable form. The principal advantage of this in IMSL's setting is that it permits the program to be debugged on one machine in basis deck form without preprocessing. By using the option to produce a distribution deck (with Converter statements removed) as output, IMSL meets its objective (4) and at the same time avoids encouraging users to convert a program to another machine.

The selective inclusion of one or another group of statements depending on the value of the COMPUTER parameter meets objective (2) quite adequately, but it has subjected the Converter (and equally, the Specializer) to surprisingly strong criticism. Not only is the strength of the criticism surprising, so also is its basic assumption: that the inclusion of a feature to permit explicit machine dependencies is equivalent to advocacy of salting every program with a horde of gratuitous machine dependencies.

To my knowledge, all of the mathematical software developers whose work is discussed here believe that every program should be as machine independent as possible. But in some cases (see {23} for an empirical study), failure to tailor programs to certain machine-compiler combinations may entail significant performance sacrifices (and hence violation of objective (4)). An example is provided by Aird in {1} in the program unit GGBIN, which contains the statements:

```
C$      IF (IBM) 1 LINE, 2 LINES
        A = (.5*ERFC(-RSQ1H*A))*RSPID3+CON1
C       CALL MDNOR(A,ERF)
C       A = ERF*RSPID3+CON1
        IF (A .LE. PS) GO TO 40
```

Here advantage is taken of a particularly accurate and efficient ERFC routine available in the IBM Fortran library; on other machines (without an ERFC) the IMSL routine MDNOR, which performs a related computation, must be used. The important point here is that there is a valid reason for introducing the machine dependency and it is introduced in such a way that it is clear that *both versions of the program are realizations of the same abstract program;* this concept will be elaborated further in Sections 5 and 6.

The precision conversion scan feature of the converter is an attempt to reduce the number of Converter statements which must be written to obtain both single and double precision versions of a program. (This problem is addressed in the Specializer by permitting "generic names" for functions, etc.) The difficulty of treating precision conversion with the other mechanisms is similar to that discussed at the end of Section 2.4: to specify this conceptually simple change would require Converter statements in many places throughout the program. With the precision conversion scan, however, no explicit indicators need be inserted by the programmer. (However, certain restrictions must be obeyed concerning positioning of specification statements and use of blanks, see {1}.)

The developers of the Converter recognize that the precision conversion scan is expensive, since each character of the program must be examined, rather than just those in the specially marked comments. For this reason they recommend having the Converter produce a basis deck in the new precision, so that the scan does not have to be repeated. Unfortunately, following this advice leaves one with two distinct basis decks for a program, which negates many of the advantages of maintaining a single version of the program discussed in Section 1. In this respect it would be better (although more complicated) for the Converter to produce a composite basis deck

containing appropriate Converter statements specifying both precisions.
In my estimation, the significance of the precision conversion scan
is not so much that it converts one precision to another, but rather
that its inclusion in the Converter demonstrates recognition that
some changes to a program can and should be made without explicit
markers in the source program.

The *degree of abstraction* permitted by the Converter is similar
to that of the Specializer except that versions of a program which
differ in the value of a numerical parameter (e.g., array dimension)
are not quite so easy to specify. For example, to achieve the effect
of the second example in Section 2.3, one must write:

```
C$     IF (IDIM .EQ. 10) 1 LINE, 1 LINE
       REAL A,B(10,8), WORK(150)

C      REAL A,B(20,8), WORK(500)
```

and use the precision conversion scan to obtain a double precision version.

The *knowledge of target environments* required by the Converter is
almost identical to that of the Specializer. Like it, the Converter
contains a master table of machine parameters which permit one to write
programs in terms of a model machine.

The *reliability* of the Converter approach is also quite similar
to that of the Specializer. The precision conversion scan seeks to
make precision changes reliable by obviating the need for explicit
actions by the programmer which must be distributed throughout the
program. However, the implementation of the precision scan may
cause reliability problems, since certain conventions must be
followed in order for it to work. Thus, a programmer using it

might think his program was completely converted when, say, some specification statement violating the conventions had not been converted. Such errors are notoriously difficult to detect and correct.

In my experience, the willingness of most programmers to examine a program which has been automatically converted from one form to another is inversely proportional to how complete that conversion is, precisely because the examination becomes less rewarding as fewer errors are found. Thus in order to achieve reliability in an automatic process, *we should strive to impose no restrictions or conventions;* if some are necessary the process should be designed to fail in a non-catastrophic manner when they are violated.

In summary, the Converter seems to have achieved the objectives set by its designers quite well. In particular, I think it achieves a nice balance between simplicity and flexibility, which makes it easy to learn and to use.

4. The "Master Library File System" of NAG

4.1 Setting

Although NAG is, like IMSL, a large commercial library development effort, the setting in which the Master Library File System (MLFS) was developed differs substantially from that of the IMSL Converter.

Historically, NAG was organized to coordinate the efforts of a group of British Universities in developing a library in both Fortran and Algol 60 for a single machine, the ICL 1906A; each university would contribute programs to a chapter of the library in which it had expertise, and would receive the entire library in return. As the high quality of the NAG library became known, however, there were numerous requests to

use it on other machines; NAG wisely decided that quality could only
be maintained if the adaptation to other machines were performed by
NAG. Finally, economic factors made it necessary for NAG to become
self-supporting, so that it is now a commercial library effort with
a library in use on a number of machines.

Throughout these changes the original decentralized organization
of NAG has persisted. This organization consists of *contributors* who
are experts in an area of numerical analysis and who develop programs
for the corresponding chapter of the library; *implementors*, who are
experts in a particular machine and who modify the programs in
several chapters of the library as necessary for their machine; and
the *central office staff*, who receive contributed programs, send them
to the various implementors, receive the completed implementations,
and make the library available to its subscribers. Historically,
the intersection of the set of contributors to a chapter and that of the
implementors for that chapter was usually null; in contrast, in the
setting of Krogh or IMSL the programmer contributing a program has
full responsibility for its implementation on all relevant machines.

This decentralized organization had substantial impact on the
design of the MLFS. Obviously, an approach which required the original
programmer to insert additional information for various machine versions
was unattractive. Moreover, the Central Office Staff were acutely aware
of the potential for blunders in managing the flow of contributed and
implemented programs to and from their destinations. Thus they sought
a system which would protect the integrity of the programs from such

blunders, as well as ease the problem of managing them. Above all, the system had to be applicable to the existing library; library development could not be interrupted to manually modify the programs for use with the system.

Another aspect of the NAG setting which affected the design of the MLFS is that the implementations for different machines proceed asynchronously with respect to one another. Thus the distributed version of the library for, say, the ICL 1906A may be at one edition, or mark, while that for another machine is at an earlier mark. Hence, the MLFS must preserve a history of marks of the library, see {18}.

4.2   Objectives

In light of the setting described above the objectives of the MLFS may be summarized as follows: (1) Its prime objective is to simplify the library activities carried out in the NAG Central Office in accordance with objectives (3) - (6). (2) It should be applicable to the existing library without modification. (3) It should provide information about the relative status of the various implementations and help to avoid divergence of them (i.e., assist management of NAG). (4) It should permit implementations to be at different marks (i.e., maintain a history). (5) It should protect an implementation from inadvertent influence from other implementors (i.e., provide security). (6) It should assist in revising the implementations of a program when its contributed version is updated (i.e., provide *predicted* implementations).

4.3   Approach

The approach adopted by NAG to meet the above objectives bears some resemblance to that used in the Specializer and Converter in that it involves a master file version of a program which consists of intermixed program statements and control statements. The similarity ends there, however, because in the MLFS the control statements are intended to be introduced automatically rather than by hand.

A control statement in the MLFS basically indicates to which implementations and marks of the library belong the program statements following it and preceding the next control statement of similar type. A *selection* can thus be performed on the master file to produce a version of the library for a particular implementation and mark thereof. One should note that it is quite possible for the master file to contain variant program statements which are not selected for any currently distributed mark of the library; nonetheless they are available to reconstruct an earlier mark if required.

The automatic production of control statements is accomplished in the MLFS by a program called the *anti-editor*. Its function is to compare an ordinary, executable program for a particular machine received from an implementor with the contributed version of that program (usually the ICL 1906A version). From this comparison the anti-editor produces a set of editing directives which modify control and program statements in the master file so that the program received from the implementor becomes selectable from the master file. In this process, some statements may be added to the master file, and others may be

conceptually deleted, i.e., their control records made to indicate
that they are not applicable from the current mark of that implementation
onward.

The operation of the components of the MLFS can be illustrated by
an analogue of the first example of Section 2.3 and of the second of
Section 3.3. Suppose the contributed version of a program contains
the statements:

```
      DO 100 I = 1, N
      SUM = SUM + X(I)*Y(I)
  100 CONTINUE
```

and a version returned by an IBM implementor at Mark 2.0 contains at
this point the statements:

```
      DO 100 I = 1, N
      SUM = SUM + DBLE(X(I))*DBLE(Y(I))
  100 CONTINUE
```

Then in comparing these two programs the anti-editor generates editing
directives which when applied to the master file cause it to contain:

```
              DO 100 I = 1, N
*600BG   010099Z9
*605BB        0200
              SUM = SUM + X(I)*Y(I)
*600BB   020099Z9
              SUM = SUM + DBLE(X(I))*DBLE(Y(I))
*600BG   010099Z9
              100 CONTINUE
```

The *600B control statements indicate who introduced the following program statements, the mark at which they were introduced (0100 in the first *600B card above), and the mark at which they were withdrawn (99Z9, the infinity mark, above). The letter G in the implementor field (after *600B) indicates the contributed version, and B indicates the IBM implementation. The *605B control statement indicates that the following statements are not part of certain implementations, and the mark at which they were withdrawn. Thus the control statements in the above example specify that the single precision accumulation of SUM was withdrawn at mark 0200 by the IBM implementation and replaced by the double precision accumulation; resumption of statements common to all versions is indicated by the final *600B control statement.

4.4   Assessment

Although the MLFS master library is similar in concept to the basis deck of the Converter or Specializer, it is not a program in executable form. In the context of the NAG setting this is quite understandable, for the motivations for using an executable master library are absent. That is, the NAG Central Office is neither interested in making users aware of the existence of other versions, nor does it require an executable form for debugging, since modifications to an executable program are made at remote sites and the modified program is automatically reincorporated into the master library by the anti-editor. The anti-editor also plays a pivotal role in achieving objective (2), since it automates the incorporation of the existing library into the MLFS.

It should be clear that the fairly detailed information contained in the MLFS control statements can be processed with various report generators to produce management information fulfilling objective (3). One unique type of report produced is the comparative printout showing which program statements appear in which of certain selected implementations and marks, see {18}. Such reports can be used, for example, to monitor divergence among implementations.

Objective (4) is met, of course, by recording the marks to which program statements belong.

To achieve objective (5) the MLFS requires mark and implementation identification with each anti-editing and editing run. Thereby it can ensure that only statements pertaining to that mark and implementation are modified in the master library.

The approach to achieving objective (6) is one of the most interesting features of the MLFS. A typical situation is that both the contributions to the library and a particular implementation of it are at mark n, and a contributor submits a slightly modified version of one of his programs for mark $n + 1$. Then the task is to furnish the implementor with his "predicted" implementation for mark $n + 1$, which he will correct as necessary and return for inclusion in the MLFS. This prediction is made by selecting from the master file according to the "equation":

$$(Predicted\ implementation)_{n+1}$$
$$= Implementation_n + Contribution_{n+1} - Contribution_n$$

i.e., the predicted implementation is the present one plus the changes made by the contributor in going from mark n to mark $n + 1$.

While such predictions are no doubt useful, especially in the transition to the use of the MLFS, one should note that they are made by "rote" rather than on the basis of any "understanding" on the part of the MLFS of the differences between machines. Thus they cannot be relied upon, and the programs must be carefully checked by hand. Nonetheless, they do direct the attention of an implementor to areas of the program likely to need his attention, and receiving them is vastly better than simply receiving the new contributed program. In Section 5 systems employing prediction based on an understanding of the differences between machines will be discussed. It is interesting to consider that the current NAG master library contains a wealth of empirical information about machine differences which could be extracted and catalogued to serve as a basis for developing such understanding.

The criteria of *degree of abstraction permitted, knowledge of target environment required,* and *reliability* are not very applicable to assessing the MLFS, since it is not intended that humans write its control statements. As the example of Section 4.3 illustrates, the MLFS offers essentially no abstraction facilities for making program variants depend on parameters rather than machine types. The MLFS does have the desirable property of not depending on indicators planted in the programs, but nonetheless it still requires of the implementors a detailed knowledge of the target environment. It does not make any contribution to the reliability of the implementation process, except insofar as a predicted implementation directs attention to areas of a program that may need to be modified.

In summary, the MLFS was designed to address a somewhat different set of objectives than do the Specializer and Converter. It meets these objectives quite well, greatly easing the information management problems of the NAG Central Office with no negative impact on contributors and implementors. In my opinion, the chief contributions of the MLFS are its anti-editor (a useful program in its own right) and its emphasis on maintaining a historical record of the library contents. Not only does the latter permit implementations to proceed asynchronously, but it also permits any earlier version of a program in the library to be reconstructed quite inexpensively. This capability would be useful in library maintenance (to determine if a section of a program causing a problem had been modified recently, or was soon to be modified because the problem had already been diagnosed), and in avoiding repetition of mistakes previously made and corrected, especially in implementing the library for a new machine range. Finally, the MLFS has been useful in calling attention to the role of *mechanical prediction* in developing transportable software, a topic which has been discussed by Ford {15} and which will be developed further in the following Section.

5. The "Transformation-Assisted Multiple Program Realization System" of NATS

   5.1 Setting

The setting of the NATS project and its successors, including the LINPACK project, is somewhat similar to that of NAG in being a cooperative, somewhat decentralized, mathematical software development effort. However, NATS/LINPACK differs from NAG (and IMSL) in that it is not commercial (rather its chief goal is research into methodology and tools for mathematical software development) and has concentrated on developing "systematized collections" of programs in particular areas of numerical analysis (analogous to library chapters) rather than on entire libraries.

The NATS/LINPACK project is also unique among the mathematical software developers discussed here in having substantially completed a second iteration of transportability system development. This second system, the Transformation-Assisted Multiple Program Realization (TAMPR) System, is the principal topic of this section. However, the development of TAMPR was undertaken as a result of the experience with, and the perceived limitations of, its predecessor, the Generalizer-Specializer system. Thus an understanding of it is necessary to understand the setting for TAMPR development (see also {4}).

5.1.1 Experience Gained from the Generalizer-Specializer System

The Generalizer-Specializer was developed for use with the systematized collection EISPACK then under development by NATS. The programs in EISPACK had been written in Fortran double precision arithmetic and tested on the IBM 360 computer. At the time development of the Generalizer-Specializer began, the project faced the task of converting them to run on at least five other computers.

It was clear that this conversion would be facilitated by the following properties of the programs: Like the Algol 60 procedures from which they were derived {27}, they were written in terms of a parametric model of machine arithmetic. Close attention had been paid to the Fortran constructions used, so that all conformed to the ANSI standard (or could be simply mapped to do so). Substantial effort had been expended to achieve a structural and stylistic uniformity among the programs in EISPACK. (Henceforth I will use the term "format" to describe this latter property.) Thus the principal conversion necessary was that from double to single precision, and it was natural to consider automating it in some simple-minded fashion. (Some other machine dependencies discovered in the testing of EISPACK are discussed in {26}.) The other objective set for the Generalizer-Specializer system was that described in Section 1, to enable maintaining but a single master version of the programs.

The approach adopted for the Generalizer-Specializer involved use of a non-executable master file similar to that later adopted for the MLFS. Its control statements were very rudimentary, basically indicating only that the following lines were in the single or double precision version, or both. A Specializer program was provided to select a version for a particular machine from the master file; later it was modified also to insert values for the machine arithmetic parameters.

The Generalizer program was designed to construct the master file of single and double precision versions from the IBM double precision version. It was thus a very simple example of a *mechanical predictor* (see Section 4.4) which codified an understanding of the differences between the double and single precision versions of EISPACK. Because of the properties of the EISPACK programs remarked above, it was possible for the Generalizer to employ simple character-string matching to locate type declarators, function names, and constant exponent indicators to be converted. Thus no detailed syntactic analysis of the source text was required, even though the conversion was accomplished without explicit indicators in the source text.

The Generalizer-Specializer system became operational in mid-1971 and for EISPACK it worked surprisingly well, despite a number of obvious limitations. The most important of these was that the Generalizer relied extensively for its correct operation on the known content and format of the EISPACK programs. Hence if used to convert an arbitrary program it might make some spurious changes and miss some changes that should be made. Thus it violated the principle of not imposing restrictions or conventions discussed in Section 3.4, and it would not be reliable for general use.

Another limitation of the generalizer was that its design did not readily permit the addition of new functions. In part this was because there was no *factoring of tasks*; i.e., for each program change the Generalizer was to make, the tasks of recognizing where to make it and making it were inextricably intertwined. Thus not only did the existing program changes fail to provide any guidance as to how to introduce new ones, they actually were a hindrance, because one had to check that the new program changes did not interact with the existing ones.

The continued development and testing of EISPACK also exposed additional applications for software tools, if appropriate ones could be developed. The most important of these was for automatic program formatting; the formatting of EISPACK was carried out by hand and was extremely time-consuming. A second tool needed was one which would permit the rapid specification and execution of one-time maintenance changes which had to be made to all or most of the programs in a package.

One more interesting objective for a future system arose from the mode of use which evolved with the Generalizer-Specializer. Originally it was supposed that the Generalizer would be used essentially once to produce the master file; thereafter maintenance changes would be made directly to the master file. In actual use, however, it proved more convenient to make the changes to the IBM 360 double precision version, test them, and then reapply the Generalizer to produce a new master file. (This was in part because of the empirical fact that nearly all the maintenance changes applied to all versions of a program, not just to one for a specific machine.) As a result of this mode of use (which parallels that of NAG) the non-executable master file caused no inconvenience, and an objective for future systems was to preserve this mode to the extent possible.

5.1.2 Evolution of the NATS Setting

As work on EISPACK neared completion, the participants in the NATS project were looking ahead to further study of mathematical software development methodology under conditions somewhat different from those for EISPACK. Specifically, there was an interest in developing packages in more complex areas of numerical analysis than linear algebra. Another area of interest was decentralized software development, wherein the programs in a package would be contributed by a number of persons from different locations who were not in every-day communication. Perhaps the most important aspect of this new setting was that plans were not completely formed when the development of TAMPR began in early 1973;

for example it was known that stylistic unity and formatting were desired in any new packages, but the exact form these should take was not known.

5.2 Objectives

From what had been learned through use of the Generalizer-Specializer, and taking into account the evolving NATS setting, the following objectives were initially set for TAMPR: (1) It should reduce the number of programs to be stored in master form as far as reasonably possible, by automating the derivation of several useful versions from each; at a minimum it should automate the derivation of any required machine-dependent versions and single and double precision versions. (2) It should permit a programmer to concentrate on producing well-structured programs written in terms of a model arithmetic; it should not require him to have a detailed knowledge of target environments in order to write transportable software. (3) It should accept ordinary Fortran programs, executable on some machine, and convert them to master form (see Section 5.1.1); insofar as possible, it should not require special markers in the source program in order to carry out its operations. (4) It should perform version derivation and other operations in as reliable a manner as possible; furthermore, its design should facilitate and encourage showing that these processes are reliable. (5) It should automate insofar as possible the melding of programs written by several authors into a single unified package; at a minimum it should unify the format and structure of the programs. (6) It should be flexible enough to meet as yet unknown requirements; among these being: formatting programs according to unknown specifications, transporting programs to unanticipated machines and languages, and performing unspecified package design changes.

Since the Generalizer-Specializer system was available for use with EISPACK, and since the development of TAMPR was regarded as fairly long-term research, no particular constraints were placed on the length of time required for its development, on its execution speed, or on the transportability of TAMPR itself.

5.3 Approach

The approach and techniques used in the TAMPR system differ substantially from those of the Specializer, Converter, and MLFS. Nonetheless, there are general similarities: Like these systems, TAMPR maintains a (non-executable) master file version of the programs it processes; this master version is called the *canonical abstract form* of a program. Also, TAMPR consists of a set of processors which operate on this canonical abstract form or convert to or from it, much as the processors of the MLFS operate on or convert to or from the MLFS master file version of a program.

To see what motivates this design approach, consider what might be required to meet the objectives of Section 5.2: there would have to be processors to produce machine-tailored versions of programs, to change the precision of programs, to perform maintenance changes to programs, to format programs, etc. Each of these processors would perform some analysis of the program, carry out its particular process, and then output the program. In this context *it is natural to ask whether these processors can be factored* (see Section 5.1.1). For example, can the analyses that would be performed by the processors be generalized in such a way that they can be performed simultaneously by a single processor, and can this analyzed form be stored as the master form of a program? Similarly, can the results of the particular processes the processors would carry out be generalized so that only a single processor is needed to convert them to executable Fortran programs?

Of course, the answer to both of these questions turns out to be yes! The particular generalization chosen for use in TAMPR is the canonical abstract form, which is obtained by parsing the Fortran program according to a BNF grammar and then processing it to uncover its structure and to discard a number of troublesome Fortran idiosyncracies (for further details see {7}). Fortunately, not only is canonical abstract form suitable as *input* to the particular processes, it is suitable as *output* from them as well (except, of course, from the formatting process). Using the same form for both input and output has an important consequence: the various particular processes can be cascaded (composed) if desired.

A further generalization and simplification of the design of TAMPR arises from addressing the objective of flexibility (objective (6)). Clearly one way to permit the specification of the particular processes to be deferred is to make the processors for them *programmable*. If the programmable processors are properly designed, the specification (programming) languages they implement will provide simpler and more rapid specification of processes than would writing corresponding specialized processors in conventional languages.

Thus at least two programmable processors are needed: the *formatter*, which converts from canonical abstract form to (say) executable Fortran according to a formatting specification written in Format Control Language (FCL) {12}, and the *transformation interpreter*, which carries out the other particular processes that go from abstract form to abstract form according to a specification of them written in the language of Intra-Grammatical Transformations (IGTs) {7}. As it turns out, these two processors and one more, the *recognizer*, which converts from Fortran to abstract form (itself programmed by the BNF grammar), are all that TAMPR requires. For, the structuring process which converts the program parse tree to canonical abstract form can be specified by IGTs, and the other processes required to fulfill the objectives of Section 5.2 can be realized by writing IGTs, FCL programs, or both.

One use of IGTs can be illustrated by showing how they can describe to TAMPR the construction of alternate versions of a program corresponding to those discussed in Sections 2.3, 3.3, and 4.3. Suppose the master version of a program contains:

```
       DO 100 I = 1, N
          SUM = SUM + DBLE(X(I))*DBLE(Y(I))
   100 CONTINUE
```

(This is the form appropriate for machines carrying 10 or fewer digits of precision.) To prepare a version for machines with more than 10 digits of precision, one writes a TAMPR transformation:

```
       <primary>
          {.SD. DBLE (<primary>"1") ==> <primary>"1" .SC.}
```

which is applied to the canonical abstract form before formatting it for those machines to produce:

```
       DO 100 I = 1, N
          SUM = SUM + X(I)*Y(I)
   100 CONTINUE
```

In a similar vein, consider the example of Section 3.4, and suppose the master version of the program is written for non-IBM machines:

```
CALL MDNOR(A,ERF)

A = ERF*RSPID3 + CON1

IF (A .LE. PS) GO TO 40
```

Then one supplies an IGT which optimizes this code segment for use on IBM machines:

```
<stmt tail>

{.SD.

        CALL MDNOR (<var>"1", <var>"2")

        <var>"3" = <expr>"1" { ? <primary> { <var>"2" } ? };

        <stmt tail>"1"

==>

        <var>"3" = <expr>"1" { ? <primary> { (.5*ERFC(-RSQ1H*<var>"1")) } ? };

        <stmt tail>"1"

    .SC.}
```

(Here ; represents the end of a Fortran statement.) This transformation locates *any instance* (in a program) of a call to MDNOR followed by an assignment statement whose right hand side contains a reference to the value computed by MDNOR, and optimizes it for the IBM 360. For the above program, the result is:

```
A = (.5*ERFC(-RSQ1H*A))*RSPID3 + CON1

IF (A .LE. PS) GO TO 40
```

One should note that as written, the above transformation assumes that the value of the variable ERF is not used later in the program. (More details on IGTs and the Fortran grammar can be found in {7}, while FCL is discussed in {12}.)

5.4 Assessment

There are two features of the TAMPR approach which have a large impact on its usefulness. One of these is the provision of special-purpose process-specification languages; the other is the ability to perform most operations without the assistance of special markers in the input program text. Of course, these two features are largely motivated by objective (6), and especially by the need to make package design changes. Such changes by definition occur late in the preparation of a package, after the programs have been written; therefore they cannot be performed by a processor which requires a long development time or special markers in the program. For similar reasons the two features are useful in achieving objective (5).

What is more interesting is the impact of these two features on fulfilling the other objectives. Their effect is to encourage one to think of specifying the derivation of versions of programs as processes *independent of the particular programs to which they apply,* rather than as components of the process of writing programs. This *factoring* of processes means that an expert can specify *the process of deriving* a particular version once and for all, after which that specification can be applied to numerous programs whose authors need at most only a very general awareness of its details. Thus the TAMPR approach is capable of a *high degree of abstraction* and requires only minimal *knowledge of the target environment*, thereby fulfilling objectives (2) and (3).

The *reliability* of program derivation processes is also enhanced by specifying them separately from programs and in a high level language. Separate specification means that they can be specified, tested, and debugged once; thereafter their correct operation is not dependent on the specification of the derivation process being correctly inserted in each program and at each location where required. Moreover, the fact that a derivation process is applied to programs without special markers means that the process itself must be analyzed carefully so that it does not apply where to do so would introduce an error. Such analysis, and the generalization and simplification which often accompany it, lead to increased reliability. For example, in examining the second IGT of Section 5.3 one fairly quickly becomes aware that a variable set by MDNOR is being deleted from this part of the program, and the effect of doing so must be considered; I think one is not nearly so conscious of this in the formulation in Section 3.4. Thus the two features of programmality and absence of markers contribute to fulfilling objective (4).

On the negative side, the TAMPR approach is somewhat costly because of its generality. Clearly the recognition of program constructions to be modified will be more costly in the absence of special markers delimiting them. In addition, many of the program derivation processes that can be specified in TAMPR turn out to be global in nature, so that the whole program must be examined; this increases processing time and memory requirements. And, of course, programmable processors are probably not so efficient as would be a set of special purpose ones. However, in my opinion the most important question about the processes carried out by TAMPR is: Are they reliable? Since most of the aforementioned sacrifices of efficiency buy increased reliability, they are well expended.

Certain aspects of the TAMPR approach, broadly related to economies of scale, make it more attractive in the context of library or package development than in a setting like Krogh's. For example, preparing the specification of a general program derivation process requires, typically, more effort than would inserting specialized specifications in a single program. Thus it may only be in the context of development of a package that the process is applied to enough programs to recover the extra cost of its development. In Section 6 the economies of scale inherent in the use of program derivation processes, or program *realization functions* as I prefer to call them, will be elaborated further.

Two other limitations of TAMPR may be cited: the lack of facilities to handle truly machine-dependent programs, and its nontransportable implementation. By truly machine-dependent programs I mean those in which nearly the entire body of the program varies from machine to machine and that variation cannot be codified in a way which permits it to be readily computed. Such programs are exemplified by the special function routines in FUNPACK {11}. As TAMPR was originally conceived {5}, the abstract form would have included provision for alternate machine versions; however, that provision has never been implemented and now is unlikely to be. In the context of library development, this lack may be a blessing, for the generality attained by being forced to avoid truly machine-dependent programs pays important economic dividends; however, in the setting of an individual numerical analyst the lack might be more keenly felt. In any event, progress is being made toward special function approximations whose machine-dependent versions would be mechanically predictable by sophisticated realization functions {17, 24}.

The nontransportable implementation of TAMPR is a consequence of the research aspect of the project. Languages were chosen for their ability to support fairly rapid, highly understandable, and readily modifiable implementation of the system. Nothing in the general approach is known to preclude constructing a transportable implementation in (say) Fortran, however.

In summary, the TAMPR system was designed to meet a number of objectives which were set as the result of experience with an earlier system; it seems to have met these well. Probably the most important contribution of TAMPR has been to encourage thinking in terms of the general program derivation processes (realization functions) discussed in the next section. Realization functions may be seen to generalize a number of concepts present in the other systems, notably: the idea of abstraction, from the Specializer; the concept of making changes without explicit markers in a program, from the Converter; and the concept of mechanical prediction with understanding, from the MLFS and Generalizer-Specializer.

Another benefit of the abstraction provided by the realization function approach to program transportability (and also, to a certain extent, by the machine parameter abstraction in the Specializer and Converter) is that *it places no a priori limitations on the spectrum of machines to which a program can be transported*. So long as programs are written according to a model of machine arithmetic of the type discussed by Brown {9}, a simple realization function may be written to transport them to a new machine obeying the model without modifications to the programs themselves. Thus the realization function approach provides all the benefits of the pure portability approach used for PORT {16}. But the realization function approach provides important benefits beyond those achievable by pure portability: (1) It makes possible transporting programs to machines whose Fortran implementations violate the PFORT Fortran subset, or on which Fortran is not implemented at all. (2) And it makes possible deriving from a single master version of a program more than just the program versions needed for transportability, as discussed in the next section.

6.   The Theme

In comparing and contrasting the four transportability systems, I have attempted to show that many of the apparent differences between them result from addressing the same problem in somewhat different settings, and that, in fact, they have substantial commonality of objectives and approach. At the least this commonality includes the idea of maintaining a master version of a set of programs, the concept of reducing the number of programs in the master version of the set where possible while maintaining the size of the set itself constant, the idea of producing variants of the master version of a program at least semi-automatically, and the goal of implementing all of these ideas reliably.

6.1 Realization Functions and the Future Evaluation of Transportability Systems

The central concept that I want to develop in this section is that of constructing *multiple realizations* of a single program by means of *realization functions*. This concept has its genesis in the observation that, as formulated above, none of the elements common to the several systems have anything to do with either "transportability" or "machine versions." This observation is remarkable only because our preoccupation with transportability and machine versions has prevented those of us developing transportability systems from perceiving the true generality of the problem we are attempting to solve.

The contrast between these two points of view is elucidated in the following two diagrams. When one concentrates on transportability, he is lead to a "horizontal" viewpoint:

in which the emphasis is on constructing members of a set of equivalent programs for different machines from another (not distinguished) member of the set. From the viewpoint of realization functions, the picture is a vertical one:

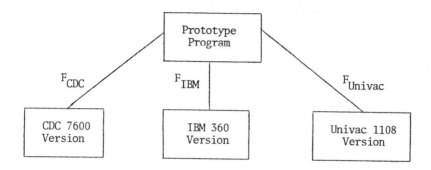

wherein several (possibly) machine-specific program realizations are constructed by appropriate realization functions from a distinguished *prototype program*. In this context the prototype program is a machine-independent program written in terms of a model of machine arithmetic, and the realization functions tailor, or optimize, it as necessary for a specific machine. Of course, in some situations no tailoring is necessary, and the (result of applying the identity realization function to the) prototype program can be used directly.

The really interesting realization functions are not those associated with transportability, however, but rather those which permit deriving more substantively different realizations from a single prototype program. Probably the simplest of these are S and D, the realization functions which construct single and double precision realizations of a prototype, respectively. (Frequently one or the other of these can be chosen to be the identity function provided the prototype programs contain no full precision *algorithm-dependent* constants.) Another important realization function is R, which realizes a prototype program written for the complex domain as a program for the real domain. It can be viewed as the process of tailoring a complex program for the case of real input data. (Typical implementations of R place constraints on the set of programs to which they apply, see {8}.) A number of other data- and program-tailoring realization functions suggest themselves.

Part of the power of realization functions comes from the ability to *compose* them. Thus if p is a complex prototype program, $S(p)$, $S(R(p))$, $D(R(p))$, and on some machines, $D(p)$, may be useful realizations. In the LINPACK package, these four realizations, plus another pair of each (obtained by replacing calls to Basic Linear Algebra modules by in-line Fortran, see {7, 8}) are potentially of interest. By this means the development and maintenance of a package containing, say, 200 programs is reduced to that of about 25 programs and three realization functions.

It is precisely this kind of *economy of scale* which is the ultimate justification for the realization function approach to mathematical software development {7}. Economies are achieved in the development of a package when the cost of developing a realization function and applying it to the programs in a package is less than that of developing the corresponding realizations by hand; of course, if several realization functions can be composed, any economies are further multiplied.

6.2   Some Open Questions Regarding Program Realizations

From this point of view, it is natural to ask: How can we maximize the potential economies of scale inherent in constructing multiple program realizations? This is still largely an open question requiring further investigation. In my view, much of the answer depends on obtaining the answers to two other open questions: How should we specify realization functions? How do we design and express prototype programs so as to have the maximum potential to yield useful realizations? For, deciding upon a method of specifying realization functions should enable one to assess what are the most sophisticated realization functions that can be specified. Similarly, answering the last question should encourage one to obtain his prototype programs in the most general form possible.

The TAMPR system is beginning to supply one possible set of answers to the first question, namely, the transformational notation discussed in Section 5.3; other approaches are certainly possible and need to be assessed. Experience with TAMPR has already raised some interesting subproblems. For example, in specifying a realization function one often faces a choice of having it perform a detailed analysis of a program to determine if a particular realization can be constructed or, alternatively, requiring some external indication whether it should be applied. (For example, how does a realization function determine that a QR algorithm cannot be straightforwardly converted from complex to real arithmetic?) An important set of open questions is thus associated with how much program analysis is possible and necessary to adequately specify realization functions. Another subproblem is whether there are needed realization functions which require a full symbolic computation facility for their specification; an example of such a realization function might be one that needed to compute approximation coefficients.

A number of especially interesting lines of investigation arise in trying to answer the last question above, about maximizing the potential of a program to yield realizations. This would seem to require that the prototype program contain as much easily recoverable information about the programmer's actual intent as possible. Since it is certainly the case that the programmer knows vastly more about the algorithm he is implementing than he is permitted to express in existing programming languages, new languages or language extensions may be necessary to represent this information explicitly in a prototype program. Some work on such languages and extensions has been done recently {9, 22}, but since it has been directed toward aiding pure portability, its applicability to specifying more general prototype programs remains to be evaluated. In my view, even more important than work on language extensions is work on generalized models, and, for want of a better term, "programming philosophies," which permit and encourage a more abstract conception of a program. For only if the programmer thinks of his program in the most general form is there real hope of maximizing the number of its realizations. Finally, a further open question is whether the most general prototype version of a program can always reasonably be made to be executable.

Paradoxically, as the generality and the clarity of specification of a prototype program are increased, so too is the possibility of automatically constructing highly optimized machine-dependent realizations of it. The general flavor of this possibility is somewhat like that of the second example of Section 5.3, and may be further understood by considering a linear algebra program in which the programmer has specified quite clearly where double precision accumulation of inner products would be beneficial, necessary, or unnecessary for the correct operation of the program. Such a prototype program could be realized using ordinary double precision if it only were available, but on some machines this prototype could also be realized using double precision accumulation (with attendant improvements in efficiency) if it were available via special language extensions.

Another example which illustrates the benefits of clarity of expression is related to one discussed by Cody {10}. He points out that the program fragment:

        IF (X .EQ. 0.0E0) Error Exit
        F = 1.0E0/X

can still result in a zero-divisor or overflow fault on a number of machines. He points out that a suitable replacement for the test on some machines is:

        IF (ABS(X) .LT. XMIN) Error Exit

while on others:

        IF (ABS(X)*XSCALE .LT. XMINS) Error Exit

is required. (Here XMIN is chosen so that 1.0E0/XMIN exists in the machine, XSCALE is 2.0^t where t is the number of bits in the significand of a floating point number, and XMINS is XMIN*XSCALE.) Converting the test against zero to one of these forms would require substantial analysis of the program. However, if the programmer were provided a small language extension which permitted him to write in the prototype program:

IF (RECIPROCAL NOT REPRESENTABLE (X)) Error Exit

F = 1.0E0/X

(and if he employed a programming philosophy which encouraged him to do so), it would be a fairly simple matter to realize this prototype optimally on each particular machine.

6.3 Machine Models and Realization Functions

The above discussion elucidates an interesting contrast between the realization function approach and the pure portability approach to mathematical software transportability. They both have the same point of departure: namely, advocacy of writing general prototype programs in terms of a model of machine arithmetic. From there they proceed in quite opposite directions: The pure portability approach uses the general prototype program to construct a single program which runs on all machines -- a kind of least common denominator program which is unlikely to be fully efficient on any machine. In contrast, if gains in efficiency warrant it (see {23}), the realization function approach can use the general prototype to construct a number of versions of the program, each tailored to be efficient on a particular machine, without sacrificing the advantages of developing and maintaining only a single version of the program.

It is worthwhile considering the above statements a bit more deeply, since doing so sheds some light on the role of realization functions in achieving reliable (correct) programs. Consider the effect of the model of arithmetic in constructing (purely) portable programs: its role is to abstract the arithmetic properties of a class of machines so that a program which operates correctly in terms of the model will operate correctly on any machine of the class. Even if not mentioned explicitly, there are axioms which characterize the model and which would be used to prove programs written in terms of it correct, if such proofs were to be carried out. In the realization function approach, the model plays the same role, but a prototype program is permitted to undergo various types of transformations, provided that they preserve the validity of the axioms of the model.

Thus realization functions may be characterized as preserving the axioms on which a proof of correctness of a prototype program is based (even if that proof has not been carried out) *while optimizing some aspect of the program other than correctness.*

This view of realization functions as preserving the meaning, or semantics, of a program is quite general, and applies to all of those considered in this paper, as discussed in {8}. Frequently, of course, the aspect of the program which is optimized is execution speed, as for example, when the realization function R optimizes a complex prototype program for use with real data.

6.4 Conclusion: Realization Functions and Reliability

Realization functions thus constitute a method of obtaining reliable programs. Programs may be shown to have *ab initio reliability,* by proof of correctness or exhaustive testing, or they may be shown to possess *induced reliability*, by showing they were reliably derived (by realization functions) from programs known to be reliable. Hence realization functions offer economies of scale not only in developing and maintaining a software package, but in testing it (or proving it correct) as well, by permitting such efforts to be concentrated on a few prototype routines. Once established, their reliability is inherited by all their realizations, which themselves require only minimal testing to guard against gross blunders. In this process, of course, the reliability of the realization functions is of paramount importance; hence I think potential answers to the open questions discussed in this section must be evaluated in terms of their contribution to emulating the paradigm:

CORRECT REALIZATION FUNCTIONS

APPLIED TO

CORRECT PROTOTYPE PROGRAMS

YIELD

CORRECT REALIZATIONS

Acknowledgements:

I am most indebted to Fred Krogh, Tom Aird, Brian Ford, and Stephen Hague for assistance in preparing and correcting the descriptions of their respective systems (for errors of detail, however, I accept responsibility) and for general discussions regarding software transportability. In addition, I wish to thank Jorge Moré, Barbara Kerns, Jim Cody, Burton Garbow, and Wayne Cowell for discussions and comments which have helped to improve the clarity of this paper.

REFERENCES

1. Thomas J. Aird, The Fortran Converter User's Guide, International Mathematical and Statistical Libraries, Houston, TX, January 1976.

2. Thomas J. Aird, The IMSL Fortran Converter: An Approach to Solving Portability Problems, These Proceedings.

3. Thomas J. Aird, Edward L. Battiste, and Walton C. Gregory, Portability of Mathematical Software Coded in Fortran, to appear in ACM Transactions on Mathematical Software.

4. James M. Boyle, Portability Problems and Solutions in NATS, Proceedings of the Software Certification Workshop (Granby, Colorado, August 1972), Wayne Cowell, Ed., Argonne National Laboratory, 1973, 80-89.

5. James M. Boyle and Kenneth W. Dritz, An Automated Programming System to Facilitate the Development of Quality Mathematical Software, Information Processing 74, North-Holland Publishing Company, 1974, 542-546.

6. James M. Boyle, Kenneth W. Dritz, Oleg B. Arushanian, and Yuri V. Kuchevskiy, Program Generation and Transformation -- Tools for Mathematical Software Development, Proceedings of IFIP Congress 77, North-Holland Publishing Co., (to appear).

7. James M. Boyle and Marilyn Matz, Automating Multiple Program Realizations, Proceedings of the M.R.I. International Symposium XXIV: Computer Software Engineering, Polytechnic Press, Brooklyn, N.Y., 1977.

8. James M. Boyle, An Introduction to the Transformation-Assisted Multiple Program Realization (TAMPR) System, Numerical Analysis in Fortran, XVII, Moscow University Press, 1976 (Russian); in J. R. Bunch, Ed. Cooperative Development of Mathematical Software, Tech. Report, Dept. of Mathematics, Univ. of Calif., San Diego, CA, 1977 (English).

9. W. S. Brown, A Realistic Model of Floating-Point Computation in Mathematical Software III, John R. Rice, Ed., Academic Press, New York, 1977, (to appear).

10. W. J. Cody, Robustness of Mathematical Software, Proceedings of the Ninth Interface Symposium on Computer Science and Statistics, Harvard University, 1-2 April 1976, 76-78.

11. W. J. Cody, The FUNPACK Package of Special Function Subroutines, ACM Transactions on Mathematical Software, 1, 1, (March 1975), 13-25.

12. Kenneth W. Dritz, An Introduction to the TAMPR System Formatter, Numerical Analysis in Fortran, XVII, Moscow University Press, 1976 (Russian); in J. R. Bunch, Ed. Cooperative Development of Mathematical Software, Tech. Report, Dept. of Mathematics, Univ. of Calif., San Diego, CA, 1977 (English).

13. Kenneth W. Dritz, Multiple Program Realizations Using the TAMPR System, These Proceedings.

14. J. J. DuCroz, S. J. Hague, and J. L. Siemieniuch, Aids to Portability Within the NAG Project, These Proceedings.

15. Brian Ford and Stephen J. Hague, Some Transformations of Numerical Software (in preparation).

16. Phyllis A. Fox, PORT -- A Portable Mathematical Subroutine Library, These Proceedings.

17. Wayne L. Fullerton, Portable Special Function Routines, These Proceedings.

18. Stephen J. Hague and Brian Ford, Portability -- Prediction and Correction, Software Practice and Experience 6, 1976, 61-69.

19. Fred T. Krogh, A Method for Simplifying the Maintenance of Software which Consists of Many Versions, Section 914, Memorandum 314, Jet Propulsion Laboratory, Pasadena, CA, 15 September 1972.

20. Fred T. Krogh and Shirley A. Singletary, Specializer User's Guide, Section 914, Memorandum 404, Jet Propulsion Laboratory, Pasadena, CA, February 1976.

21. Fred T. Krogh, Features for FORTRAN Portability, These Proceedings.

22. Michael Malcolm, Critique of a Talk on Software Portability, Los Alamos Scientific Laboratory, Los Alamos, New Mexico, May 1976.

23. B. N. Parlett and Y. Wang, The Influence of the Compiler on the Cost of Mathematical Software, ACM Transactions on Mathematical Software 1, 1, (March 1975), 35-46.

24. J. L. Schonfelder, The Production and Testing of Special Function Software in the NAG Library, These Proceedings.

25. Brian T. Smith, James M. Boyle, and William J. Cody, The NATS
     Approach to Quality Mathematical Software, in Software for
     Numerical Mathematics, D. J. Evans, Ed., Academic Press,
     New York, 1974, 393-405.

26. Brian T. Smith, Fortran Poisoning and Antidotes, These Proceedings.

27. James H. Wilkinson and Christian Reinsch, Handbook for Automatic
     Computation, Volume II, Linear Algebra, Part 2, Springer-Verlag,
     New York, Heidelberg, Berlin, 1971.

FEATURES FOR FORTRAN PORTABILITY*

Fred T. Krogh

ABSTRACT

This paper briefly summarizes experience with a Fortran
preprocessor that simplifies the task of writing efficient
and transportable mathematical software. The main part of
the paper lists features in the preprocessor and other
features that would have been useful to have.

*This paper presents the results of one phase of research carried
out at the Jet Propulsion Laboratory, California Institute of
Technology, under Contract NAS7-100, sponsored by the National
Aeronautics and Space Administration.

A.   INTRODUCTION

With the assistance of Shirley Singletary we have developed at JPL a Fortran preprocessor, "The Specializer," which provides a means for a programmer to indicate differences in the Fortran code for different computing environments or for different areas of application of the code. Transformations to be made are indicated in specially flagged Fortran comments as described in {1}, except for a few minor modifications and additions.

We have used the Specializer in writing a package for solving ordinary differential equations ($\approx$ 5500 lines of Fortran code) and on a nonlinear least squares package. Altogether we have had experience on 6 different machines including two mini-computers. Because of poor access, weak motivation, and frustration at the lack of documentation for ways in which the compiler was substandard, we did not get the ODE package to execute on one of the minicomputers. Except for that problem, we have found specialized code reasonably easy to get running on a new machine, and the process is such that the more different machines the code is running on, the more likely it is to run on any new machine without modification.

The Specializer has two main drawbacks: it is a little clumsy to learn and use, and the Specializer comments obscure the general flow of a program. As a result of our experience and our overall biases we summarize below features that we believe to be desirable for a person interested in writing mathematical software that executes efficiently in a wide variety of computer environments. We assume that a preprocessor is used to map the portable representation of the code into Fortran, since any other choice adversely impacts either the portability or the efficiency of the resulting code.

For a good overview of other work done in this area, see {2}.

B.   FEATURES

1.   The preprocessor should be portable.

We would not use the words "portable" or "transportable" for code that requires a non-portable preprocessor in order to be passed from a user on one machine to a user on another machine.

2.   The preprocessor should allow the use of small assembly modules for certain critical functions. (E.g., I/O)

Without this, the preprocessor will not see the use that it ought to have.

3.   It should be simple to specify conversion to commonly used machines.

For example, to get code for a Univac 1108 using {1}, one simply writes .MACH. = 'U1108'.

4.   It should be easy to convert long floating point constants or fractions of the form (N/D) where N and D are floating point integers, to correctly rounded floating point constants of the proper precision and type.

5.   The type specifications for variables whose type changes with the base precision for the code should be easy to indicate.

We now replace .TYPE. by INTEGER,REAL,DOUBLE PRECISION, or COMPLEX depending on whether 'I', 'S', 'D', or 'C' was last assigned to .IMP. This makes it easy to use mixed precision in a code.

6. Access to environmental and execution parameters should be provided.

> For example, overflow limits could be indicated by .OVFLO., .IOVFLO., .SOVFLO., .DOVFLO., where the value replacing .OVFLO. is one of the other three values and depends on .IMP., and the other values get replaced by the overflow limit for integer, and single and double precision arithmetic respectively. Other examples: .NIN.=nominal input unit, .NCHAR.=number of characters in a word, .LIST. indicates the listing option. Many environmental parameters should have their values determined by the value last assigned to .MACH.

7. The user should be able to define his own parameters in order to indicate how different versions of a code are to differ.

> This is particularly important if code is to be tailored to fit the application.

8. It must be possible to indicate code that is only to be included under certain conditions.

9. It should be possible to display parameter values and to indicate default values.

> For example, .MACH. :: 'IBM 360' could be used to indicate that the code currently applies to an IBM 360, and if the code is run through the preprocessor without defining .MACH., such a statement defines .MACH. to have its old value. (And if .MACH. had been assigned a new value, that value would replace the 'IBM 360'.)

10. It should be possible to indicate different values to be assigned to a parameter, depending on the values of other parameters, and to have parameter expressions evaluated.

11. It is convenient to have access to an executable basis deck.

>By this, we mean a deck that executes properly on a given machine, but still contains all the information required to convert the code to other versions.

12. Some provision for storing Hollerith data in data statements is highly desirable.

13. Certain special functions are of real assistance in doing a conversion.

>For example, we have used the function AFORM(n) to generate the proper format to output n characters. Thus AFORM(14) is replaced by 3A4,A2; 2A6,A2; or A10,A4 depending on whether the machine has 4,6, or 10 characters/word.

14. A facility to insert, ahead of a deck, control cards that depend on the program name is a useful feature on some systems.

15. It is very useful to be able to define global parameters and global Fortran variables as well as local ones.

>Fortran named common is inconvenient because: any change requires changes in more than one program, and type specifications are not a part of the common block declaration. It is desirable to have global variables

declared global in one way and made available in a
program unit in another way. The Fortran COMMON
statement does not give this kind of structure.

16. Generic function names simplify the writing of code that
is to execute in single and double precision.

17. The code that is written should look very much like Fortran,
or better yet like Fortran with the addition of control
structures that facilitate structured programming.

18. It is useful to allow as part of the code the specification
of an error message that is to be printed by the preprocessor
if parameters satisfy specified conditions.

19. A facility to define and invoke macros is extremely desirable.
Such macros ought to be Fortran-like in appearance and easy
to understand.

C. CONCLUDING REMARKS

The Specializer was written to be portable, with the assumption
that Fortran integers would be represented by at least 32 bits. Features
3 to 14 are available in the Specializer (but not exactly as described
above) and have been found useful. The remaining features we have
wished we had, particularly a macro facility which would be of use
in many ways.

REFERENCES

1. Krogh, Fred T., "A Language to Simplify Maintenance of Software Which Has Many Versions." Section 914 Internal Computing Memorandum No. 360, April 18, 1974. Jet Propulsion Laboratory, Pasadena, California 91103.

2. Boyle, James M., "Mathematical Software Transportability Systems - Have The Variations A Theme?" This Proceedings.

THE IMSL FORTRAN CONVERTER:

AN APPROACH TO SOLVING PORTABILITY PROBLEMS

T. J. Aird

ABSTRACT

IMSL produces and supports a library of Fortran subroutines in the areas of mathematics and statistics and currently serves approximately 400 subscribers. In order to solve the problems involved in the production and maintenance of this library, IMSL has developed a Fortran preprocessor, called a Converter. The Converter provides a practical, cost effective, solution to the IMSL portability problems. This approach allows compilation and execution in one computer-compiler environment without Converter processing, and local, non-portable, compiler features may be used whenever such usage improves performance in a measurable way.

1. Introduction

IMSL produces and supports a library of Fortran subroutines in the areas of mathematics and statistics and currently serves approximately 400 subscribers. The library is available on the computers of seven manufacturers. The specific computer-compiler environments served by IMSL are as follows:

IBM 360/370 series - Fortran IV {9},

UNIVAC 1100 series - Fortran V {13},

Control Data 6000/7000 and Cyber 70/170 series - Fortran Extended {5} and Fortran {6},

Honeywell 6000 series - Fortran Y {12},

Xerox Sigma series - Extended Fortran {14},

Digital Equipment series 10 - Fortran 10 {7}, and

Burroughs 6700/7700 series - Fortran {4}.

The problems involved in the production and maintenance of this library on seven computer ranges have motivated IMSL to devote time and resources to the study of portability. Most of the work has been done during the past three years and IMSL has been aided in this work by the support of the National Science Foundation under Grant GJ-40044. Details of the IMSL study are reported in the references {1}, {2}, {8}, and {15}. This paper summarizes material from those references and discusses usage of the new techniques at IMSL.

2. Portability Concepts and Basic Assumptions

The approach that IMSL has taken in solving its portability problems is based on the desire to provide a subscriber with codes that perform as well as possible in the intended environment. In order to attain this goal, it is necessary that codes developed for a given computer-compiler environment be allowed to utilize local, possibly non-portable, features whenever such usage improves performance in a measurable way. Of course, IMSL avoids unnecessary computer dependent code so that the effort to produce a library for a new environment will not be made more difficult. This approach implies that few codes in the IMSL library are portable across the seven environments. That is, most codes from one of the environments require some changes before they will compile and execute properly in another environment. For the majority of codes in the IMSL library, the differences between versions are small and usually 90% or more of the lines are common to all versions. There are extremes and some codes are portable (all lines common) across all seven environments while others (e.g., special function codes) have such great differences that each version is treated separately. Some examples are given here to help illustrate the points made above.

A.   IMSL subroutines that perform statistical computations may be required to compute a sum of many (several thousand) terms. When these codes execute on IBM and Xerox computers in single precision (approx. 6 digits of precision) it is desirable to accumulate the sum in double precision (approx. 16 digits of precision) to avoid a possible unfavorable accumulation of rounding errors. When the same codes are executed on CDC computers in single precision (approx. 14 digits of precision) it is not necessary to use intermediate double precision calculations. In fact, it is desirable to avoid the use of double precision due to the extra computer time required.

In these situations, IMSL's development procedure is to use double precision only in those environments where it is determined to be necessary.

Some compilers allow an extension to the form of ANSI-allowable {3} subscripts. For example, X(IP(J)) is allowed by IBM compilers. But this is not an extension of the Fortran language across all seven environments and specifically it is not allowed by the Univac compiler. IMSL does not use the more general form in order to avoid a potential portability problem.

It is often necessary to access constants that are environment dependent. The largest representable real number, and the base and precision of floating point numbers are examples of such constants. The maximum usable number of digits in the representation of decimal constants varies with computer and compiler. Some compilers issue diagnostic messages when decimal constants exceed the usable length (CDC Fortran Extended) and some invoke double precision when the number of digits exceeds the allowable length for single precision numbers (Burroughs). IMSL always restricts decimal constants to be within the limits set by the compiler in order to avoid these problems.

In summary, IMSL codes differ from environment to environment when there is a need for such differences, but unnecessary computer dependencies are avoided. The PFORT verifier {11} is a great aid in flagging non-portable constructions. Statements flagged by the verifier are examined to determine if the non-portable construction is necessary.

IMSL has developed and uses a Fortran preprocessor, called a Converter, to aid in solving its portability problems. Programmers using the Converter to write codes for operation in multiple environments are assumed to have knowledge of each of these environments. A programmer's manual {8} provides information for the seven environments listed in Section 1. Also, a master table used by the Converter provides values for computer specific constants (CONSTANTS) and other information about those environments (PARAMETERS) that aid the programmer.

3.   The Fortran Converter

The Fortran Converter is a program which performs automatic conversion of Fortran programs and subprograms from one computer-compiler environment to another. The present version of the Converter has built-in features which aid in the conversion between environments given in Section 1. The set of environments can be extended easily to include a new environment but this does not imply that codes developed for other environments will automatically operate correctly in the new environment.

The Converter operates on a set of program units which compile and execute correctly in one environment, by making changes which cause them to compile and execute correctly in another environment - the target environment. Input to the Converter is a basis deck (a set of program units that contain Converter instructions), and output is another basis deck or a distribution deck, either of which operates correctly in the target environment. A distribution deck is produced by deleting all Converter instructions from the basis deck. IMSL subscribers receive distribution decks, and basis decks are used to aid in-house development and maintenance. Converter usage by IMSL is transparent to subscribers. The Converter obtains information concerning conversion from three sources: Converter instructions contained in the basis deck, options selected for conversion, and a master table. Converter instructions are distinguished by the characters C$ in columns 1 and 2. The symbol $ can be redefined. Instructions are written in columns 7 through 72, are free format and follow the Fortran convention for continuation. Exemplification of Converter instructions is given later in this section. The main function of the OPTIONS statement is to select the target environment: COMPUTER, PRECISION and whether or not the Converter SCAN option (to convert from single precision to double precision) is to be used. For example, the following statement selects IBM double precision as the target environment and the SCAN option is selected:

C$ OPTIONS,COMPUTER=IBM,PRECISION=DOUBLE,SCAN

The Converter has access to a master table and Section 6 shows a subset of this table. Whenever the COMPUTER option is set, the master table is examined by the Converter. Certain OPTIONS are selected as they are specified in the table. Also, PARAMETERS and CONSTANTS are assigned values from the table, correct for the target environment. After the master table has been examined, processing of the basis deck continues. OPTIONS may be selected, and PARAMETERS and CONSTANTS may be defined or redefined by placing Converter statements in the basis deck.

Other Converter statements are illustrated through examples which follow. The Converter User's Guide {2} gives complete definitions and explanations, and Section 7 of this paper gives an example of a complete IBM basis deck.

Example 3.1

Assume that a single precision code is intended to deal with large input vectors and must accumulate a sum in double precision whenever the number of significant digits in single precision arithmetic is less than 10. The following Converter statement is used:

```
C$      IF (SSIG10.LT.10) 1 LINE
        DOUBLE PRECISION SUM

          .
          .
          .

        SUM = 0.0
        DO 10 I=1,N
   10 SUM = SUM + X(I)
```

The C$ IF Statement causes the line following it to be deactivated (by placing a C in column 1) whenever the number of significant digits in single precision is 10 or greater. The Converter obtains this number, for the selected target environment, from the PARAMETERS section of the master table. When a distribution deck is requested (by placing DISTRIBUTION in the options list) inactive lines are removed. Other forms of the C$ IF statement allow convenient reference to blocks of code for activation or deactivation. These forms are discussed in {2}.

Example 3.2

A code may be required to use double precision arithmetic in the IBM and XEROX environments and single precision arithmetic in other environments. The Converter SCAN option is intended to be of aid in such a situation. At IMSL, a single precision code may be developed using statements such as:

```
REAL       X,Y,Z,ONE,P85

DATA       ONE/1.0/,P85/.85/

X = ABS(Y+2.5)
```

When the SCAN option is selected, the Converter replaces these lines as
follows:

```
DOUBLE PRECISION X,Y,Z,ONE,P85

DATA                ONE/1.0D0/,P85/.85D0/

X = DABS(Y+2.5D0)
```

The SCAN option causes changes to be made in declaration statements by
changing from REAL to DOUBLE PRECISION, COMPLEX to COMPLEX*16, and
COMPLEX to DOUBLE COMPLEX, as required by the target environment.
Decimal constants are changed from single precision form to double
precision form and single precision intrinsic and basic external
function names are changed to their double precision counterparts.
Further flexibility is available through Converter statements that
allow the automatic precision conversion option to be overridden
temporarily. There are certain restrictions that apply to the
SCAN option and these are explained in the User's Guide {2}. In
particular, blanks are treated as significant characters. As an
example, the variable SI N is not changed to DSI N. Also, R EAL
is not changed to DOUBLE PRECISION. But, IMSL does not use blanks
within variable names and this restriction does not present a problem.
Conversion from double precision to single precision is done when
PRECISION = SINGLE and SCAN are selected on an OPTIONS statement.

Example 3.3

Computer environment constants such as the largest representable real number, and the base and precision of floating point arithmetic are available through the CONSTANTS section of the master table. Their names are SINFP,SRADIX, and SEPS respectively (see Section 6). The following statement causes the Converter to generate a DATA statement with the correct (for the target environment) value for SRADIX (and the Fortran variable R)

C$ DATA R/SRADIX/

The C$ DATA statement is followed in a basis deck by a DATA statement correct for the operating environment. In the IBM environment, one would use:

C$ DATA R/SRADIX/

       DATA R/16.0/

The basis deck using these statements is operable in the IBM environment. For the CDC environment the Converter would change these to the following:

C$ DATA R/SRADIX/

       DATA R/2.0/

Through the master table, the Converter provides access to a large set of computer environment constants. The table can be augmented to include additional constants and constants for additional environments. Also, Converter CONSTANTS can be defined or redefined within a basis deck. The SCAN option also handles automatic conversion between single precision CONSTANTS whose names begin with S and double precision CONSTANTS whose names begin with D.

The C$ DATA statement also allows decimal constants to be used as values. These constants may be specified to any number of digits and the Converter rounds them to the correct number of digits for the target environment.

4.   Converter Usage in the IMSL Environment

The Fortran Converter presented here and in {1}, {2} and {15} has been found to be a practical aid in solving IMSL portability problems. Converter usage does not cause degradation of code quality and it simplifies code production and maintenance. At the present time, about 50 IMSL subroutines (of approx. 350 in the Library) exist in basis deck form and the Converter is operating in all seven environments listed in Section 1. Most new IMSL codes are being developed as basis decks and old codes are being transformed to basis deck form whenever maintenance of them is required.

The most frequently used Converter features are the precision conversion (SCAN option) and the C$ DATA statement to reference computer environment constants. The ability to compile and execute a basis deck in one environment, without conversion, is very important for IMSL development work. A code may be compiled and executed many times during its development stage and the ability to do this without the Converter is a cost saving feature. Most work at IMSL is done initially in the IBM environment and, consequently, IBM basis decks are developed.

The simplicity of the Converter statement and its usage makes it attractive to programmers. Most Converter statements are self-explanatory and easily understood with a small amount of training and study. Complicated basis decks are discouraged by IMSL and the separate deck development approach is preferred whenever the number of Converter statements would exceed 20% of the number of lines of a code. The Converter's strength is also its weakness. There are times when a more flexible form of the C$ IF statement (to allow AND and OR conditions) should be desirable.

Further Converter features may be desirable in environments slightly different from that of IMSL. If Hollerith strings are used frequently, features to handle them automatically (see {10} for discussion of such a feature) would be important. Also, in some situations a macro processing capability would be very desirable {10}. The IMSL Converter is most likely to expand in these directions if expansion is required. However, for the present IMSL environment, the current set of Converter features seems adequate and its simplicity of learning and usage is very desirable.

Converter usage by IMSL is also aided by an automatic maintenance procedure for basis decks. When a code must change in all seven environments, an IMSL programmer need only make changes to the basis deck. When testing is completed, the basis deck is used as input to a source compare program which automatically prepares the maintenance letter and the accompanying correction deck that IMSL sends to its subscribers. Thus, maintenance of a basis deck is much simpler, and less costly for IMSL, than maintenance of one code with seven separate versions.

5.   Summary

The Fortran Converter developed and used by IMSL provides a practical, cost effective, solution to the portability problems associated with the production and maintenance of a subroutine library

for several environments. The simplicity of the Converter is both a
strength and weakness. Basis deck development is easy to learn and
codes are not unnecessarily complicated by Converter statements.
However, more flexibility in certain Converter features and some
new features may be desirable. The ability to compile and execute
a basis deck in one environment, without Converter processing, is a
feature that is important to IMSL in reducing computer costs. A
most important feature of this approach to portability problems is
that it allows local, non-portable, compiler features to be used
whenever such usage improves performance in a measurable way.

6. A Subset of the Master Table

```
C$    OPTIONS,     ISIGS(IBM) = 7,     ISIGD(IBM) = 16,
C$   *             ISIGS(XEROX) = 7,   ISIGD(XEROX) = 16,
C$   *             ISIGS(UNIVAC) = 9,  ISIGD(UNIVAC) = 18,
C$   *             ISIGS(HIS) = 9,     ISIGD(HIS) = 18,
C$   *             ISIGS(DEC) = 9,     ISIGD(DEC) = 18,
C$   *             ISIGS(CDC) = 14,    ISIGD(CDC) = 29,
C$   *             ISIGS(BGH) = 11,    ISIGD(BGH) = 22
C$    PARAMETERS   MASTER TABLE
C     /
C     / IMSL LIBRARY NUMBER
C     / -------------------
C     /
C        LIB      (IBM   ) = 1
C        LIB      (XEROX ) = 1
C        LIB      (UNIVAC) = 2
C        LIB      (HIS   ) = 2
C        LIB      (DEC   ) = 2
C        LIB      (CDC   ) = 3
C        LIB      (BGH   ) = 3
C     /
C     / NUMBER OF SIGNIFICANT DIGITS - SINGLE PRECISION
C     / ----------------------------------------------
C     /
C        SSIG10   (IBM   ) = 6.3
C        SSIG10   (XEROX ) = 6.3
C        SSIG10   (UNIVAC) = 8.1
C        SSIG10   (HIS   ) = 8.1
C        SSIG10   (DEC   ) = 8.1
C        SSIG10   (CDC   ) = 14.4
C        SSIG10   (BGH   ) = 11.1
C     /
C     / NUMBER OF SIGNIFICANT DIGITS - DOUBLE PRECISION
C     / ----------------------------------------------
C     /
C        DSIG10   (IBM   ) = 16
C        DSIG10   (XEROX ) = 16
C        DSIG10   (UNIVAC) = 18
C        DSIG10   (HIS   ) = 19.3
C        DSIG10   (DEC   ) = 18.6
C        DSIG10   (CDC   ) = 28.6
C        DSIG10   (BGH   ) = 22.8
C     /
C     / BASE OF FLOATING POINT REPRESENTATION
C     / -------------------------------------
C     /
C        RADIX    (IBM   ) = 16
C        RADIX    (XEROX ) = 16
C        RADIX    (UNIVAC) = 2
C        RADIX    (HIS   ) = 2
C        RADIX    (DEC   ) = 2
C        RADIX    (CDC   ) = 2
C        RADIX    (BGH   ) = 8
C     /
C     / NUMBER OF CHARACTERS PER WORD
C     / -----------------------------
C     /
C        NCHAR    (IBM   ) = 4
C        NCHAR    (XEROX ) = 4
C        NCHAR    (UNIVAC) = 6
C        NCHAR    (HIS   ) = 6
C        NCHAR    (DEC   ) = 5
C        NCHAR    (CDC   ) = 10
C        NCHAR    (BGH   ) = 6
```

```
C$    CONSTANTS    MASTER TABLE

C     /
C     /
C     /C O M P U T E R   E N V I R O N M E N T   C O N S T A N T S
C     /
C     /
C     /   BASE OF FLOATING POINT REPRESENTATION
C     /   ------------------------------------
C     /
C         IRADIX   (IBM   ) = 16/
C         IRADIX   (XEROX ) = 16/
C         IRADIX   (UNIVAC) = 2/
C         IRADIX   (HIS   ) = 2/
C         IRADIX   (DEC   ) = 2/
C         IRADIX   (CDC   ) = 2/
C         IRADIX   (BGH   ) = 8/
C         SRADIX   (IBM   ) = 16.Ø/
C         SRADIX   (XEROX ) = 16.Ø/
C         SRADIX   (UNIVAC) = 2.Ø/
C         SRADIX   (HIS   ) = 2.Ø/
C         SRADIX   (DEC   ) = 2.Ø/
C         SRADIX   (CDC   ) = 2.Ø/
C         SRADIX   (BGH   ) = 8.Ø/
C         DRADIX   (IBM   ) = 16.ØDØ/
C         DRADIX   (XEROX ) = 16.ØDØ/
C         DRADIX   (UNIVAC) = 2.ØDØ/
C         DRADIX   (HIS   ) = 2.ØDØ/
C         DRADIX   (DEC   ) = 2.ØDØ/
C         DRADIX   (CDC   ) = 2.ØDØ/
C         DRADIX   (BGH   ) = 8.ØDØ/
C     /
C     /   THE LARGEST POSITIVE INTEGER
C     /   ----------------------------
C     /            •
C         IINFP    (IBM   ) = Z7FFFFFFF/2**31-1
C         IINFP    (XEROX ) = Z7FFFFFFF/2**31-1
C         IINFP    (UNIVAC) = Ø377777777777/2**35-1
C         IINFP    (HIS   ) = Ø377777777777/2**35-1
C         IINFP    (DEC   ) = "377777777777/2**35-1
C         IINFP    (CDC   ) = ØØØØ7777777777777777B/2**48-1
C         IINFP    (BGH   ) = ØØØØ7777777777777/8**13-1
C     /
C     /   THE LARGEST POSITIVE REAL
C     /   -------------------------
C     /
C         SINFP    (IBM   ) = Z7FFFFFFF/16**63(1-16**-6)
C         SINFP    (XEROX ) = Z7FFFFFFF/16**63(1-16**-6)
C         SINFP    (UNIVAC) = Ø377777777777/2**127(1-2**-27)
C         SINFP    (HIS   ) = Ø376777777777/2**127(1-2**-27)
C         SINFP    (DEC   ) = "377777777777/2**127(1-2**-27)
C         SINFP    (CDC   ) = 37767777777777777777B/2**1Ø22(2**48-1)
C         SINFP    (BGH   ) = ØØ777777777777777/8**63(8**13-1)
C         DINFP    (IBM   ) = Z7FFFFFFFFFFFFFFF/16**63(1-16**-14)
C         DINFP    (XEROX ) = Z7FFFFFFFFFFFFFFF/16**63(1-16**-14)
C         DINFP    (UNIVAC) = Ø377777777777,
C    *                        Ø777777777777/2**1Ø23(1-2**-6Ø)
C         DINFP    (HIS   ) = Ø376777777777,
C    *                        Ø777777777777/2**127(1-2**-63)
C         DINFP    (DEC   ) = "377777777777,
C    *                        "777777777777/2**127(1-2**-62)
C         DINFP    (CDC   ) = 37767777777777777777B,
C    *                        37167777777777777777B/2**1Ø22(2**48-2**-48)
C         DINFP    (BGH   ) = ØØ777777777777777,
C    *                        Ø7777777777777777/8**32767(8**13-8**-13)
```

```
C     /
C     /   THE SMALLEST POSITIVE REAL NUMBER SUCH THAT
C     /   ETA AND -ETA CAN BOTH BE REPRESENTED
C     /   ----------------------------------------
C     /
C         SETA    (IBM   ) = Z00100000/16**-65
C         SETA    (XEROX ) = Z001000000/16**-65
C         SETA    (UNIVAC) = 0000400000000/2**-129
C         SETA    (HIS   ) = 0400400000001/2**-128(2**-1+2**-27)
C         SETA    (DEC   ) = "000400000000/2**-129
C         SETA    (CDC   ) = 00014000000000000000B/2**-975
C         SETA    (BGH   ) = 01771000000000000/8**-51
C         DETA    (IBM   ) = Z0010000000000000/16**-65
C         DETA    (XEROX ) = Z0010000000000000/16**-65
C         DETA    (UNIVAC) = 0000040000000,
C     *                      00000000000000/2**-1025
C         DETA    (HIS   ) = 0400400000000,
C     *                      0000000000001/2**-128(2**-1+2**-63)
C         DETA    (DEC   ) = "000400000000,
C     *                      "000000000000/2**-129
C         DETA    (CDC   ) = 00014000000000000000B,
C     *                      00000000000000000000B/2**-975
C         DETA    (BGH   ) = 01771000000000000,
C     *                      07770000000000000/8**-32755
C     /
C     /
C     /   NUMBER OF CHARACTERS PER WORD
C     /   -----------------------------
C     /
C         NCHAR   (IBM   ) = 4/CHARACTERS PER WORD
C         NCHAR   (XEROX ) = 4/CHARACTERS PER WORD
C         NCHAR   (UNIVAC) = 6/CHARACTERS PER WORD
C         NCHAR   (HIS   ) = 6/CHARACTERS PER WORD
C         NCHAR   (DEC   ) = 5/CHARACTERS PER WORD
C         NCHAR   (CDC   ) = 10/CHARACTERS PER WORD
C         NCHAR   (BGH   ) = 6/CHARACTERS PER WORD
C     /
C     /   INPUT UNIT
C     /   ----------
C     /
C         NIN     (IBM   ) = 5/
C         NIN     (XEROX ) = 5/
C         NIN     (UNIVAC) = 5/
C         NIN     (HIS   ) = 5/
C         NIN     (DEC   ) = 5/
C         NIN     (CDC   ) = 5LINPUT/
C         NIN     (BGH   ) = 5/
C     /
C     /   OUTPUT UNIT
C     /   -----------
C     /
C         NOUT    (IBM   ) = 6/
C         NOUT    (XEROX ) = 6/
C         NOUT    (UNIVAC) = 6/
C         NOUT    (HIS   ) = 6/
C         NOUT    (DEC   ) = 6/
C         NOUT    (CDC   ) = 6LOUTPUT/
C         NOUT    (BGH   ) = 6/
C     /
C     /
```

```
C       /
C       /   THE SMALLEST POSITIVE REAL
C       /   --------------------------
C       /
C
C           SETAP    (IBM    ) = ZØØ100000/16**-65
C           SETAP    (XEROX  ) = ZØØ100000/16**-65
C           SETAP    (UNIVAC) = 0000400000000/2**-129
C           SETAP    (HIS    ) = 0400400000000/2**-129
C           SETAP    (DEC    ) = "ØØ0400000000/2**-129
C           SETAP    (CDC    ) = ØØ014000000000000000ØB/2**-975
C           SETAP    (BGH    ) = 01771000000000000/8**-51
C           DETAP    (IBM    ) = ZØØ10000000000000/16**-65
C           DETAP    (XEROX  ) = ZØØ10000000000000/16**-65
C           DETAP    (UNIVAC) = 0000040000000,
C      *                        0000000000000/2**-1025
C           DETAP    (HIS    ) = 0400400000000,
C      *                        0000000000000/2**-129
C           DETAP    (DEC    ) = "000400000000,
C      *                        "000000000000/2**-129
C           DETAP    (CDC    ) = ØØ014000000000000000000B,
C      *                        0000000000000000000000B/2**-975
C           DETAP    (BGH    ) = 01771000000000000,
C      *                        07770000000000000/8**-32755
C       /
C       /   THE SMALLEST POSITIVE REAL NUMBER SUCH THAT
C       /   1.Ø+EPS .GT. 1.Ø  .AND.  1.Ø .GT. 1.Ø-EPS
C       /   -------------------------------------------
C       /
C
C           SEPS     (IBM    ) = Z3C100000/16**-5
C           SEPS     (XEROX  ) = Z3C100000/16**-5
C           SEPS     (UNIVAC) = 0147400000000/2**-26
C           SEPS     (HIS    ) = 0714400000000/2**-27
C           SEPS     (DEC    ) = "147400000000/2**-26
C           SEPS     (CDC    ) = 16414000000000000000B/2**-47
C           SEPS     (BGH    ) = 013Ø1000000000000/8**-12
C           DEPS     (IBM    ) = Z3410000000000000/16**-13
C           DEPS     (XEROX  ) = Z3410000000000000/16**-13
C           DEPS     (UNIVAC) = 0170640000000,
C      *                        0000000000000/2**-59
C           DEPS     (HIS    ) = 06Ø6400000000,
C      *                        0000000000000/2**-62
C           DEPS     (DEC    ) = "104400000000,
C      *                        "000000000000/2**-61
C           DEPS     (CDC    ) = 15614000000000000000B,
C      *                        15Ø10000000000000000B/2**-95
C           DEPS     (BGH    ) = 01451000000000000,
C      *                        0000000000000000/8**-25
C       /
C       /   LN(RADIX), NATURAL LOG OF RADIX
C       /   -------------------------------
C       /
C           SNLGRX   (IBM    ) = 2.7725887222397812376689284B6/
C           SNLGRX   (XEROX  ) = 2.7725887222397812376689284B6/
C           SNLGRX   (UNIVAC) = .6931471805599453Ø941723212146/
C           SNLGRX   (HIS    ) = .6931471805599453Ø941723212146/
C           SNLGRX   (DEC    ) = .6931471805599453Ø941723212146/
C           SNLGRX   (CDC    ) = .6931471805599453Ø941723212146/
C           SNLGRX   (BGH    ) = 2.Ø794415416798359282516963644/
C           DNLGRX   (IBM    ) = 2.7725887222397812376689284B6DØ/
C           DNLGRX   (XEROX  ) = 2.7725887222397812376689284B6DØ/
C           DNLGRX   (UNIVAC) = .6931471805599453Ø941723212146DØ/
C           DNLGRX   (HIS    ) = .6931471805599453Ø941723212146DØ/
C           DNLGRX   (DEC    ) = .6931471805599453Ø941723212146DØ/
C           DNLGRX   (CDC    ) = .6931471805599453Ø941723212146DØ/
C           DNLGRX   (BGH    ) = 2.Ø794415416798359282516963644DØ/
```

```
C      /MATHEMATICAL CONSTANTS
C      /
C      /
C      / E AND RATIONAL FUNCTIONS OF E
C      / -----------------------------
C      /
C      /           E
C        SE           = 2.7182818284590452353602874713/E
C        DE           = 2.7182818284590452353602874713D0/E
C      /
C      / PI AND RATIONAL FUNCTIONS OF PI
C      / ------------------------------
C      /
C      /           PI
C        SPI          = 3.14159265358979323846264338310/PI
C        DPI          = 3.14159265358979323846264338310D0/PI
C      /
C      / CONSTANTS INVOLVING NATURAL LOGARITHMS
C      / --------------------------------------
C      /
C      /           LN(2)
C        SNLG2        = .6931471805599453094172321214/LN(2)
C        DNLG2        = .6931471805599453094172321214D0/LN(2)
C      /           LN(10)
C        SNLG10       = 2.30258509299404568401799145470/LN(10)
C        DNLG10       = 2.30258509299404568401799145470D0/LN(10)
C      /
C      / CONSTANTS INVOLVING COMMON LOGARITHMS
C      / -------------------------------------
C      /
C      /           LOG(2)
C        SCLG2        = .30102999566398119521373889472/LOG(2)
C        DCLG2        = .30102999566398119521373889472D0/LOG(2)
C      /           LOG(E)
C        SCLGE        = .43429448190325182765112891891/LOG(E)
C        DCLGE        = .43429448190325182765112891891D0/LOG(E)
C      /
C      / CONSTANTS INVOLVING SQUARE ROOTS
C      / --------------------------------
C      /
C      /           SQRT(.5)
C        SSQ1H        = .70710678118654752440084436210/SQRT(.5)
C        DSQ1H        = .70710678118654752440084436210D0/SQRT(.5)
C      /           SQRT(2)
C        SSQ2         = 1.4142135623730950488016887242/SQRT(2)
C        DSQ2         = 1.4142135623730950488016887242D0/SQRT(2)
C      /
C      / OTHER CONSTANTS
C      / ---------------
C      /
C      /           EULER'S CONSTANT
C        SEULC        = .57721566490153286060651209008/EULER'S
C        DEULC        = .57721566490153286060651209008D0/EULER'S
C      /
C$     END MASTER TABLE
```

7. An Example of an IBM Basis Deck

```
C$      IF (DISTRIBUTION) OPTIONS FOR ZQADC
C$      OPTIONS,
C$    *    PRECISION(IBM) = DOUBLE, SCAN(IBM),
C$    *    PRECISION(XEROX) = DOUBLE, SCAN(XEROX),
C$    *    PRECISION(UNIVAC) = SINGLE, SCAN(UNIVAC) = 0,
C3    *    PRECISION(HIS) = SINGLE, SCAN(HIS) = 0,
C$    *    PRECISION(DEC) = SINGLE, SCAN(DEC) = 0,
C$    *    PRECISION(CDC) = SINGLE, SCAN(CDC) = 0,
C$    *    PRECISION(BGH) = SINGLE, SCAN(BGH) = 0
C$LBL OPTIONS
C       SUBROUTINE ZQADC  (A,B,C,ZSM,ZLG,IER)                           ZQDC0010
C                                                                        ZQDC0020
C-ZQADC-----------------------------------------------------------------ZQDC0030
C                                                                        ZQDC0040
C    FUNCTION              - FIND THE ROOTS OF THE QUADRATIC EQUATION     ZQDC0050
C                            A*Z**2+B*Z+C = 0.0, WHERE THE               ZQDC0060
C                            COEFFICIENTS A, B, AND C ARE COMPLEX        ZQDC0070
C                            NUMBERS.                                    ZQDC0080
C    USAGE                 - CALL ZQADC(A,B,C,ZSM,ZLG,IER)              ZQDC0090
C    PARAMETERS   A        - COEFFICIENT OF THE QUADRATIC EQUATION (INPUT).ZQDC0100
C                 B        - COEFFICIENT OF THE QUADRATIC EQUATION (INPUT).ZQDC0110
C                 C        - COEFFICIENT OF THE QUADRATIC EQUATION (INPUT).ZQDC0120
C                            (NOTE - A, B, AND C MUST BE DECLARED TYPE   ZQDC0130
C                            COMPLEX.)                                   ZQDC0140
C                 ZSM      - ROOT OF THE QUADRATIC EQUATION (OUTPUT).    ZQDC0150
C                 ZLG      - ROOT OF THE QUADRATIC EQUATION (OUTPUT).    ZQDC0160
C                            (NOTE - ZSM AND ZLG MUST BE DECLARED TYPE   ZQDC0170
C                            COMPLEX.)                                   ZQDC0180
C                            FOR THE ROOTS ZSM AND ZLG THE FOLLOWING     ZQDC0190
C                            CONDITION HOLDS - CABS(ZSM) .LE. CABS(ZLG)  ZQDC0200
C                 IER      - ERROR PARAMETER                            ZQDC0210
C                            WARNING (WITH FIX)                         ZQDC0220
C                              IER = 65, IMPLIES A=B=0.0                 ZQDC0230
C                                IN THIS CASE, THE LARGE ROOT,          ZQDC0240
C                                ZLG = SIGN(FINITY,-B), AND             ZQDC0250
C                                THE SMALL ROOT, ZSM = -ZLG, WHERE      ZQDC0260
C                                FINITY = LARGEST NUMBER WHICH CAN BE    ZQDC0270
C                                REPRESENTED IN THE MACHINE.            ZQDC0280
C                              IER = 66, IMPLIES A=0.0                   ZQDC0290
C                                IN THIS CASE, THE LARGE ROOT,          ZQDC0300
C                                ZLG = SIGN(FINITY,-B), WHERE           ZQDC0310
C                                FINITY = LARGEST NUMBER WHICH CAN BE    ZQDC0320
C                                REPRESENTED IN THE MACHINE.            ZQDC0330
C$      IF(DOUBLE) 1 LINE, 1 LINE                                       ZQDC0340
CC    PRECISION            - DOUBLE                                     ZQDC0350
CC    PRECISION            - SINGLE                                     ZQDC0360
C$      IF (LIB.EQ.1) 1 LINE, 1 LINE                                    ZQDC0370
CC    REQD. IMSL ROUTINES - UERTST,VXPADD                              ZQDC0380
CC    REQD. IMSL ROUTINES - UERTST                                     ZQDC0390
C    LANGUAGE             - FORTRAN                                     ZQDC0400
C-----------------------------------------------------------------------ZQDC0410
C    LATEST REVISION      - FEBRUARY 7, 1975                           ZQDC0420
C                                                                        ZQDC0430
        SUBROUTINE ZQADC  (A,B,C,ZSM,ZLG,IER)                           ZQDC0440
C                                                                        ZQDC0450
        DIMENSION          AA(2),BB(2),CC(2),ZSA(2),ZLA(2)              ZQDC0460
C$      (SINGLE/DOUBLE) 1 LINE                                          ZQDC0470
        DOUBLE PRECISION   DR,DI,D,D1                                   ZQDC0480
        REAL               HALF,ZERO,FINITY,AA,BB,CC,ZSA,ZLA,AR,BR,CR,  ZQDC0490
       1                   ZSR,ZLR,AI,BI,CI,ZSI,ZLI,TEMP                ZQDC0500
        COMPLEX            A,B,C,ZSM,ZLG,A0,B0,C0,ZS,ZL                 ZQDC0510
        EQUIVALENCE        (AA(1),A0),(BB(1),B0),(CC(1),C0),(ZSA(1),ZS),ZQDC0520
       1                   (ZLA(1),ZL),(AA(1),AR),(BB(1),BR),(CC(1),CR),ZQDC0530
       2                   (ZSA(1),ZSR),(ZLA(1),ZLR),(AA(2),AI),        ZQDC0540
       3                   (BB(2),BI),(CC(2),CI),(ZSA(2),ZSI),(ZLA(2),ZLI)ZQDC0550
```

```
C$     DATA                 FINITY/SINFP/                                     ZQDC0560
       DATA                 FINITY/Z7FFFFFFF/                                 ZQDC0570
C$     (SINGLE/DOUBLE) 4 LINES                                               ZQDC0580
C$     DATA                 RADIX/SRADIX/,                                    ZQDC0590
C$    1                     RNLGRX/SNLGRX/                                    ZQDC0600
       DATA                 RADIX/16.0/,                                      ZQDC0610
      1                     RNLGRX/2.772589/                                  ZQDC0620
       DATA                 ZERO/0.0/,HALF/0.5/                               ZQDC0630
       IER = 0                                                               ZQDC0640
C                                         PUT THE COEFFICIENTS IN TEMPORARY TO ZQDC0650
C                                         SAVE EXECUTION TIME.                ZQDC0660
       A0 = A                                                                ZQDC0670
       B0 = -B                                                               ZQDC0680
       C0 = C                                                                ZQDC0690
C                                         CHECK FOR A=ZERO OR C=ZERO.         ZQDC0700
       IF(AR .NE. ZERO .OR. AI .NE. ZERO) GO TO 5                            ZQDC0710
       IER = 65                                                              ZQDC0720
       ZL = CMPLX(FINITY,ZERO)                                               ZQDC0730
       ZS = ZL                                                               ZQDC0740
       IF(BR .EQ. ZERO .AND. BI .EQ. ZERO) GO TO 35                          ZQDC0750
       IER = 66                                                              ZQDC0760
       ZS = C0/B0                                                            ZQDC0770
       GO TO 35                                                              ZQDC0780
     5 IF(CR .NE. ZERO .OR. CI .NE. ZERO) GO TO 10                           ZQDC0790
       ZS = CMPLX(ZERO,ZERO)                                                 ZQDC0800
       GO TO 30                                                              ZQDC0810
C                                         SCALING TO AVOID OVERFLOW OR        ZQDC0820
C                                         UNDERFLOW. SCALE THE COEFFICIENTS   ZQDC0830
C                                         SO THAT A*C IS APPROXIMATELY ONE.   ZQDC0840
C                                         THE SCALE FACTOR CSQRT(A*C) FITS    ZQDC0850
C                                         THIS REQUIREMENT BUT MAY CAUSE      ZQDC0860
C                                         OVERFLOW OR UNDERFLOW IN THE        ZQDC0870
C                                         SCALING PROCEDURE.                  ZQDC0880
C                                         LET AMAX1(ABS(AR),ABS(AI)) BE       ZQDC0890
C                                         REPRESENTED BY RADIX**IA AND LET    ZQDC0900
C                                         AMAX1(ABS(CR),ABS(CI)) BE           ZQDC0910
C                                         REPRESENTED BY RADIX**IC.           ZQDC0920
C                                         THE SCALE FACTOR, SCALE, IS DEFINED ZQDC0930
C                                         BY THE FOLLOWING FORMULA:           ZQDC0940
C                                         SCALE=RADIX**IS, WHERE              ZQDC0950
C                                         IS=ENTIER((IA+IC+1)/2) AND          ZQDC0960
C                                         ENTIER IS THE MATHEMATICAL GREATEST ZQDC0970
C                                         INTEGER FUNCTION.                   ZQDC0980
C$     IF (DOUBLE) 3 LINES, 2 LINES                                          ZQDC0990
C  10 IS = (ALOG(AMAX1(ABS(SNGL(AR)),ABS(SNGL(AI))))+                        ZQDC1000
C     1      ALOG(AMAX1(ABS(SNGL(CR)),ABS(SNGL(CI))))+RNLGRX)/               ZQDC1010
C     2      (RNLGRX+RNLGRX)                                                  ZQDC1020
    10 IS = (ALOG(AMAX1(ABS(AR),ABS(AI)))+ALOG(AMAX1(ABS(CR),ABS(CI)))+      ZQDC1030
      1      RNLGRX)/(RNLGRX+RNLGRX)                                         ZQDC1040
       SCALE = RADIX**IS                                                     ZQDC1050
C                                         IF THE SCALE FACTOR .LE.            ZQDC1060
C                                         DEPS*MAX(ABS(BR),ABS(BI))           ZQDC1070
C                                         DO NOT SCALE THE COEFFICIENTS.      ZQDC1080
C$     IF (DOUBLE) 7 LINES, 5 LINES                                          ZQDC1090
C      D1 = DMAX1(DABS(BR),DABS(BI))                                         ZQDC1100
C      CALL VXPZRO                                                           ZQDC1110
C      CALL VXPADD(D1)                                                       ZQDC1120
C      CALL VXPADD(DBLE(SCALE))                                              ZQDC1130
C      CALL VXPADD(-D1)                                                      ZQDC1140
C      CALL VXPSTO(D)                                                        ZQDC1150
C      IF (D .EQ. ZERO) GO TO 25                                             ZQDC1160
       TEMP = AMAX1(ABS(BR),ABS(BI))                                         ZQDC1170
       D1 = DBLE(TEMP)                                                       ZQDC1180
       D = D1+SCALE                                                          ZQDC1190
       D = D-D1                                                              ZQDC1200
```

```
        IF (SNGL(D) .EQ. ZERO) GO TO 25                          ZQDC1210
C                               IF MAX(ABS(BR),ABS(BI)) .GE.     ZQDC1220
C                               DEPS*SCALE FACTOR THEN SCALE     ZQDC1230
C                               BØ. OTHERWISE SET BØ = ZERO.     ZQDC1240
CS      IF (DOUBLE) 6 LINES, 3 LINES                             ZQDC1250
C       CALL VXPZRO                                              ZQDC1260
C       CALL VXPADD(D1)                                          ZQDC1270
C       CALL VXPADD(DBLE(SCALE))                                 ZQDC1280
C       CALL VXPADD(DBLE(-SCALE))                                ZQDC1290
C       CALL VXPSTO(D)                                           ZQDC1300
C       IF (D .NE. ZERO) GO TO 15                                ZQDC1310
        D = D1+SCALE                                             ZQDC1320
        D = D-SCALE                                              ZQDC1330
        IF (SNGL(D) .NE. ZERO) GO TO 15                          ZQDC1340
        BR = ZERO                                                ZQDC1350
        BI = ZERO                                                ZQDC1360
        GO TO 20                                                 ZQDC1370
     15 BR = (BR/SCALE)*HALF                                     ZQDC1380
        BI = (BI/SCALE)*HALF                                     ZQDC1390
     20 AR = AR/SCALE                                            ZQDC1400
        AI = AI/SCALE                                            ZQDC1410
        CR = CR/SCALE                                            ZQDC1420
        CI = CI/SCALE                                            ZQDC1430
C                               SOLVE AØ*Z**2-2.Ø*BØ*Z+CØ=ZERO   ZQDC1440
CS      IF (DOUBLE) 14 LINES, 5 LINES                            ZQDC1450
C       DI = 2.ØDØ*BR                                            ZQDC1460
C       CALL VXPZRO                                              ZQDC1470
C       CALL VXPMUL(BR,BR)                                       ZQDC1480
C       CALL VXPMUL(-BI,BI)                                      ZQDC1490
C       CALL VXPMUL(-AR,CR)                                      ZQDC1500
C       CALL VXPMUL(AI,CI)                                       ZQDC1510
C       CALL VXPSTO(DR)                                          ZQDC1520
C       CALL VXPZRO                                              ZQDC1530
C       CALL VXPMUL(BI,DI)                                       ZQDC1540
C       CALL VXPMUL(-AI,CR)                                      ZQDC1550
C       CALL VXPMUL(-AR,CI)                                      ZQDC1560
C       CALL VXPSTO(DI) *                                        ZQDC1570
C       ZS = DCMPLX(DR,DI)                                       ZQDC1580
C       ZS = CDSQRT(ZS)                                          ZQDC1590
        DR = DBLE(BR)**2                                         ZQDC1600
        DI = DBLE(BI)*(2.ØDØ*DBLE(BR))                           ZQDC1610
        ZS = CMPLX(SNGL(((DR-DBLE(BI)**2)-DBLE(AR)*DBLE(CR))+DBLE(AI)*  ZQDC1620
     1      DBLE(CI)),SNGL((DI-DBLE(AI)*DBLE(CR))-DBLE(AR)*DBLE(CI)))  ZQDC1630
        ZS = CSQRT(ZS)                                           ZQDC1640
C                               CHOOSE THE SIGN OF ZS SUCH THAT  ZQDC1650
C                               CABS(B)=AMAX1(CABS(B+ZS),CABS(B-ZS)). ZQDC1660
CS      IF (DOUBLE) 5 LINES, 1 LINE                              ZQDC1670
C       CALL VXPZRO                                              ZQDC1680
C       CALL VXPMUL(ZSR,BR)                                      ZQDC1690
C       CALL VXPMUL(ZSI,BI)                                      ZQDC1700
C       CALL VXPSTO(TEMP)                                        ZQDC1710
C       IF (TEMP .LE. ZERO) ZS = -ZS                            ZQDC1720
        IF(DBLE(ZSR)*DBLE(BR)+DBLE(ZSI)*DBLE(BI) .LE. ZERO) ZS = -ZS  ZQDC1730
        BØ = BØ+ZS                                               ZQDC1740
C                               PERFORM THE FINAL COMPLEX OPERATION  ZQDC1750
C                               FOR THE ZEROS.                   ZQDC1760
     25 ZS = CØ/BØ                                               ZQDC1770
     30 ZL = BØ/AØ                                               ZQDC1780
     35 ZSM = ZS                                                 ZQDC1790
        ZLG = ZL                                                 ZQDC1800
   9000 CONTINUE                                                 ZQDC1810
        IF(IER .NE. Ø) CALL UERTST(IER,6HZQADC )                 ZQDC1820
   9005 RETURN                                                   ZQDC1830
        END                                                      ZQDC1840
```

REFERENCES

1. Aird, T. J., Battiste, E. L., and Gregory, W. C., "Portability of Mathematical Software Coded in Fortran," Submitted to ACM TOMS, 1976.

2. Aird, T. J., "The Fortran Converter User's Guide," International Mathematical and Statistical Libraries, Inc., Houston, Texas, 1975.

3. American National Standard Fortran - ANSI X3.9-1966, American National Standards Institute, New York, NY 1966.

4. Burroughs B6700/B7700 Fortran Reference Manual, 5000458, Burroughs Corporation, 1974.

5. Control Data Fortran Extended Version 4 Reference Manual, 60305600, Control Data Corporation, 1974.

6. Control Data Fortran Reference Manual, 60174900, Control Data Corporation, 1972.

7. DEC System 10 Fortran-10 Language Manual, DEC-10-LFORA-B-A, Second Edition, Digital Equipment Corporation, 1974.

8. Gregory, W. C., "Converter Portable Fortran Programmer's Manual," International Mathematical & Statistical Libraries, Inc., Houston, Texas, 1976.

9. IBM System/360 and System/370 Fortran IV Language, GC28-6515-10, Eleventh Edition, International Business Machines Corporation, 1974.

10. Krogh, F. T., "A Language to Simplify Maintenance of Software Which Has Many Versions," Computing Memorandum No. 360, Jet Propulsion Laboratory, 1974.

11. Ryder, B. G., "The Fortran Verifier: User's Guide," Computing Science Technical Report #12, Bell Telephone Laboratories, 1973, Revised 1975.

12. Series 6000 Fortran Compiler, BJ67, Rev. 0, Honeywell Information Systems, Inc., 1972.

13. Sperry Univac 1100 Series Fortran V, UP-4060 Rev. 2, Sperry Rand Corporation, 1974.

14. Xerox Extended Fortran IV, 90 09 56F, Xerox Corporation, 1975.

15. IMSL Fortran Converter, Distribution Guide, IMSL, Inc., Houston, Texas, 1976.

AIDS TO PORTABILITY WITHIN THE NAG PROJECT

J. J. Du Croz, S. J. Hague, and J. L. Siemieniuch

ABSTRACT

We discuss the increasing relevance of portability aids to NAG in its efforts to develop a multi-machine mathematical software library. In particular we outline some of the utilities employed so far and give an informal guide to a FORTRAN transformer (APT) currently under development.

1.   Introduction

When first faced with the problem of implementing its library on other machines, NAG {1} adopted what has been termed the purely corrective approach to portability. Software originally written for the ICL 1906A was modified, as necessary, on each specific machine range. The changes were generally carried out in a disciplined way, manually in some cases and by varying degrees of automation in others. All, however, were made by each implementation group, independently, after receiving the basic version from NAG Central Office.

The principal automatic aid developed by the Central Office at this stage was the Master Library File System (MLFS) {2}. This is essentially a source-text management system for identifying and storing the changes made by each implementation group and for updating each implementation as the basic version of the library is itself updated from one mark to the next.

We have been fortunate in that a sufficient number of competent people have been available to perform the task of implementation, but this labour-intensive approach has obvious disadvantages:

- it is time-consuming and often tedious;

- it may lead to duplication of effort;

- it may lead to unnecessary divergence among implementations;

- despite close coordination, incorrect or inappropriate changes
  may be introduced.

The need to overcome these disadvantages has been the motive for NAG's growing interest in automatic aids to portability. We place equal stress on economy (less manpower required) and reliability (greater control over the manner of implementation and the type of changes made).

NAG's *current* approach to portability of the library, as embodied in the current Mark 5, involves a combination of <u>prediction</u> and <u>correction</u>: we attempt to predict most of the changes required (and to effect them automatically) and we expect that implementors will still need to make a small number of further corrections (usually by hand). The predictive phase includes:

(a)   <u>Standardisation</u> of the basic version so that firstly many changes are no longer required (the library becomes more <u>portable</u>) and secondly it becomes easier to effect the necessary changes by simple automatic means (the library becomes more <u>transportable</u>);

(b)   <u>Transformation</u> of the standardised basic version to produce a predicted source-text for a specific implementation.

Automatic aids are equally important in both these processes.

Section 2 discusses standardisation, especially in connection with the recently completed exercise of bringing the whole of the existing library up to the new standards. Section 3 discusses transformation, with particular emphasis on an experimental transformer program which is already being used for precision conversion in the FORTRAN library.

2.   Aids to Standardisation

2.1 Language Restrictions

We can eliminate some implementation changes by ensuring that the basic version of the library is written in a suitably restricted subset of the relevant programming language (NAG is concerned with FORTRAN, Algol 60 and Algol 68). Ideally this subset will be the intersection of all the dialects of the language available on the target machines,

(but of course it may happen that some of these machines present
mutually conflicting dialects). Strict adherence to the ANSI FORTRAN
(1966) standard, for instance, is an obvious first step and many good
compilers will detect most, though not all, violations of these
standards. For FORTRAN, Ryder {3} has defined a more restricted
and more portable subset, called PFORT; and even more important,
produced a program, the PFORT verifier, to check for conformity with
PFORT. (It is a lost cause to try to impose any standards on a large
body of software without some automatic means of checking the process!)
The PFORT verifier, being written in PFORT, if itself highly portable,
(except for special routines for packing and unpacking character strings).

2.2 Machine-Dependent Constants

In numerical software, constants are a frequent source of implementation-
dependence. The constants may be mathematically defined, but because of
compiler restrictions, on many machines only a limited number of significant
digits may be specified; (ideally all compilers would accept, and handle
correctly, any number of significant digits, so that by specifying a
sufficiently large number of digits, the basic version could satisfy
all implementations). Other constants, such as the relative machine
precision or the largest representable integer, are intrinsically
machine-dependent and different values must be supplied for each
implementation.

NAG's current approach is to make all the machine-dependent
constants and the most common mathematical constants, such as π,
accessible through function or procedure calls, e.g., X01AAF(X)

returns the best approximation to the value of π. This removes
implementation-dependencies from the body of the source-text, and
relegates them to the single occurrences of the constants within
those functions. (The overhead of an extra function-call is
usually negligible; where it is not, as in special function
routines, a different mechanism is used.)

Another advantage of using those functions, which can easily
be written in machine-code, is that we avoid any problems of inexact
conversion of a decimal constant into machine-representation (e.g.,
some compilers on the IBM 360 only allow 7 digits, but this is not
always enough to specify a floating-point number to full machine
accuracy).

An alternative mechanism for handling all numerical constants
is currently being investigated (see Section 3); but in any event the
first problem is to locate all occurrences of numerical constants in
order to check that they are handled correctly, and here an automatic
aid, though desirable, is on the whole lacking; the program SOAP {4}
does locate constants in Algol 60 programs, but in a rather unhelpful
manner.

2.3 Imposing Further Standardisation

It is of course an extra advantage if we can automatically modify
the source-text to make it conform to our standards. As an example, NAG
requires that in FORTRAN subroutines all variables are explicitly declared -
not to meet any compiler restrictions, but to facilitate the production
of double precision versions (see Section 3). We have therefore developed
an automatic procedure to insert type declarations where necessary (actually
a system macro, using the output from a compiler). We also require that

declarations occur in a certain order: this is required by certain

compilers, e.g., one FORTRAN compiler will accept

        INTEGER N, IA(N)

but not

        INTEGER IA(N), N

N and IA being parameters of a subroutine. The required re-ordering

has been effected automatically. (When such compiler restrictions

are uncovered, the question arises whether to impose a further

constraint on the basic version of the library, or to make the necessary

changes only for the affected implementation; but in either case the

text must be modified, and this should be done automatically if possible.)

2.4 Formatting The Source-Text

It is now standard NAG practice to pass all source-text through an

appropriate formatting program, e.g.:

POLISH {5} for FORTRAN;

ALGLIST {6} or SOAP {4} for Algol 60;

SARA {7} for Algol 68.

Such programs, in addition to improving the appearance and legibility

of the code, aid portability by making subsequent transformations

easier to effect. Such transformations may well be of an ad hoc nature:

for example, an implementor may wish temporarily to change all statements

of a certain well-defined type to overcome a compiler bug. Or, in order

to implement a version of the library which uses Large Core Memory on

the CDC 7600, the implementor needs to insert LEVEL 2 declarations for

all array parameters for all routines in the library; if the basic

source-text is in a suitably standardised layout, with the declarations

occurring in a standard order, this can be achieved by a single editing

pass.

In practice there is no hard and fast line between formatting and the imposition of standardisation: all the above programs perform some simple standardisations, e.g., POLISH ensures that all DO-loops terminate with a CONTINUE-statement and gathers all FORMAT statements together at the end of a program segment.

2.5 Future Approach to Standardisation

The previous sections have emphasized that all aspects of what Gentleman has called "the software factory" are important to achieve portability with reasonable ease and reliability. Also we have had to remain "open for business during alterations," so some of our present tools and practices are frankly not suited for general appraisal.

However, armed with the experience gained during the recent revision of the whole library, we now aim to develop a more integrated suite of automatic aids, which will ensure that the basic version of the library meets all our standards. These aids must not only be transportable, but also well-documented and easy to use, so that they can be made available to our contributors.

It should be obvious that the various aids described (language-checkers, formatters, standardisers) must all perform a similar syntactic analysis of the source-text, so there is considerable potential for greater integration.

3. Transformations

We now consider some of the transformations which we need to perform on the standardised source-text, in order to implement the library on different machine-ranges.

First, a simple example: in the Algol 60 version of the library
we need to convert the representations of Algol basic symbols, e.g.,
'GE' to $\geq$. Implementors for various machine-ranges have written
programs to do this.

The Algol 68 library faces a more serious problem: two distinct
dialects of the language - Algol 68S and Algol 68R - with different
syntax. Maybrey {8} has written a program to convert from one to
the other.

But the most extensive transformation is the conversion of the
FORTRAN library from single to double precision, or vice versa. The
contributed library and several implementations use single precision
as the basic precision; implementations for the IBM 360 and ICL 2900,
for example, use double precision as basic. This category of change
accounted for over 70% (see { 9 }) of all changes made in implementing
Mark 4 of the library. Therefore at NAG Central Office an Automatic
Precision Transformer (APT) is being developed, to automate the change.
An experimental version has been implemented by Siemieniuch, and is already
being used to produce a double precision version of Mark 5 of the Library.

It may be regarded as having two components:

    SOPT: Software Precision Transformer, which changes language
               constructs between single and double precision;

and HAPT: Hardware Precision Transformer, which changes machine-
               specific values and can make other changes which are
               specific to one or more implementations.

These are in fact combined in a single one-pass program, but the
facilities of SOPT are all that we need to handle the NAG Library in

its present form, so we give an informal guide to SOPT first, then discuss briefly the action of HAPT. Further details of the APT system including language restrictions and processor limitations are to be found in Siemieniuch {10}.

3.1 SOftware Precision Transformation (SOPT)

Consider a section of single-precision contributed source-text:

```
C=E
      SUBROUTINE D01AGY(N2, P, W, MAXSO, M, W1)
C     NAG COPYRIGHT 1975
C     MARK 5 RELEASE
      INTEGER IN, N2, I, IM, M, MAXSO, NM, N
      REAL PI, AN, AI, W1(129), P(65), W(129), D01AGZ, X01AAF
      COMMON /AD01AG/ N
      PI = X01AAF(PI)
      AN = FLOAT(N)
      IN = N2 - 1
      DO 20 I=1,IN,2
        IM = I*M + 1
        P(IM) = COS(PI*FLOAT(I)/AN)
   20 CONTINUE
      IN = MAXSO/2 + 1
      DO 40 I=IN,N,2
        W1(I) = 0.0
        AI = FLOAT(I)
        W1(I+1) = 1.0/(AI*AI-1.0)
   40 CONTINUE
```

etc.

The initial comment informs SOPT that this is a single-precision version and that the character = is used to mark special comment cards that are to be processed by SOPT. Transformation to double-precision produces this (changes in italics):

```
C=D   SOFTWARE TRANSFORMATION BY SOPT MK 5.1
      SUBROUTINE D01AGY(N2, P, W, MAXSO, M, W1)
C     NAG COPYRIGHT 1975
C     MARK 5 RELEASE
      INTEGER IN, N2, I, IM, M, MAXSO, NM, N
      DOUBLE PRECISION PI, AN, AI, W1(129), P(65), W(129), D01AGZ,
     * X01AAF
      COMMON /AD01AG/ N
      PI = X01AAF(PI)
      AN = DBLE(FLOAT(N))
      IN = N2 - 1
      DO 20 I=1,IN,2
```

```
            IM = I*M + 1
            P(IM) = DCOS(PI*DBLE(FLOAT(I)))/AN)
   20 CONTINUE
      IN = MAXSO/2 + 1
      DO 40 I=IN,N,2
         W1(I) = 0.0DO
         AI = DBLE(FLOAT(I))
         W1(I+1) = 1.0DO/(AI*AI-1.0DO)
   40 CONTINUE
```

etc.

This example includes most categories of change, namely:

- changes in type declarations, e.g.,

    REAL → DOUBLE PRECISION

- changes to the syntax of real constants

$$\left.\begin{array}{c} 1.0 \\ 1.0E0 \end{array}\right\} \rightarrow 1.0D0$$

- changes to calls of intrinsic and basic external functions

    e.g.,   ABS( )   →   DABS( )
            FLOAT( ) →   DBLE(FLOAT( ))
            IFIX( )   →   IFIX(SNGL( ))

the fourth category of systematic change occurs in FORMAT statements

        Ew.d → Dw.d

these do not occur in routines but do occur in test programs.

Assuming the source-text to be transformed conforms to ANSI(1966) standards, we require of SOPT that the transformed text remains ANSI-compatible. This means, for example, that SOPT cannot transform constructs of type COMPLEX; the required double precision equivalent, if it exists, depends upon the specific implementation, and so is handled by the HAPT component of the transformer.

All the changes in the above examples are <u>reversible</u>; we can transform the double precision text back to the original single precision. The original program may, however, involve mixed precision where an alternative to the base precision is required for some elements of the code, for example the double precision accumulation of an inner product of two single precision vectors. If SOPT took no account of this mixed precision, it would lead to an irreversible transformation, as in this example:

```
      DOUBLE PRECISION SUM
      REAL A(N),B(N)
      SUM = 0.0D0
      DO 20 I = 1,N
   20 SUM = SUM + DBLE(A(I))*DBLE(B(I))
```

This would be transformed into:

```
      DOUBLE PRECISION SUM
      DOUBLE PRECISION A(N),B(N)
      SUM = 0.0D0
      DO 20 I = 1,N
   20 SUM = SUM + (A(I))*(B(I))
```

If we now restore that double precision text back to single precision, we would have:

```
      REAL SUM
      REAL A(N),B(N)
      SUM = 0.0E0
      DO 20 I = 1,N
   20 SUM = SUM + (A(I))*(B(I))
```

i.e., the mixed precision arithmetic has been lost. (Of course, if in a particular implementation, quadruple precision facilities existed, the HAPT component might be used to transform $0.0D0 \rightarrow 0.0Q0$, DBLE $\rightarrow$ QLEN, etc.). The property of reversibility is desirable because library codes are not only implemented in both precisions, but may be contributed in either. For example, during the development of a new routine, it may be tested in double precision on an IBM 360/370 and in single precision on a CDC 7600. To overcome the problem typified by the above example, SOPT checks

for the occurrence of constructs in the alternate precision, e.g.,
(DOUBLE PRECISION types and constants, DBLE, etc., within a single
precision source). If a potential irreversibility is found, SOPT
flags its presence by imbedding processor information code into the
text for future processing. Thus, SOPT in transforming the mixed
precision innerproduct would generate this text:

```
C=P01
        DOUBLE PRECISION SUM
        DOUBLE PRECISION A(N),B(N)
C=P01
        SUM = 0.0D0
        DO 20 I = 1,N
C=P0101
C   20 SUM = SUM + DBLE(A(I))*DBLE(B(I))
     20 SUM = SUM + (A(I))*(B(I))
```

i.e., special SOPT comments of the form Pnm or Pn are planted. If
the above text were transformed back to single precision, SOPT leaves
unchanged the n lines following a Pn comment and 'activates' the n
comment lines and 'de-activates' the next m lines after encountering a
Pnm flag (m, n are two digit integers). There is also a facility Cnm
to allow the programmer to plant processor flags which have an effect
similar to that of Pnm. By using these devices, SOPT can transform
from one precision to the other and back again, producing (computationally)
identical text.

3.2 HArdware Precision Transformation (HAPT)

The second function of the transformer relates to machine-specific
constants and constructs which are implementation-dependent.

HAPT can be used to embed the correct value of a machine-dependent
constant for a particular implementation. Of the several techniques
considered, the following was found to be the simplest and most effective:

certain machine-dependent constants were chosen and given reserved names. The value of each constant for all implementations under consideration is held in decimal representation (both single and double precision for non-integer constants) in a data file. The occurrence of a reserved name in a specially flagged DATA statement simply initiates a character-by-character copy from the data file to the appropriate position in the DATA value list. The form of flag (planted by the programmer) is C=D, e.g., the occurrence of

C=D

      DATA PI/ 0.31415926536E+1/

in a single precision program (with PI adjusted for an ICL 1900 machine), would be transformed to

C=D

      DATA PI/ 0.3141592653589793238462643383D+1/

for a CDC 7600 double precision program. The transformation is reversible as with SOPT and indeed can be applied to produce any of the implementation values for which the data file caters.

The C=D flag can now have an argument which is a constant expressed to a maximum of 40 decimal digits with an exponent to a maximum of 5. The action of the processor on encountering a C=D is to replace all the DATA item values, whose names do not match those of the reserved names, by the argument (rounded to the correct number of significant figures for the desired implementation). The argument is given an E- or D-exponent depending upon the DATA item value(s) being replaced.

For the ICL 1900 implementation, for example, the value of $\sqrt{\pi}$ is contained in PISQ (and EPS is a reserved name):

C=D 1772453850905516027298167(+1)

      DATA PISQ,EPS/0.17724538509E+1, 0.72759576142E-11/

Transforming this into an IBM 360/370 implementation, we obtain

C=D 1772453850905516027298167(+1)

     DATA PISQ,EPS/0.1772453850905516D+1, 0.2220446049250313D-15/

For implementation-dependent coding which is not handled by SOPT or the value-substitute facility of HAPT, there is a record selection mechanism which allows the programmer to plant implementation-dependent records, e.g., in preparing an ICL 1900 program, the following records might occur:

```
C=+ 1 01
      MASTER PROGRAM
C=+ 1 07
C      PROGRAM TEST(OUTPUT, TAPE6=OUTPUT)
C=- 1 01 07
C      THIS IS THE FIRST RECORD FOR ALL BUT 01 AND 07
```

$$\vdots$$

In this example, the MASTER record is 'active' in implementation 01 (the ICL 1900 implementation number). If we use HAPT to generate implementation 07 (CDC 6000/7000), the MASTER statement is 'de-activated', and the PROGRAM statement is 'activated'. For any other implementation, neither of these statements is active. The general form of the selection device is

$$C{=}{+} \; n \; a_1 \; a_2 \; \ldots \ldots a_{22}$$

where n is the number of records to be activated (+) or not (-), if the desired implementation number is one of the set $a_1, \ldots, a_{22}$ (22 is the maximum number permitted). To avoid interaction with SOPT, the occurrence of a sequence of p records concerned with selection must be preceded by a $C{=}P_p$, i.e., effectively switching off SOPT for the next p records. In the above example, the record C=P06 must precede the text shown.

4. Summary

We have indicated the increasing importance attached by NAG to mechanical aids in its pursuit of a single library for many machines. The APT program described in Section 3 represents a significant advance in our approach to FORTRAN transformation. The production version of APT, when installed, may vary in detail from the present experimental program, but not in essence; all those changes that can be safely automated should be performed mechanically and other deviations which can be predicted should be catered for in advance.

The problems of source text transformation at the level described in this paper are clearly soluble in theory, if not yet completely overcome in practice. A problem of a deeper complexity concerns transformations not within the bounds of FORTRAN (one dialect to another) but from one language to another. In NAG's case, there are at least three language versions; Algol 60, Algol 68 and FORTRAN (and possibly BASIC too). What are the implications of developments in one language version for the others? Language converters (usually from FORTRAN to Algol 60 or Algol 68) do exist, but there are obvious limitations because many constructs have no counterpart in the other language. Most compiling systems still cannot provide acceptable mixed (high-level) language facilities. Given that the demand for these different language versions continues, one question we face is to what extent mechanical aids can alleviate the burden of developing and maintaining several language versions of a library on many machines.

REFERENCES

1. The NAG Annual Reports (1973/4) and (1974/5), available from the NAG Central Office, 13 Banbury Road, Oxford, U.K.

2. Richardson, M. G., and Hague, S. J. The Design and Implementation of the NAG Master Library File System. *Software P. and E.* In Press (1976).

3. Ryder, B. G. The PFORT Verifier. *Software P. and E.* 1 (1974), pp. 359-378.

4. Scowen, R. S., Allin, D., Hillman, A. L., and Shimell, L. SOAP - A Program Which Documents and Edits Algol 60 Programs. Computer J. 14 (1971), pp. 133-135.

5. Dorrenbacher, J., Paddock, D., Wisneski, D., and Fosdick, L. D. POLISH, A FORTRAN Program to Edit FORTRAN Programs. Dept. of Computer Science, University of Colorado, Report #CU-CS-050-74 (1974).

6. Oxford University Computing Laboratory, Computing Service Handbook, Part 5, Chapter 2.

7. Woodward, P. Skeleton Analyser and Reader for Algol-68R ("SARA"). Algol 68 paper, available from R.S.R.E., Malvern (U.K.) (1975).

8. Maybrey, M. A. On Using Two Forms of Algol 68. In *Proceedings of the Conference on Experience with Algol 68 - Liverpool, 1975,* ed. Leng P., Liverpool University Press (1975).

9. Bentley, J., and Ford, B. On the Enhancement of Portability Within the NAG Project - A Statistical Survey. This Proceedings.

10. Siemieniuch, J. L. APT - Automatic Precision Transformation to Enhance the Portability of Numerical Software Coded in FORTRAN. To appear (1976).

MULTIPLE PROGRAM REALIZATIONS USING THE TAMPR SYSTEM

Kenneth W. Dritz

ABSTRACT

Elsewhere in these Proceedings, Boyle has
presented an overview of the issues relating to
transportability of mathematical software libraries and has surveyed four of the systems, including
TAMPR, which have been designed to assist in automating the development, conversion, and maintenance
of such libraries. This report considers in more
detail the architecture of TAMPR and briefly discusses typical applications. It concludes with
a discussion of our experiences and plans for the
future.

This report is organized as a survey of the
historical and operational aspects of TAMPR. It
stops short of supplying full details and examples
but cites other publications in which they may be
found.

Introduction

At a basic conceptual level the operation of TAMPR can be described
as the application of a realization function to a prototype program to
obtain a particular realization of that program. Realization functions
of particular interest are those semantics-preserving functions that
yield programs tailored to specific machine/compiler environments;
equally useful are those that accomplish certain optimizations, such
as integration of subroutines as in-line code and change of data
representation. The power of realization functions is due to their
generality and the possibility of composing them (applying several
in sequence) to achieve reliable composite transformations.

Although the overall effect of a realization function is a source-
to-source transformation, the process of its application differs greatly
from conventional string-oriented manipulations such as text editing or
macro expansion. Basically, TAMPR processing occurs in three steps:

(1)   A Fortran source program is converted by a recognizer
       into an abstract form which is represented internally
       essentially as a parse tree in the TAMPR version of a
       Fortran grammar.

(2)   Intragrammatical transformations (IGT's) are applied to
       the abstract form. These are operations on trees which
       preserve their grammaticality.

(3)   The transformed tree is converted back into source form
       by a formatting process.

Because transformations are applied to trees whose components are
syntactically labeled, spurious changes are easily avoided. The

reliability of the overall process derives partly from the confidence that transformations will be applied only in the intended contexts and partly from the intragrammaticality restriction, which guarantees at least syntactically legal Fortran programs as output.

To simplify the description and application of realization functions as IGT's, the abstract form is first converted to what we have defined as a canonical abstract form by the application of specialized structure clarification transformations. One effect of these is to reduce the complexity of the program by substituting simpler, context-free constructs for context-dependent ones while preserving semantic content. An example is the unraveling of nested DO loops which share the same final statement; they are replaced by properly nested loops, each with its own unique terminator. A related effect is the replacement of certain uses of the GOTO statement by the equivalent context-free structured construct without labels, i.e., IF-THEN, IF-THEN-ELSE, DO-WHILE, loop-with-exit, etc., for which purpose the underlying grammar is actually that of a structured extension of Fortran. Besides permitting the simplification of transformations relating to control flow, such as those to effect certain optimizations across loops, these structure clarification transformations ultimately enhance the formatter's ability to reveal the hierarchical structure of the program by indentation. These are the major features of TAMPR depicted in Figure 1.

Details and examples of IGT's may be found in {5, 8}.

The TAMPR System Formatter

When we began the design of the TAMPR system in 1973 we had only an incomplete idea of the formatting rules we wanted to employ. Hence, to facilitate experimentation with different formatting strategies as they evolved we decided to make the formatter programmable. This turned out to be a wise decision because at the current time we are still increasing the sophistication of our formatting strategies. Another, hardly less important, reason for endowing the formatter with pro-grammability was the expectation that many types of realization functions could be implemented with it alone, avoiding the need (in these cases) for IGT's.

Formatting specifications are supplied to the formatter in the form of an FCL (Format Control Language) program. The formatter is organized as a translator/interpreter for FCL; it applies the given FCL program to a canonical abstract form input to obtain a concrete Fortran source program as output. A particular FCL program suffices for the translation of a whole class of Fortran programs from abstract to concrete form. Our "standard" FCL program produces executable, or real, Fortran as output while another produces "structured Fortran" intended for the human reader. Note that FCL is powerful enough to provide either of these results from the same abstract form. Other FCL programs will be mentioned later.

The constituents of the language, FCL, and the Formatter's rules for interpreting an FCL program are described in {2}. For purposes of this overview, the following abbreviated account of FCL and the behavior of the formatter will suffice.

(1)   The essential feature of the system is that formatting action specifications are, in general, syntax-directed; that is, they may be related to specific phrases in the TAMPR grammar (and thus to specific subtrees in the abstract form).

(2) The simplest formatting actions are built into the formatter
and are provided automatically. In general, the majority
of formatting requirements are met by these built-in actions.

(3) When the desired formatting actions for a particular phrase
in the TAMPR grammar differ from the built-in actions,
explicit formatting actions for that phrase are included
in the FCL program. Thus, the size of an FCL program is
determined by the extent to which the desired formatting
actions deviate from the simple, built-in actions.

(4) Explicit formatting action specifications for a particular
phrase may be made dependent on the context in which that
phrase appears in the abstract form tree. This is achieved
by nesting the specification for the context-dependent
action inside another specification, which serves to
define the context.

(5) Textual output fragments may be produced directly from the
terminal nodes in the abstract form, or they may be computed
from data obtained from terminal nodes and/or from constants
in the FCL program itself. The latter capability is particularly useful in the handling of localized cosmetic alterations,
such as those involved in generating real Fortran from the
(structured) canonical abstract form or in Fortran tailored
to a specific compiler; it can also be used to "generate"
additional comments in the output.

(6) Facilities are provided to re-introduce low-level stylistic
elements, such as indentation and spacing, which were
discarded by the recognizer. Particular attention has been

devoted to the problem of line overflow and continuation
with the result that the FCL programmer may achieve some
reasonably sophisticated effects in this area (see {2}).

Current Applications of TAMPR

Although the TAMPR system had its origin in the NATS project,
where it was motivated by the need to automate the more routine
adaptations required to transport systematized collections of numerical
software from one machine to another {1, 3}, its fundamental role has
since been enlarged to include the derivation of related members of
such a collection from a single prototypical member {5, 9}. Examples
of both types of application are cited in this section, along with one
or two others that are more distantly related to these purposes but
are well within the capabilities of the system.

The majority of the problems that can be anticipated when a Fortran
program is transported from one implementation of Fortran (the "source"
version) to another (the "target" version) usually arise when at least
one of the two versions is not an implementation of the Standard {4}.
These problems have several different origins:

(1)   In the source version, an extension not implemented in
      the target version has been used (e.g., the IMPLICIT
      statement or generalized subscripts).

(2)   A feature common to both versions has been used, but in
      the target version certain restrictions (of either a
      qualitative or quantitative nature) not applicable to
      the source version apply. For example, many implemen-
      tations contain restrictions on the placement or ordering

of specification statements (see {6}). Restrictions on
the type of statement permitted as the last statement in
the range of a DO loop are of a similar qualitative nature.
Quantitative restrictions abound, such as the maximum DO
loop nesting depth, etc.

(3)   In some cases, features common to both versions are given
different interpretations. For example, some compilers
associate the non-Standard extension X**Y**Z from the
left while others associate it from the right (see {6}).

Problems such as these (with the exception of certain quantitative
restrictions, such as the maximum number of array dimensions) can be
solved by a process of localized reduction to, or substitution of,
equivalent language components -- a process made possible by the
redundancy that exists within Fortran. By "localized" we mean
isolated in the sense that the required or suggested changes are
uncoupled from other features used in the program and especially
from a deep semantic understanding of the program or the algorithm
it represents. Changes that can be characterized in this way lend
themselves readily to automation in systems such as TAMPR*.

Another class of transportability problems relates to differences
such as word length, normalization, and rounding strategy in the
underlying machines. Some specific problems in this class can be

---

*"Localized" must not be construed to imply that the portion of the
program requiring examination to effect such a change is limited.
For example, a TAMPR transformation to replace IMPLICIT statements
by additional type specifications obviously requires an examination
of the entire program unit.

solved by localized substitutions and are thus amenable to mechanization with TAMPR; for example, TAMPR can systematically make a change in precision from single to double to regain what is lost in the move to a machine with a shorter word length. On the other hand, substitution of a different convergence test, or perhaps an entirely different series approximation, requires more semantic reasoning about programs than TAMPR is currently capable of performing.

Most of our practical experience with transportability transformations was gained during the translation of a large program containing many subroutines from CDC Fortran to IBM Fortran. Among the specific changes handled by TAMPR, we note the following.

(1) Multiple statements per line were reduced to single statements per line.

(2) Multiple assignments were reduced to simple assignments.

(3) Hollerith format items in the form *text* were changed to nHtext; commas were inserted between format items.

(4) Logical IF's with labels for true and false branch targets were replaced by logical IF's containing GOTO statements.

(5) Variable names were shortened to six characters. The new names were chosen manually but inserted mechanically.

Some of the problems we encountered in this exercise are mentioned later.

At the present time, enhancements of TAMPR are being guided by the emerging requirements for its use in the LINPACK software development effort {7}. The experience gained in this application (which is described in more detail below) led, in 1975, to conceiving the system as directed toward the production of multiple realizations of programs from a single prototype version and to the adoption of the name TAMPR (Transformation-Assisted Multiple Program Realization system) to emphasize that concept. Thus, LINPACK occupies a dominant niche in the recent evolution of the TAMPR system.

Three sets of transformations have so far been designed to assist in the automatic derivation of some members of the LINPACK collection from others.

One set converts a program that uses real (i.e., single precision) arithmetic to one that uses double precision. Very little work is involved. One change involves substituting double precision counterparts for the names of single precision intrinsic and external functions. Another requires that "REAL" be changed to "DOUBLE PRECISION" in type specification statements. The final one consists of attaching "D0" to real constants expressed without an exponent or changing "E" to "D" in those containing an exponent. (Clearly, if one contemplates using a program as a prototype for constructing realizations in single and double precision then he must write any constants needed using enough digits to yield the required accuracy in double precision.) These transformations, which are discussed in more detail in {8}, can be wholly carried out in the formatting process since they have no substantial context dependency; thus, they do not require IGT's.

The second set of LINPACK transformations is designed to convert a program using complex arithmetic to one that uses real arithmetic. Such a conversion is only meaningful (with respect to preservation of semantics) in conjunction with certain assumptions, discussed in {9}, concerning the magnitudes of the imaginary parts of complex variables that could arise in a computation presented with pure real arguments. In general, an analysis of a potential prototype is required to determine its suitability for this type of conversion. The transformations are specifically concerned, in part, with changing complex constants with zero imaginary parts to the corresponding real constants, changing COMPLEX type specifications to REAL type specifications, changing LINPACK subroutine names, and simplifying expressions containing the intrinsic functions REAL, CONJG, and CMPLX (with zero second argument). These simplifications, and a few others not described here, yield several optimizations with the potential for eliminating redundant assignments. For further discussion, see {8}.

The third set of LINPACK transformations has as its goal the replacement of calls on the Hanson, Lawson, and Krogh Basic Linear Algebra modules {10} by inline code. For at least some machines this is a desirable optimization. The implementation of the BLA transformations for six of the BLA's used in LINPACK is covered in detail in {5, 8}. Here we shall merely remark that the IGT's for BLA removal are capable of specializing the replacement code for several typical special cases which are detected by analyzing the arguments of the BLA call along with the dimension information for the arrays appearing among them.*

The three sets of LINPACK transformations can be applied in various combinations to yield compound transformations. There are clearly eight different compound transformations, as shown in Figure 2. Their use in

---

*This is another example of a transformation using globally obtained information to effect a localized change.

LINPACK permits the cost of developing about two dozen prototypical core routines (in complex, single precision form with BLA calls) to be spread over a much larger collection of about two hundred members.

Finally, we shall describe a current application of TAMPR which is rather distinct from all the rest: it is being used as a component in an experimental program verification system. The goal of the IGT's in this case is to reduce a (suitably restricted) Fortran program to a sequence of assertions characterizing the effects of each statement on the program's variables. For this purpose, and also to accommodate user-supplied assertions relating the program's outputs to its inputs (and perhaps to state other facts), the TAMPR Fortran grammar has been extended to include an ASSERT statement and appropriate logical operators and quantifiers. The FCL program for this application sees only these ASSERT statements; the "verification conditions" it produces as output consist of a sequence of formulas in the first-order predicate calculus. (A variation of this FCL program produces its output in "clause form" suitable for direct input to a resolution-based automatic theorem prover.) The programmability of the formatter is used to advantage here to reformat certain subexpressions employing infix operators in prefix (i.e., function-argument) form.

Results of Experience with TAMPR

From the earliest days of its availability, TAMPR has played the dual role of experimental (or developmental) system and production system. Concurrent with its present uses in LINPACK development, in program verification studies, and in transportability of programs, it is being evaluated for its effectiveness in solving the problems for which it

was designed, and it is undergoing further development aimed at
increasing its effectiveness and broadening its scope. It has
reached its current level of capabilities only through repeated
additions to the original conceptual framework. For instance,
initially we were without concrete plans for handling comments
in a program undergoing transformation. Similarly, we had only
vague ideas for handling the line overflow problem during formatting.
Thus we concentrated on making progress in areas we understood well.
As we gained experience with the system we found ourselves in a much
better position to appreciate the remaining problems, to contemplate
solutions, and to test them in the system. Frequently we found the
flexibility inherent in the programmability of the system allowed us
to at least approximate some new techniques and evaluate them before
committing ourselves to new built-in features. And, looking backward,
we can now confirm our original faith in the value of a modular,
structured, and clean design; the original investment in designing
generality and expandability into the system has indeed been justified.

The remainder of this section will be devoted to a discussion of
some of the unsolved problems or deficiencies catalyzing our continuing
development of TAMPR. Most of the items in this discussion are
characterizable either as functional enhancements or as revisions that
enhance some property of the system without providing new function.
The one to be discussed first, however, seems to be separable from
the others.

At the outset of the CDC to IBM source program conversion effort described earlier we found that the recognizer component of TAMPR, which produces the abstract form, was not prepared to recognize some of the syntactical constructs by which CDC Fortran differs from "TAMPR Fortran" as it stood then. It was necessary to revise our grammar. This process was at least partly aided by our use of the XPL compiler generator system in the construction of the recognizer. Nevertheless, the peculiar lexical aspects of Fortran complicate both the grammar and the largely *ad hoc* lexical analyzer to the extent that changes such as these severely strain the recognizer, making it the least responsive to change of any TAMPR component.

Many types of transformations that work at the syntactical level of expressions produce the need to perform arithmetic on constants present in those expressions {11}. This capability is currently lacking in the IGT interpreter; at best, one can define transformations for specific cases (including specific constants) that are likely to arise as the result of other transformations. We plan in the future to introduce a functional notation into the syntax of IGT's. This might be used to perform arithmetic on constants, generate new identifiers, and so forth.

While the features of the formatter for dealing with the line overflow and continuation problem have been generally satisfactory and have, in fact, permitted the achievement of some sophisticated -- almost surprising -- results, they are still not as general as we believe may be necessary. Currently, line breaks and continuations can be made to occur at points in the output text that have some particular relationship to the phrase structure of that text. We plan to introduce lower-level facilities to permit break and continuation points to be related to absolute column or tab positions.

A number of efficiency and usability improvements are planned in various components of the system. A major improvement of this type for the IGT interpreter deals with the mechanisms for constraining the applicability of certain transformations. We currently use a cumbersome marking method which we hope to replace with applicability predicates attached to transformations in the manner of Loveman {12}. The formatter needs to be optimized to use less main memory and to permit processing a "batch" of abstract forms against a single FCL program in one run. Finally, while the use of XPL, LISP, and PL/I in the development of TAMPR was consistent with its status as a research project, it would now seem highly desirable to be able to share our experience with others by producing a portable version of TAMPR.

Future Plans

Several investigations are planned for TAMPR in the future. In the direction of additional applications we expect, of course, to study more transportability transformations. Transformations to improve the efficiency of programs is also on the agenda. Another attractive possibility is the automatic instrumentation of programs for testing or measurement. And we have been contemplating interlanguage translation, for instance from Fortran to PL/I. This involves much more than simple syntactic substitutions; interesting semantic problems are posed, for instance, by the distinct ways in which dimension information is passed from one routine to another in these two languages. It would seem that successful transformations to handle this problem will have to do a considerable amount of "reasoning about programs." (We would also expect to be hampered by the IGT intragrammaticality restriction in a language translation application, or be faced with enlarging the TAMPR grammar by effectively merging those for two different languages.)

We also intend to study further several theoretical questions relating to TAMPR and to program transformation systems in general. Two such are the following: What properties make programs useful as prototypes for multiple program realizations? How do we prove that a transformation is correct (i.e., preserves correctness)? Loveman's "post-assertions" for transformations {12} will probably be instrumental in a rigorous treatment of the correctness problem.

Conclusions

All of the uses of TAMPR described here have in common the motivation of economy. The use of multiple program realizations in conjunction with package development yields economies at several points. To take an example from LINPACK, less human effort is involved in the creation of twenty-five prototypes and three realization functions than in the creation of the whole package of about two hundred routines. Economies are achieved by the enhanced reliability of the program generation process when realization functions are used, because they (having once been proved sound) transfer the correctness of a prototype to the program realized from it. Further economies are achieved by extending this reliability to the maintenance aspect, by performing the maintenance on the "documentation version," a well-structured prototype, and then obtaining the necessary "executable versions," which most likely gain efficiency at the expense of clarity and structure, by transformation.

Automated systems can sometimes hinder rather than help. To be free of potential interference with the user's goals they must be sufficiently general and flexible, and they must be well-behaved. By the former we mean that they must permit the user to achieve his intent without constantly circumventing limitations or simulating non-existent capabilities, and this is exactly what has led us to develop programmable components in our system. By the latter we mean that the meaning as well as the implementation of the transformation process is so well defined and constrained that the user can be confident that the system will carry out his intent, as described, without producing unexpected side effects or spurious results. In TAMPR this has been achieved largely through the use of abstract forms.

421

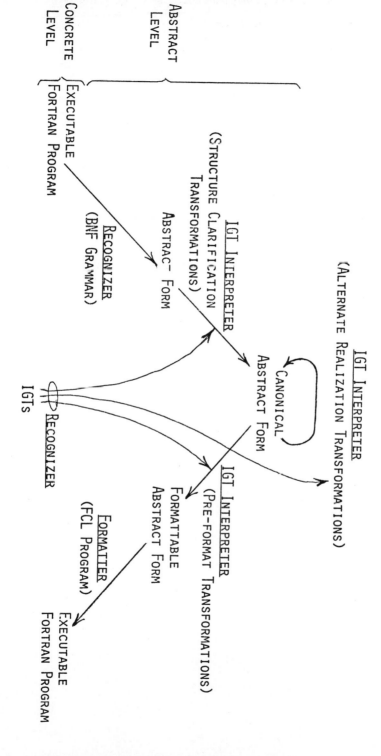

FIGURE 1

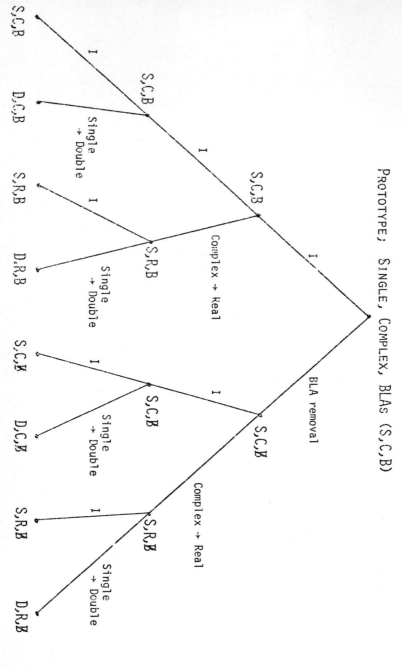

CONSTRUCTION OF ALTERNATE LINPACK REALIZATIONS

PROTOTYPE; SINGLE, COMPLEX, BLAS (S,C,B)

REFERENCES

1. James M. Boyle, Mathematical Software Transportability Systems --
    Have the Variations a Theme? (These Proceedings).

2. Kenneth W. Dritz, An Introduction to the TAMPR System Formatter,
    in James R. Bunch (ed.), Cooperative Development of Mathematical
    Software, Technical Report, Dept. of Mathematics, Univ. of Calif.
    at San Diego, San Diego, Calif., 1977.

3. James M. Boyle and Kenneth W. Dritz, An Automated Programming
    System to Facilitate the Development of Quality Mathematical
    Software, Information Processing 74, North-Holland Publishing
    Co., 1974, 542-546.

4. Brian T. Smith, Fortran Poisons and Antidotes, (These Proceedings).

5. James M. Boyle and Marilyn M. Matz, Automating Multiple Program
    Realizations, Proceedings of the M.R.I. International Symposium
    XXIV: Computer Software Engineering, Polytechnic Press, Brooklyn,
    NY, 1977.

6. Eva Könberg and Ingemar Widegren, Fortran Dialects -- A Selection,
    Report C-1500, National Defense Research Institute, Stockholm,
    Sweden, Feb. 1973.

7. The LINPACK Prospectus and Working Notes, Applied Mathematics
    Division, Argonne National Laboratory, Argonne, Illinois 60439.

8. Michael E. Frantz, Automated Program Realizations: BLA Replacement
    and Complex to Real Transformations for LINPACK, Report ANL 76-55,
    Argonne National Laboratory, Argonne, Illinois 60439 (To appear).

9. James M. Boyle, An Introduction to the Transformation-Assisted
    Multiple Program Realization (TAMPR) System, in James R. Bunch
    (ed.), Cooperative Development of Mathematical Software, Dept.
    of Mathematics, Univ. of Calif. at San Diego, San Diego, Calif.
    1977.

10. R. J. Hanson, F. T. Krogh, and C. L. Lawson, A Proposal for
    Standard Linear Algebra Subprograms, Technical Memorandum
    33-660, Jet Propulsion Laboratory, Pasadena, Calif. Nov. 1973.

11. T. A. Standish, D. C. Harriman, D. F. Kibler, and J. M. Neighbors,
    The Irvine Program Transformation Catalogue, Dept. of Information
    and Computer Science, Univ. of Calif. at Irvine, Irvine, Calif.
    92717, Jan. 1976.

12. David B. Loveman, Program Improvement by Source-to-Source
    Transformation, JACM 24, 1, Jan. 1977, 121-145.

SOFTWARE DESIGN TO FACILITATE PORTABILITY

Order and simplification are the first steps toward the mastery of a subject.

— *Thomas Mann*

The three papers in this section address program design questions which bear on portability. Both large applications programs and basic function subroutines are represented. J. L. Schonfelder discusses the strategy and tactics of the NAG Library project in providing special function software across various machine ranges. L. Wayne Fullerton shows how earlier approaches can be modified to produce absolutely portable special function routines, requiring no changes across machine ranges. He analyzes the benefits of this approach in terms of accuracy, reliability, and maintainability. The costs are examined in terms of storage and execution time.

The paper by N. Victor and M. Sund is written from the perspective of the designers and users of statistical software for medical data applications. It deals with a number of design considerations related to the portability of such software with special emphasis placed on the use of standardized data interfaces.

THE PRODUCTION AND TESTING OF SPECIAL FUNCTION SOFTWARE
IN THE NAG LIBRARY

J. L. Schonfelder*

ABSTRACT

A summary of the major machine ranges and arithmetics involved in the NAG Library project is given, thus setting the context and the scope of the portability problems involved. The system that has been evolved to produce the NAG special functions software is described in outline and some idea of the resulting compromises made to gain transportability is included. The actual form of the software is illustrated by use of a concrete example for the function tan(x) (S07AAF). The problem of providing portable testing software is also discussed and a distinction is made between validation and certification; validation being the verification of "correct" functioning of each routine pathway at least once, certification being an attempt to provide an accurate and extensive measurement of the routine's accuracy profile and/or error behaviour. The approach to providing portable certification software being adopted for future marks of the NAG Library is outlined and illustrated by example.

*Much of the preliminary work on certification testing was done while the author was a visiting lecturer at the Basser Dept. of Computer Science, University of Sydney, Australia.

1. INTRODUCTION

The NAG Library caters for a wide range of machines and hence a wide range of arithmetics. In order to give some idea of the breadth of this range, and hence some idea of the scope of portability problems that we are faced with in providing special function software in this context, the following table shows the major machines involved and a brief summary of their arithmetic characteristics:

MACHINES	PRECISION	XMAX	XMIN	BASE
ICL 1900	$11.1D \rightarrow 11.4D*$ $22.3D \rightarrow 22.6D$	5.8×10^{76}	4.3×10^{-78}	2
IBM 360/370	$6.0D \rightarrow 7.2D$	7.2×10^{75}	5.3×10^{-79}	16
ICL System 4 } ICL 2900	$15.7D \rightarrow 16.8D*$			
CDC 6000/7000	$14.1D \rightarrow 14.5D*$ $28.6D \rightarrow 28.9D*$	1.3×10^{322}	3.1×10^{-294}	2
Univac 1100	$7.8D \rightarrow 8.1D$ $17.8D \rightarrow 18.1D*$ $11.4D \rightarrow 12.0D*$	1.7×10^{38} 9.0×10^{307} 4.3×10^{68}	1.5×10^{-39} 2.8×10^{-309} 8.8×10^{-47}	2
Burroughs 6700	$22.9D \rightarrow 23.8D$	1.9×10^{29603}	1.9×10^{-29581}	8
DEC PDP 10	$8.1D \rightarrow 8.4D*$	1.7×10^{38}	1.5×10^{-39}	2

The precisions were calculated as the log to base ten of the relative error possible due to the finite mantissa size ($-\log_{10}$ (rep.err.)). This quantity has a range of values depending on which end of the normalization range the number occupies. The range is quite large for the base 8 and 16 arithmetics because of the "wobbling word length" phenomenon about which we will say more later. XMAX and XMIN give the approximate ranges for which representations without over or under flow are possible. The starred precisions are those for which current implementations are available or in the process of being produced.

Thus we have eight major active machine ranges covering five different precisions from roughly 8D to 18D and, if in the future, both double and single precision implementations were required on all machines this would grow to ten ranging from 7D to 29D. Coupled with this are number ranges which vary from 10^{-47} to 10^{+68} at the smallest to 10^{-29581} to 10^{+29603} at the largest. To add to the labour and to the complications the library is required in both Algol 60 and FORTRAN. (We shall not say much if anything about the Algol version since the FORTRAN has the wider application but the fact that both versions have been required has had a not insignificant effect on design decisions). It should also be pointed out that in the near future Algol 68 implementations will also be required which will add to the problem.

An obvious approach to the problem would be to eliminate machine dependencies altogether. This could be done by setting the accuracy required at the lowest attainable on any machine, 8D say and the ranges at the most stringent for any machine. This approach would give highly portable software but would hardly be adequate on any machine. Therefore, portability being clearly out of the question the aim was and is to provide software, designed at least, to give full machine accuracy on each implementation and to utilize as far as is practicable the whole representation range on each machine. However, even though such software is necessarily highly machine dependent, the aim is to design and produce it in such a way that an implementation on machines other than the parent one can be produced easily without the need for specialist function approximation knowledge on the part of the implementor. That is, the aim was and is to produce transportable software.

The system that has been evolved for producing software of this type was described in some detail in the papers of references {1} and {2}. In fact the evolution of the system can be seen quite clearly by comparing the earlier papers {1} with the later {2}. This evolutionary process is still continuing but I think the overall structure and approach is now fairly well established. In Section 2 of this paper we shall give an outline of the system and illustrate some of the difficulties and partial solutions by way of examples. Section 3 will contain a discussion of the problem of portable stringent testing of such software and again the approach adopted for the NAG Library will be outlined and illustrated.

2. PRODUCING THE NAG SPECIAL FUNCTION ROUTINES

The linch-pin of the system for producing the routines and the testing material is a multiple precision arithmetic facility, MLARITHA {3} written in Algol 68. This allows one to write programs working to arbitrary accuracy with almost the same ease as one would write the same programs working at standard floating point precisions. Using this package a suit of programs for generating Chebyshev expansions to any accuracy has been developed.

The Chebyshev expansion was chosen as the basic approximating technique, since its truncations provide reasonably efficient approximants, being close to the mini-max polynomials. Suitable approximants at each of the various precisions can be obtained easily by simple changes in truncation order and the Chebyshev expansions are fairly easy to generate. The Chebyshev expansions have to be generated once only, provided this is done to sufficient accuracy, say 30D. This is a highly significant fact. True mini-max rational approximants would provide greater efficiency but to provide for all precisions involved in this way would require generating a sizeable subset of the Walsh array. This would have involved a vast expenditure of machine and human time both of which are in somewhat limited supply. Also the Chebyshev expansions would probably be produced initially to provide suitable starting approximations for a min-max rational approximation program {4}. At some future stage, if it becomes necessary, the system that has been developed could be modified to make use of rational approximants without too much change.

Once an algorithm has been designed and the expansions generated and checked the expansion coefficients are stored in a file on the machine. Each expansion is stored with identifying information, its maximum order and precision and a unique index to be used for retrieval. At this point the actual writing of the routine begins. The aim in writing the routines is to produce a routine which is "correct" for the parent machine and which can be converted into one that is "correct" for any of the other implementations by a simple and obvious editing job. As much of the code as possible is therefore made machine independent. Certain code such as that evaluating the actual expansion and any code introducing constants that do not have an "exact" representation for all machines is duplicated. The unwanted code is rendered inoperative by turning it into comments. This is done by introducing a suitable prefix. Precision dependent code is prefixed by Cdd where dd is the number of decimal figures of the intended precision involved in any constants.

Machine dependent rather than strictly precision dependent code is included in a different manner but is identifiable and can be made operative or not by removing or adding comment prefixes.

Isolated constants are fed in via DATA statements. These include mathematical constants such as π, or machine constants setting failure thresholds or other switch points. For any function these constants are few in number but are highly specific to the function. The behaviour of the routine is frequently highly sensitive to the precise choice of these constants. These constants are included by hand. However, the expansion code involves a large number of constants arising from the coefficients of the truncated Chebyshev expansions. Each precision requires a different truncation and a different rounding of the coefficients. Writing this code in its many duplicated versions with the attendant vast number of multidigit constants is not possible by hand. Nor is it desirable, as the frequency of error in hand transcription of large numbers of multidigit constants is extremely high. Therefore expansion evaluation code is included in the routines mechanically by machine processing.

A routine is written initially without evaluation code. At the point in the routine where expansion evaluation code is to be included a special trigger record is inserted. This trigger record contains information such as the retrieval index of the expansion to be used, the variable names used as argument and result and the expansion evaluation technique code. This incomplete routine is processed by a program, which constructs and inserts the required expansion evaluation code automatically. This process and the evaluation techniques employed were discussed at length in (2) and I shall not repeat that discussion here. This method of dealing with the expansion evaluation code is now quite well tried and is efficient and reliable.

Most of the problems in introducing a new implementation arise in choosing the threshold and range constants, etc. One difficulty occurs because of the existence of non-binary machines. The Hex and octal arithmetic machines suffer from the phenomenon of "wobbling word length" described very clearly by Cody {5}. In effect "wobbling word length" is always present in any floating point arithmetic system. If x is given by

$$x = f\beta^r$$

where β is the arithmetic base

r is the exponent and

f is the normalized mantissa, ($f \neq 0$)

then the relative error in x is given by

$$\delta = \left| \frac{\bar{f} - f}{f} \right|$$ where $\bar{f}$ is the finite length approximation to f, and

the difference between f and $\bar{f}$ is bounded by β^{-t} or $\frac{1}{2}\beta^{-t}$ depending on whether the system truncates or rounds to t digits. Therefore

$$\delta \leq \frac{\beta^{-t}}{|f|} \quad \text{or} \quad \delta \leq \frac{\frac{1}{2}\beta^{-t}}{|f|} \; .$$

Now f is usually normalized so that $\frac{1}{\beta} \leq |f| < 1$;

therefore the relative error due to representation varies from $\beta^{-t}(\frac{1}{2}\beta^{-t})$ to $\beta^{-t+1}(\frac{1}{2}\beta^{-t+1})$. That is, there is a ratio of β between best and worst bounds. Obviously this effect is minimized for a binary base, $\beta=2$, but is quite large for an octal base, $\beta=8$ or worse still for a Hex base, $\beta=16$. In these latter bases the bad cases result in the first few most significant bits being unset; possibly two bits in octal and three in Hex. It is hardly profitable to try and maximize significance on a binary machine by very complex choice of constants to ensure best possible normalization since at best we would get a factor of 2 and the extra complications in arithmetic would probably lose it again anyway. However for Hex and octal machines the possible loss of significance due to this effect can be very serious, particularly if badly normalized constants are involved in sensitive calculation steps. For example, in the algorithm shown below for the tangent function the value $\frac{\pi}{2}$ is needed. This constant is very badly normalized on both octal and Hex machines but $\frac{2}{\pi}$ is reasonably well normalized on both; at least on both the most significant bit is set. Thus in all routines which use a factor of $\frac{\pi}{2}$ we introduce this as division by $\frac{2}{\pi}$.

The role of such constants and the other machine dependent threshold constants is best illustrated by an example. We will look at the algorithm for tan (x). This was the algorithm used in reference {2} to illustrate the approach adopted by NAG, and the proposed Mark 6 code is included here as an appendix to illustrate the changes that have already been made since that paper appeared, some of them as a result of problems highlighted by Cody {5} and some reflecting simplifications possible due to improved portability processing available automatically at NAG Central Office {6}.

The algorithm is straightforward and well known {7}.

Let $x = N\frac{\pi}{2} + \theta$ $\qquad \theta \in \left[-\frac{\pi}{4}, \frac{\pi}{4} \right]$, N integral

$$\theta = x - \{\frac{2x}{\pi}\}\frac{\pi}{2} \text{ where } \{\frac{2x}{\pi}\} = \text{ the nearest integer to } \frac{2x}{\pi}$$

then $\tan(x) = \begin{cases} \tan\theta, & N \text{ even} \\ -1/\tan\theta, & N \text{ odd} \end{cases}$

$$\tan(\theta) = \theta\Sigma'a_r \, T_r(t), \quad t = 2(\frac{4\theta}{\pi})^2 - 1$$
$$\quad\quad\quad r=0$$

The expansion above is well known and was published to 20D by
Clenshaw {7}. For the purposes of the NAG library we regenerated it
to 30D and the expansion evaluation code was handled as I have already
described. This expansion can be evaluated with complete stability
in its converted simple polynomial form using Horner's nested multiplication
form, the necessary truncations and conversions all being done automatically.
The critical points in this algorithm affecting accuracy are the calculation
of θ and subsequently t. The range reduction is done using a constant for
$\frac{2}{\pi}$, TBPI, say to ensure good normalization on all machines as already described.

The evaluation of t requires the calculation of the expression

$$t = \frac{32}{\pi^2} \theta^2 - 1.$$

The obvious and most efficient way of dealing with this is to
include a further constant for $\frac{32}{\pi^2}$, TTBPIS, say and evaluate t as

$$T = TTBPIS*THETA*THETA-1.0$$

but again bad normalization on Hex and octal machines is a problem.
So for the Mark 6 version we use the form

$$T = 8.0*(TBPI*THETA)**2-1.0$$

thus retaining good normalization on all machines.

Also for this algorithm we have three other problems if we want
the routine to be robust. First if

$$\left| \frac{2\theta}{\pi} \right| \lesssim \sqrt{XMIN}$$

then underflow will be caused during calculation of T. However, if
$|\theta| \lesssim \sqrt{macheps}$ then to within the precision of the machine

$\tan(\theta) = \theta$. Therefore by including a constant XSQEPS $\doteq \sqrt{\text{macheps}}$ and a test for this condition, we not only prevent underflows but save on having to evaluate the expansion unnecessarily. Second we shall require a failure threshold for large arguments. For large x significance is necessarily lost due to the most significant digits of x determining only N, the number of multiples of $\frac{\pi}{2}$. This leads to a necessary loss of significance in θ and hence in $\tan(x)$. Eventually for large enough x no digits will be determined in θ and so the routine must fail. The threshold here is set to cause a failure exit when there is a danger of the result having no more than three significant figures. (In fact on the 1900 this threshold is set to avoid integer overflow in the calculation of N since this machine has a very small range of integers). Finally we need a second failure threshold to deal with the singularities in $\tan(x)$ that occur for x at an odd multiple of $\frac{\pi}{2}$. Here again the loss of significance in determining θ sets limits on the relative accuracy obtainable, and since $\tan(x) \sim \frac{-1}{\theta}$, also the absolute accuracy. Again we set this threshold to invoke a failure exit when there is a danger of less than three significant figures in the result.

These four precision dependent constants and the expansion generation code contain all the implementation dependent information required for this routine. However there are some functions for which there will be actual machine dependent code required. One such example was discussed by Cody {8}, namely Dawson's integral. For large x this function behaves like $\frac{1}{2x}$. On machines with a fractional normalization for the mantissa $\frac{1}{2x}$ is always representable when x is representable but on machines with an integer normalization, CDC and Burroughs, this is not the case.

For instance on CDC if x lies between roughly $10^{293} \rightarrow 10^{322}$ the calculation of $\frac{1}{2x}$ would cause underflow. Therefore implementations of Dawson's integral must include code to guard against such underflow when it can occur. Similar machine dependencies will occur for different functions on various machines and these must be dealt with by inclusion of the appropriate code and editing information to allow activation to be easily accomplished when needed.

This gives some idea of the system used to produce the functions for the NAG library. To give some idea of the degree of trans-portability this achieves, converting from the parent 1900 to the CDC implementation for tan(x) took about two minutes at a terminal to perform the necessary edits. To introduce a completely new implementation into the system took about ten minutes plus the running of a two second job to include the necessary expansion code. To introduce a new function into the system is a bit more difficult to estimate. The time involved varies with how difficult it is to design a suitable algorithm and generate the final necessary expansions. On average I would say about one to two man-weeks over five to ten weeks elapsed time is the order of effort involved.

3. TESTING MULTI-MACHINE SPECIAL FUNCTIONS

On the whole the testing of the routines presents as many portability problems as production. One of the requirements of the NAG library is that test programs form an essential part of the library and no implementation is considered complete until all the test programs also have been implemented and run successfully. This means that portability is a major design criterion here as well as for the routines.

Testing of the NAG software has two distinct levels, example testing and stringent testing. Example testing is designed to be mainly illustrative

of use. The example program for each routine, along with data and
results, are published as part of the routine's documentation. Such
programs are necessarily simple and work to low accuracy (4D) and are
hence essentially portable. As a first level of testing they are
useful in that gross bugs will show up but that is all. Stringent
testing is however where the real exercising of the routine takes
place. Stringent testing is to full accuracy on each machine and
it is meant to abuse as well as use the routine.

    For special function software it is possible to distinguish
two levels of testing within the stringent testing framework,
validation and certification. Validation testing might be considered
to be testing each pathway through the routine at least once and
verifying that reasonable results are produced. That is, the routine
is run with arguments that cause it to work through each of
its subranges and/or its standard range. It should be made to work
through each of its range reduction pathways. Failure conditions
should be checked to see that they are correctly detected. If possible
mathematical consistency should be checked. For instance $x = \sinh(\text{arcsinh}(x))$,
$\text{Cosh}^2 (x) = 1 + \text{Sinh}^2 (x)$, etc., are relations that could be used to verify the
consistency of the hyperbolic functions set in the library. Verification
of reasonable results frequently presents difficulties since standard tables
seldom give sufficient accuracy and hand checking gives only a negative
form of testing. In fact validation is really only a method of saying
the routine is not actually badly wrong. If the routine systematically
fails to meet the design accuracies by a wide margin this should be
detected but small random inaccuracies will escape or worse, unforeseen
failure conditions might well be missed.

Certification testing is designed to thoroughly exercise the
routine over the whole range of arguments for which it is designed
to operate, or at least over the major portion of this range, and to
provide a more or less precise measure of the accuracy actually attained.
Certification should provide testing that is sufficiently thorough that
few genuine bugs in design or implementation could remain undetected,
and it should provide good measure of the accuracy profile of the
routine on test. It should exercise the routine for a large sample
of arguments spread uniformly but not systematically over the necessary
range. For each argument it should produce a reasonably accurate (3D or
4D) measurement of the generated error due to the routine. It should do
this for each machine.

In providing certification there are three critical requirements,
each of which has attendant portability difficulties:

1.   Extra precision expected values of the function.

2.   The arguments which generate the expected values must
      be "exactly" the same as those used to exercise the
      routine on test.

3.   The calculation of the error, which involves a subtraction
      of actual and expected values of the function, must be done
      so as to produce at least 2 or 3 significant figures accuracy.

An obvious way of satisfying all three requirements is to make use
of extra-precision arithmetic. A priori routines working to self-controlled
accuracies and using extra-precision throughout could be written to generate
the expected values and the extra precision used for the crucial subtraction.
This solution has to be ruled out for two reasons. Firstly, a number of

implementations do not support an arithmetic of precision higher than that being used for the library. Secondly the problem of producing and validating extra-precision a priori routines on all machines is greater than the original testing problem we are trying to solve. High precision a priori routines are needed but these must be "off-line." One implementation can be produced and validated on one machine and used to produce argument and expected result pairs. The MLARITHA system has again been used for this purpose and a number of a priori arbitrary accuracy function routines have been produced drawing on ideas gained from conversations with Clenshaw and Olver and from Brent {9}. Requirements 2 and 3 are still problems, but 3 can be solved fairly easily by isolating the error calculation in a small subprogram which if necessary can simulate extra-precision for the subtraction.

The certification testing system which is to be used for Mark 6 onwards works as follows:

The validation testing that was the main stringent testing up to Mark 5 remains intact but verification against standard tables is replaced by certification testing. Data for the certification testing is produced by machine, using MLARITHA and the appropriate a priori routine. The a priori routines are previously validated independently and are not involved in any way with the production of the routine on test. This data in the main consists of argument expected-value pairs, the argument having 6 significant figures and the expected value at least 24. These pairs are read by a special subprogram which is the same for all programs and only has to be produced once for each implementation.

This reads the expected value in extra-precision and the argument at normal. The test program generates the actual value by calling the appropriate routine and the error is calculated by a second special subroutine. The result is printed in full, if required, and statistics of the error behaviour are accumulated. Histograms are also produced of the distribution of mean, RMS and maximum errors, and are plotted on the line printer. The scales on these histograms are related to the representation error on each machine which is also set in the data. Also if the error criterion is relative a histogram of the significance level obtained is plotted.

The following figures 1, 2, 3 and 4 show the histograms produced on the 1900 for the tan(x) routine, shown in the appendix. These were produced from 1000 arguments between -1.57 and +1.57, i.e., almost $-\frac{\pi}{2}$ to $\frac{\pi}{2}$. The expected loss of accuracy near the end points is clearly seen but away from these the errors are at worst five or six times the representation error. It is difficult to see how one could do better than this without performing all calculations in the routine in extra-precision. The sensitivity of this testing is demonstrated by figure 5 which shows the maximum errors obtained when an error of 1 in the eleventh figure is introduced into the constant $\frac{2}{\pi}$. Also for comparison the maximum error plot for the CDC implementation is shown in figure 6.

The trans-portability of the system can again be demonstrated by the fact that I converted the test program for the CDC in less than ten minutes. This is a little unfair since both 1900 and CDC FORTRAN implementations are single precision and so standard double precision is available for the extra-precision I/O and error calculation, but I should not think it would take more than an hour or so to produce equivalent subprograms using simulated extra-precision.

This technique still suffers from the fact that arguments will be slightly modified by the decimal to binary conversions on each machine. Limiting arguments to 6 figures should result in contamination that is strictly bounded by the representation errors. The labour and complexity of having to produce exact binary arguments and dealing with their I/O does not really seem justified by the improvement this would produce. Thus we are content to continue with some contamination with propagated error in our measured error. One reason for accepting this is that the routines are documented on the assumption that propagated error is always present. This is done in an effort to encourage users to be aware of the problem that floating point arithmetic is, in principle, inexact and that in all calculations the effects of propagating errors should be recognized and considered.

Given that all routines included in the library or updated at or after Mark 6 will be subjected to this form of certification testing, the reliability of the software should be considerably improved and we will be able to certify in documentation the precise level of significance that can be actually obtained when using the routines on any implementation.

4.   CONCLUSIONS

It is often said that "politics is the art of the possible." To paraphrase this cliche we might say that "producing portable software is the art of the practicable." The art is in keeping

not too many jumps behind the changes that manufacturers are inclined
to inflict on us without warning. Practicable rather than possible,
since in software almost anything is possible given enough man and
machine resources. However both machine resources and man power
are in strictly limited supply, particularly the latter.

I hope, therefore, that the picture I have painted of my view
of the practicable in this area is reasonably clear. An underlying
aim in all that I have done has been to produce something that works
now and to create a system which keeps the options open to allow future
improvements to be incorporated without major disruptions.

APPENDIX

The following is a reproduction of the FORTRAN source for the routine S07AAF. This is the Mark 6 version of the routine for tan(x) as described in the text.

```
      REAL FUNCTION S07AAF(X,IFAIL)
C     TAN(X)
C
      REAL X,ERR,FL1,FL2,TBPI,XSQEPS,SGN,THETA,T,Y
      INTEGER P01AAF
      DATA ERR/6HS07AAF/
C
C     PRECISION DEPENDENT CONSTANTS
C08   DATA FL1,FL2,TBPI,XSQEPS
C08   A/1.0E5,1.0E-5,6.3661977E-1,1.0E-4/
      DATA FL1,FL2,TBPI,XSQEPS
      A/1.0E7,1.0E-9,6.36619772368E-1,1.0E-6/
C14   DATA FL1,FL2,TBPI,XSQEPS
C14   A/1.0E11,1.0E-11,6.3661977236758E-1,1.0E-7/
C16   DATA FL1,FL2,TBPI,XSQEPS
C16   A/1.0D13,1.0D-13,6.366197723675813E-1,1.0E-8/
C18   DATA FL1,FL2,TBPI,XSQEPS
C18   A/1.0D15,1.0D-15,6.36619772367581343E-1,1.0E-9/
C
C     ERROR 1 TEST
      IF(ABS(X).GT.FL1) GO TO 10
C     RANGE REDUCTION
      SGN=1.0
      IF(X.LT.0.0) SGN=-SGN
      N=INT(X*TBPI+SGN*0.5)
      THETA=X-FLOAT(N)/TBPI
      N=MOD(IABS(N),2)
C     ERROR 2 TEST
      IF(N.EQ.1.AND.ABS(THETA).LT.ABS(X)*FL2) GO TO 20
C     SMALL THETA TEST
      Y=1.0
      IF(ABS(THETA).LT.XSQEPS) GO TO 5
C     EXPANSION ARGUMENT
      T=8.0*(TBPI*THETA)**2 -1.0
C
C     EXPANSION (0007) EVALUATED AS Y(T)  --PRECISION 08E
C08   Y=  +1.1173014E+0+T*(  +1.3386247E-1+T*(  +1.8924537E-2+
C08   AT*(  +2.7010438E-3+T*(  +3.8561954E-4+T*(  +5.5077467E-5+
C08   BT*(  +8.2048199E-6+T*(  +1.1781590E-6)))))))
C
C     EXPANSION (0007) EVALUATED AS Y(T)  --PRECISION 12E
      Y=  +1.11730141051E+0+T*(  +1.33862473672E-1+
      AT*(  +1.89244955238E-2+T*(  +2.70103244160E-3+
      BT*(  +3.85828287625E-4+T*(  +5.51184605550E-5+
      CT*(  +7.87409839245E-6+T*(  +1.12350071991E-6+
      DT*(  +1.60455379111E-7+T*(  +2.42925841226E-8+
      ET*(  +3.48825813626E-9))))))))))
C
```

```
C       EXPANSION (0007) EVALUATED AS Y(T)  --PRECISION 14E
C14   Y=  +1.1173014105142E+0+T*(  +1.3386247366625E-1+
C14   AT*(  +1.8924495521240E-2+T*(  +2.7010325492100E-3+
C14   BT*(  +3.8582831712490E-4+T*(  +5.5117857920090E-5+
C14   CT*(  +7.8739725241234E-6+T*(  +1.1248781710765E-6+
C14   DT*(  +1.6069812517338E-7+T*(  +2.2915132958756E-8+
C14   ET*(  +3.2724838584129E-9+T*(  +5.0089133229408E-10+
C14   FT*(  +7.1924759283372E-11)))))))))))))
C
CINC1 EXPANSION (0007) EVALUATED AS Y(T)  --PRECISION 16E
CINC1 EXPANSION (0007) EVALUATED AS Y(T)  --PRECISION 18E
    5 S07AAF=THETA*Y
      IFAIL=0
      IF(N.EQ.1) S07AAF=-1.0/S07AAF
      RETURN
C     ERROR 1 EXIT
   10 S07AAF=0.0
      IFAIL=P01AAF(IFAIL,1,ERR)
      RETURN
C     ERROR 2 EXIT
   20 IF(THETA.EQ.0.0) THETA=-1.0
      S07AAF=-SIGN(1.0/(X*FL2),THETA)
      IFAIL=P01AAF(IFAIL,2,ERR)
      RETURN
      END
```

A few points to note are:

1. No account is now taken of double-precision explicitly; all code is single precision. Conversion of constants and intrinsic functions to D form is to be done automatically by NAG central office.

2. The code intended for the byte machines is included as 16 significant figure precision, since most compilers in use on these machines perform their decimal → Hex conversions by truncating to 16 decimal figures first. Therefore better results will be obtained by rounding constants to 16 digits than including the potentially more accurate 17 digits. In fact some compilers actually flag 17 digits as an error. A similar situation occurs for CDC machines which could potentially use 15 digits but in fact will only convert 14.

3. The trigger records for 16 and 18 digits are included but the code for this precision has not yet been inserted. They show the method of introducing a new precision.

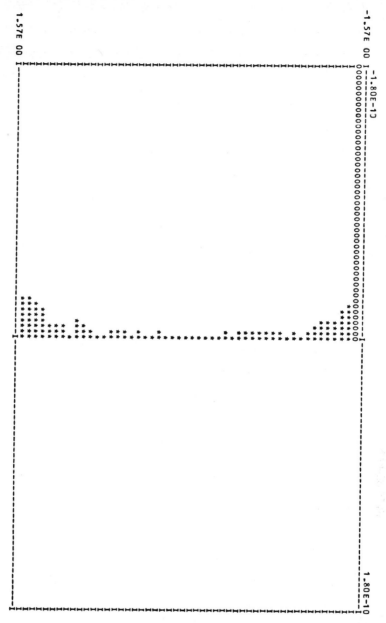

HISTOGRAM OF MEAN ERROR PER BIN

1* = (0.5 - 1.0) REP.ERR.

MEAN ERROR FOR S07AAF ON IGL 1906A

FIGURE 1

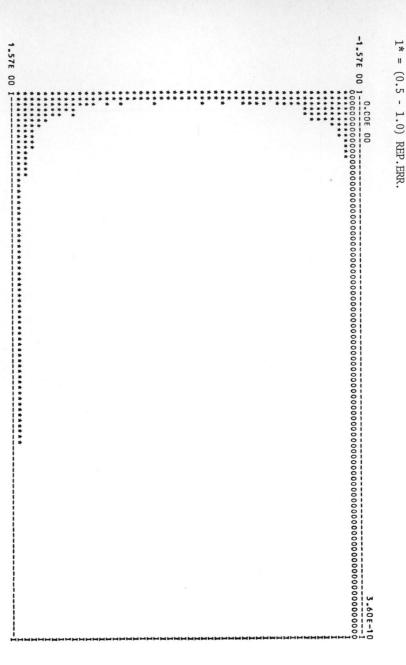

HISTOGRAM OF RMS ERROR PER BIN

1* = (0.5 - 1.0) REP.ERR.

FIGURE 2

ROOT-MEAN-SQUARE ERROR FOR S07AAF ON ICL 1906A

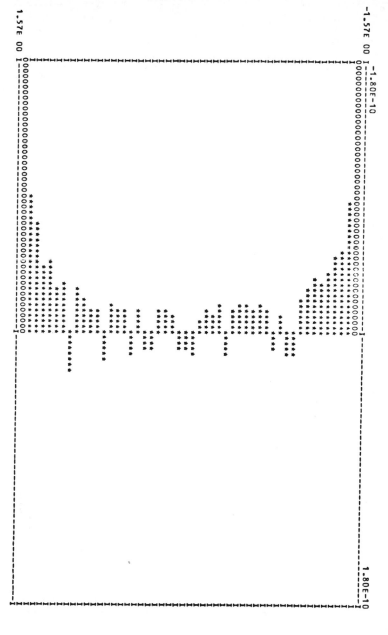

HISTOGRAM OF MAX ERROR PER BIN

1* = (0.5 - 1.0) REP.ERR.

FIGURE 3

MAXIMUM ERROR FOR S07AAF ON ICL 1906A

HISTOGRAM OF MINIMUM SIGNIFICANCE
HIGHEST ATTAINABLE SIGNIFICANCE = 1.14E 01
25* = 1 SIGNIFICANT FIGURE

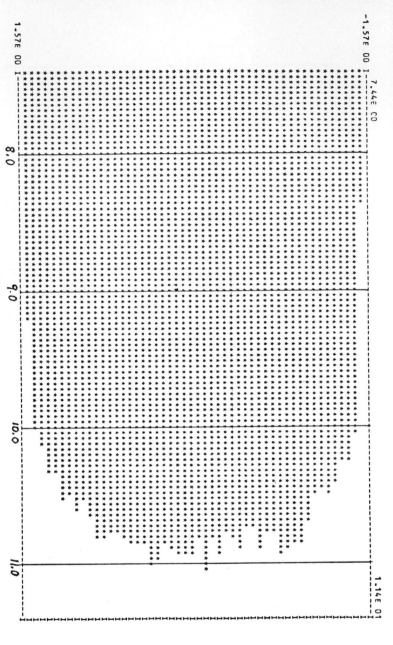

FIGURE 4

Plot of - Log$_{10}$ (Max relative error) for S07AAF as run on the 1906A

Extra scale lines show the number of significant figures achieved

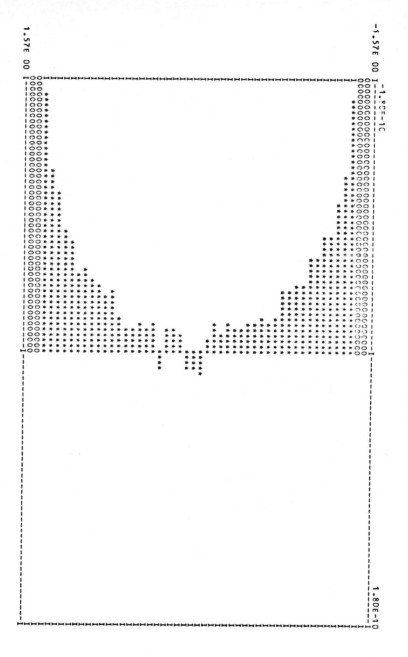

HISTOGRAM OF MAX ERROR PER BIN
1* = (0.5 - 1.0) REP.ERR.

FIGURE 5

Maximum Errors per bin for S07AAF with an error of 1 in eleventh figure of $\frac{2}{\pi}$ constant, c.f. Figure 3.

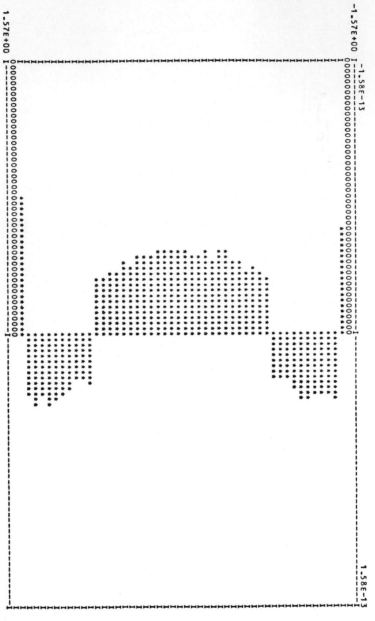

HISTOGRAM OF MAX ERROR PER BIN
1* = (0.5 - 1.0) REP.ERR.

MAXIMUM ERROR FOR S07AAF on CDC 7600

FIGURE 6

REFERENCES AND FOOTNOTES

1. J. L. Schonfelder, "Special Functions in the NAG Library,"
    Software for Numerical Mathematics, D. J. Evans (Ed.),
    Academic Press (1974).

    J. L. Schonfelder, "The NAG Library and its Special Functions
    Chapter," International Computing Symposium 1973, A. Gunther
    et al. (Eds.) North Holland (1974).

2. J. L. Schonfelder, "The Production of Special Function Routines
    for a Multi-Machine Library," Software-Practice and Experience
    6, 71-82 (1976).

3. J. L. Schonfelder and J. T. Thomason, "Applications Support by
    Direct Language Extension - An Arbitrary Precision Arithmetic
    Facility in Algol 68," International Computing Symposium (1975),
    Gelenbe and Potier (Eds.) North Holland (1975).

4. C. W. Clenshaw, "Rational Approximations for Special Functions,"
    Software for Numerical Mathematics, D. J. Evans (Ed.), Academic
    Press (1974).

5. W. J. Cody, "An Overview of Software Development for Special
    Functions," Numerical Analysis Dundee (1975), Dold and Eckmann
    (Eds.), Lecture Notes in Mathematics, Springer-Verlag (1976).

6. Duplicated Code to deal with single-double precision conversion
    is no longer required as this switch is performed by automatic
    processing in the NAG central office.

7. C. W. Clenshaw, Mathematical Tables, Vol. 5, National Physical
    Laboratory, HMSO London (1962).

8. W. J. Cody, "The FUNPACK Package of Special Function Subroutines,"
    ACM Trans. Maths. Software, 1, 13-25 (1975).

9. R. P. Brent, "A Fortran Multiple-Precision Arithmetic Package,"
    (Private Communications).

PORTABLE SPECIAL FUNCTION ROUTINES

L. Wayne Fullerton

1.   Introduction and Motivation

1.1 Introduction

"Portable" has become a catchword that conveniently means whatever an author desires it to mean. A portable routine can be one that is just slightly more difficult to rewrite than to modify and implement in a new environment. Or it may be a routine that is required to work correctly without modification on any computer in any country. I intend a portable special function to work without modification on any major computer in the United States and Canada. My definition therefore coincides with the usage of "portable" by Fox {7}.

One might regard the construction of portable special function routines to be a much easier task than writing, say, portable differential equation solvers, because, after all, special functions are well under-stood and the routines are generally small. But differential equation solvers are relatively high-level routines far removed from the machine; portability is achieved in part simply by requiring (portable or non-portable) routines such as SQRT and EXP to be available. Special function routines, on the other hand, are low level and close to the machine; it is accordingly more difficult to produce more-or-less machine independent special function routines than to write portable differential equation solvers. In fact, it is not at all obvious that portable special function routines can be written, because the FORTRAN standard allows considerable flexibility in the implementation of so-called standard features.

1.2 Objectives in Software Production

A goal in producing a routine of any kind, portable or not, should be to minimize the effort required to write, maintain and use the routine. The total effort or cost is a complicated non-linear function of many parameters, some of which may not yet be identified and some of which are time dependent. For example, the cost of software relative to hardware has dramatically increased in the past decade. Therefore we no longer view an execution speed improvement of ten percent with the same awe we once did, because we suspect the attendant software cost exceeded the benefit likely to have accrued from the ten percent "improvement." But there is another problem. Even among the identified parameters that are time independent, the cost is difficult to measure, and so no consensus on these parameters' importance is likely to be obtained. For example, how do we measure the cost of maintaining an unreadable code when we cannot agree what "readable code" is? (Is FORTRAN even readable?) Consider another example of a difficult-to-measure parameter: What is the cost of referencing $\tan^{-1}x$ by ATAN(x), but $\cos^{-1}x$ by ARCOS(x)? How many users have mistakenly referenced ARTAN and ACOS, and how much time have they devoted to looking up the names in their manuals just to be sure? The problem of minimizing total effort undoubtedly does not have a unique solution. If it does, we probably will not agree precisely what that solution is.

The solution I adopt is a moderate one. It seems superfluous to argue that no single feature of a code should be weighed highly to the exclusion of all others. But we have all heard immodest claims such as "My code is best because it is fastest," or maybe "My code is best because it uses the least storage," or perhaps it is most accurate, or most portable, and so on. We all surely know there is not just one important consideration in designing and evaluating a code. Unfortunately, a claim to greatness seems to be lacking when no single feature of a program or library is best. When compromises must be made, they also must be justified. Moreover, the impossibility of measuring precisely the value of the tradeoffs inevitably leads to arguments about the quality of the program that the author wanted to be best. I subject myself to such arguments in Section 2.

1.3 Motivation

A fundamental motivation for producing a portable function library at Los Alamos is that we have roughly three kinds of computers, three operating systems, and three separate libraries. How much does it cost us to maintain three separate libraries? I am sure the cost is non-trivial. Much more significant, I believe, is the cost to our users of switching among these various libraries with different characteristics. The availability of a single high quality portable library would substantially reduce the cost of maintenance and would reduce the effort required of our programmers to use the routines.

The motivation for producing the library at this time is that we are testing a new super-computer that has little software. We have asked whether we should build yet a fourth library. We have decided to consolidate now while the expense is relatively small. In so doing, we eliminate a portion of our dependence on the computers we now own, because the same routines can work on any brand of super-computer as well as on sufficiently standard mini-computers.

Another reason for building a single special function library is that these portable routines can be made more reliable than their non-portable cousins. If the routines work on more than one machine, we can afford to spend some extra time making them especially reliable. Testing a routine on one machine gives us confidence that the same routine will also work correctly on other computers. If a routine has been used for a year on one computer, we have increased confidence in its reliability. Again, this helps to reduce the effort of users when the routine is moved to a new machine. If they encounter an error, they are not led to suspect a brand new library routine. Since they are using an old and trusted library routine, they are properly encouraged to look at their own program for the error.

Minimizing the total effort associated with any routine is still another important reason for producing portable routines. I confess to laziness of a special type. If I must write a Gamma function routine this year, I do not also want to guarantee having to write one next year when a new computer arrives. Nor do I want to guarantee having to tinker with a preprocessor.

2. General Design Considerations

Another phrase for "design considerations" might be "value judgements" in view of the difficulty of measuring costs associated with various design parameters. Nonetheless, I shall outline in this section a number of considerations that entered in the design of a portable special function

library. I will call upon limited statistics when they are available,
but too often I will merely appeal to your intuition and hope your
experience has not been too different from mine.

2.1 Language

The only language widely implemented and used by the scientific
community is FORTRAN. Although there are many dialects, a sufficiently
small subset of FORTRAN is portable. It is not obvious that the portable
subset of FORTRAN is large enough to permit special function routines to
be written; I maintain it is, although we shall consider a few problem
areas later. The subset of FORTRAN that is portable is called PFORT
by Ryder {13}. Bell Laboratories has shown that it is possible
to move large codes without change to many different environments.
Just as important, Bell Laboratories has made available several tools
to assist the development of portable routines. One of these tools is
the PFORT verifier {13}. Another is a package of utilities for
portably handling error conditions and obtaining machine constants
{7, 3, 8}. The widespread use of FORTRAN in the scientific community,
the evidence of the portability of properly written FORTRAN programs and
finally the availability of tools that define "properly written" all
suggest FORTRAN as the only language currently suitable for portable
routines.

2.2 Library Lifetime

When grossly incorrect routines can survive ten years, it is easy
to imagine that reliable portable routines might survive forever. Such
an outlook, however, ignores the evolution of computer hardware and
computer science. It is likely that we shall all eventually have most

elementary functions available as hardware instructions. Many of my routines will then become obsolete. Before that happens, we may find machine constants to be defined as compiler parameters. My routines would then be easier to write, and certainly they would be more readable. When Waite's vision {17} of a portable compiler becomes reality, my surviving routines can be coded in whatever language the compiler implements, and the routines will be portable. I will be disappointed in the progress of computer science if my routines survive a decade.

2.3 User Community

As noted earlier, the portable routines are intended to work on all major computers in the United States and Canada. To design routines to work on any larger subset of computers may make it impossible to write portable routines, and in any event requires much more information than I have. Furthermore, routines of foreign origin usually require a local "priest" to assist the user community, and the routines must compete with versions written by local experts. Someday someone may write programs for the world but not this year or next.

2.4 Storage

Inadequate computer memory no longer is the primary concern of programmers that it once was. A memory size of 50,000 words is now considered small. Also, virtual storage and paging computers are common. If a programmer uses 50 words of storage so that sensible error messages may be printed rather than cryptic octal error numbers, we applaud his efficient usage of storage. And if another 50 words must be sacrificed to make the routine portable, few programmers are upset. Of course, programmers still write segmented programs because the programs would otherwise require too much storage, and these programmers always object when a larger routine is substituted for a smaller one. But there are too many other important competing considerations to allow us to weigh these programmers' objections infinitely highly.

2.5 Speed

Execution speed, like storage, has been a traditionally important design consideration. I should like to argue that differences of 50% are relatively unimportant. A few users, of course, will have very limited computing budgets, and they may argue that the Airy function should be especially fast so that their whole program might be speeded up. But all design considerations are subject to such special pleading. We, the builders of program packages, must sit in judgement of all such arguments.

If we wish to optimize, then we should globally optimize. The programs that require the biggest share of the computing budget should be singled out for optimization; subprograms that consume only a small fraction of the budget are not worth optimizing even if they are inefficient. Unfortunately, statistics that enable us to evaluate the importance of the execution speed of special functions are scarce.

Bailey and Jones {1} have monitored the usage of mathematical library routines at Sandia Laboratories for an eight-month period. Their usage statistics do not include the elementary special functions (such as EXP , ALOG, SQRT,...) but this is appropriate: Most installations will already have fast and accurate versions of these routines and portable versions are not likely to be used. Bailey and Jones' statistics were based on the number of times each routine was loaded rather than on the percentage of execution times the routine consumed. Nonetheless, the statistics are useful. They found the three most heavily used mathematical areas to be numerical quadrature, zeroes of polynomials and Fourier transforms. Bessel functions were fourth in rank. Although each loading of a Bessel function accounted for many executions, we still would expect numerical quadrature and Fourier transform routines to consume more time. All hyperbolic functions combined ranked only tenth in usage, well behind typically time-consuming routines like ordinary differential equation solvers. Even the relatively elementary Gamma functions ranked only eighteenth in usage. As further evidence of the infrequent usage of special functions, only twelve percent of all jobs loaded a mathematical library routine of any kind. One can argue that these statistics are typical for most computing facilities, but I would expect the higher mathematical functions to be more heavily used at a scientific research laboratory than, say, at a university.

Additional evidence bearing on the relative unimportance of the execution speed of special functions was obtained incidentally in connection with an optimization project carried out at Los Alamos Scientific Laboratory {12}. Of course, modifying algorithms can yield the greatest gains, but it is difficult enough even to understand 20 to 50 thousand statement programs, let alone obscure algorithms! The standard optimization procedure, therefore, consisted of monitoring a program during several typical runs and printing execution time statistics at the end of each run. These statistics included the percentage of the total time consumed by each subprogram. Typically, a few small low level subroutines would be identified as the real time consumers, and they would be recoded. Even though individual programs were monitored, the statistics do provide reasonably accurate estimates of the laboratory-wide usage of special functions, because fewer than a

dozen programs consume about seventy five percent of the computing time.
Members of the optimization project estimate that all elementary functions
consume less than ten percent of the computing time; I think seven or
eight percent is a good estimate. If the three lowest level special
functions EXP, ALOG, and SQRT are excluded, then the percentage drops
below five percent. Thus, a difference of 50% in the execution time
of a Bessel function becomes quite insignificant.

These arguments should not be construed as arguments against the
production of hardware special functions. There are a number of good
reasons for producing hardware functions, one of which is greatly
increased execution speed. A factor of 100 improvement in speed would
lead to many new applications that are now impractical. We have all
witnessed the dramatically increased utilization of Fourier transform
techniques that resulted from the fast Fourier transform. Similar
effects can be expected from hardware special functions that are hundreds
of times faster than current software. Factors of 1.5, however, will
not lead to new algorithms. It is not worth a day of programmer time
to save a few minutes of computer time each year.

2.6 Accuracy

Arguments that routines should be accurate to the last bit have
considerable merit. But arguments that routines must be accurate to
the last bit fail to take into account how scientists use computers.
Generally they use single precision until they observe a suspicious
result, then they switch to double precision. Scientists do not
compute with only one or two guard bits; they often have at least two
or three guard decimal digits. Indeed, on some machines they may work

with ten guard digits. A number of users at Los Alamos use fast
half-precision routines, because they decided it was important for
the six-digit routines to be three times faster than the nearly
full-accuracy versions. While we may debate the wisdom of their
decision, we cannot dispute their need for only six digits of accuracy.
When they are lucky to measure temperatures and pressures to three
significant figures, six digit accuracy is double precision.

IBM provoked the wrath of many numerical analysts when they
introduced the 360 series of computers. There are two problems.
The first is the small word size. The other problem is that normalized
hexadecimal numbers can have as many as three leading binary zeroes; thus,
the hardware limits the maximum precision to a variable number of bits,
21 to 24 bits. Even if an answer contains 24 bits it may be accurate
only to 21 bits, because an intermediate result may have contained
three leading binary zeroes. When IBM designed the 360, they decided
that the low precision would be suitable for some applications and,
moreover, that the leading insignificant zeroes would produce no serious
loss of accuracy. Who am I to argue? Why should I insure that some
values of the Fresnel integral be accurate to 23 bits when other values
cannot be more accurate than 21 bits? (An unsatisfactory answer is
that round-off errors propagate somewhat less rapidly.) If a user
really needs results accurate to 23 bits, then he should use double
precision at a cost in speed of 20%. Indeed, many numerical analysts
advocate using double precision all the time on IBM machines.

It should be considered an indictment of current languages and
hardware that readability, portability, speed and storage all suffer
greatly whenever highly accurate routines are written.

The discussion above should not be regarded as a license to ignore all that has been done in the past decade to produce accurate routines. Many recent authors of portable and non-portable libraries have successfully written highly accurate routines. Cody {5}, in particular, has demonstrated that very accurate routines can be written in FORTRAN. We would be remiss if we ignored some of the machine independent techniques for insuring accuracy, for they do not adversely affect the readability or portability of the routines.

2.7 Portability

This section has so far been a defense of compromise, and the reader might conclude that the defense will rest here, because "portable" is in the title of this paper. Although I certainly weigh portability more highly than other authors of function libraries, there must still be room for compromise. The fact that I merely intend the routines to work on major computing systems in the United States and Canada allows considerable room for compromise. I do not include one-of-a-kind operating systems unless the designers addressed the problem of portability. Likewise, I do not include the myriad of minicomputers whose manufacturers have not properly considered the problem of portability. Finally, the limitation to systems in the United States and Canada frees me from worrying about systems to which I have no access, about which I have no knowledge and on which my routines will never be run.

Even with these constraints, we should still be willing to compromise. A routine once was considered inferior if it was not faster than the speed of light. Later, a routine was considered inferior if it was not accurate to the very last bit. We now should be careful to prevent a similar unreasonable attitude developing toward portability. Consider Cody's famous example,

$$\text{IF (B .NE. 0.0) C = A/B}$$

On CDC hardware the expression can result in a zero divide, because the logical expression tests for a true zero while the divide unit checks only for a zero biased exponent. Of course, this problem could be corrected in the compiler, but it is not. If my routines fail on some machines, because the above expression is improperly executed, some users may choose to remove the portable label from my routines. I suggest, though, that they write the manufacturer. Only when we demonstrate an unwillingness to accommodate each computer's weaknesses will manufacturers and compiler writers understand how expensive the defects really are. We must emphasize the similarities of computers, not their differences.

2.8 Conclusion

The problem of determining appropriate weights for design consider-ations begs for a solution. Studies such as Bailey and Jones' extended so that statistics gathering is designed into the operating system can provide part of a basis for rational decisions. But we also need other elusive statistics such as the average (manpower and computing) cost and benefit of providing, for example, an assembly language version of the Gamma function. Until such statistics are available, originators of libraries must be content with arguing about value judgements.

3.   Specific Design Considerations

3.1   Background

In the previous section we were concerned with general issues. This
section is devoted to specific design choices. Fortunately, the design
work need not be done in a vacuum, because several function libraries
have been developed to work on more than one machine. Schonfelder {14,
15, 16} has constructed a transportable library of special functions as
a part of the NAG library. Also, Cody {5} has built a set of libraries
that work on the major North American computers. Cody, Schonfelder and
I have all reached different moderate solutions to the problem of minimizing the total effort, because we have used different weights for each
of the design considerations. Cody's routines are characterized by their
accuracy and robustness. Schonfelder weighed speed and transportability
highly. I have chosen to weigh portability and my personal time two
years from now highly. Relative to Cody, I am willing to sacrifice some
accuracy and robustness. Relative to Schonfelder, I sacrifice speed
and storage.

Of the two recently built libraries, my library, FNLIB, more
closely resemble Schonfelder's. But FNLIB most closely resembles
a much earlier ALGOL library written by Clenshaw, Miller and Woodger
{4}. They chose to use Chebyshev series expansions of auxiliary
functions that were evaluated by a low level routine. But their
routines were not strictly portable, because several machine dependent
constants had to be set before they were used. These constants included
the number of expansion terms and legal argument bounds. Although a
procedure was provided for calculating the number of terms required
for a specified accuracy, the constants in each routine were modified

by hand. Schonfelder adopted the basic algorithm of Clenshaw, et al.
and added a preprocessor so that correct machine dependent constants
could be selected automatically. Later Schonfelder abandoned the use
of Chebyshev series because of the execution time overhead associated
with the Chebyshev series evaluation.

There are numerous advantages to using Chebyshev series. In the
words of Clenshaw, Miller and Woodger: "The method of Chebyshev series
expansion ... is economical, in the sense that the number of data required
to specify each function is small; the error of an approximation is easily
assessed; and it is widely applicable (any continuous function of bounded
variation has a convergent Chebyshev series expansion)." To these
advantages we may add that Chebyshev series are generally more stable than
conventional polynomials and, more important, they provide variable accuracy
portable approximations.

There is one potentially serious objection to using Chebyshev series:
They are only polynomial approximations that are nearly best in the sense
of absolute error. Even though the exponential integral is continuous
and bounded in the interval 10^{-10} to 1, no one would seriously consider
using a polynomial approximation to the exponential integral itself due
to the logarithmic singularity at 0. Moreover, one usually wants function
values that are nearly best in the sense of relative error; and if an
auxiliary function like $\frac{\sin x}{x}$ is being approximated, the optimal
weight is not close to unity (absolute error). Thus, one might object
that Chebyshev expansions are insufficiently general and that true
minimax approximations with proper weight and basis functions should
be used. This objection is primarily an academic one for the well-
understood elementary special functions, because they are easily

manipulated so that Chebyshev series approximations contain only a few terms more than the true minimax approximations. However, it will be shown in the next section that Chebyshev series are just a special case of a much more general expansion valid for almost arbitrary weight and basis functions, and these more general expansions may be used if Chebyshev expansions are unsuitable.

With Chebyshev expansions adopted as the approximation form for the elementary functions, the only remaining important choice concerns the method of including machine dependent constants in the routines. The biggest advantage of using a preprocessor is that it may be modified so that the peculiarities of any compiler in the world may be accommodated. But a preprocessor guarantees that work must be done later when a new machine arrives at some installation. It also guarantees the necessity of maintaining an extra program. Even if the preprocessor is simple to use, the chore of assembling the old input, testing the new routines, and disseminating the routines is likely to consume at least one week. Is it not far better <u>almost</u> to insure that no extra work will have to be done when a new machine arrives? If a few basic machine constants such as the largest and smallest representable numbers and the machine epsilon are localized in a low level routine, then the legal argument bounds may be calculated. Even if the bounds are non-linear functions of the basic constants, we still can easily find very good limits portably, and Section 5 contains several examples.

Fourteen years after the work of Clenshaw, Miller and Woodger, my contribution to the field of elementary special function software has been to write routines in FORTRAN rather than ALGOL and to obtain machine constants from low level utilities.

3.2 A Template for Portable Special Function Routines

Consider now in outline form a template for a portable special function of one variable. Naturally, not every routine will follow this precise prescription, but the well-defined template provides a convenient means of introducing some of the more detailed design choices.

A. Store Coefficients of the Approximation in DATA Statements

Separate single and double precision routines for each function are written. In order to permit function accuracies of 15 significant figures (50 bits), approximations accurate to 16 significant digits are stored. Double precision approximations are accurate to 31 significant figures.

From the standpoint of portability this step is the weakest. In order to write a single routine for all machines, it is necessary to store the most accurate approximation that ever will be needed. Unfortunately, the standard IBM compiler issues an error message if it encounters a number containing many significant figures. If necessary, a very simple preprocessor can be written in FORTRAN to truncate numbers that are deemed too long by the compiler. Although the processed portable routines will not themselves be portable, at least they will work with the diseased compiler. If there were many such exceptions to the portability of portable routines, one could well question the wisdom of trying to write portable rather than transportable routines, but to the best of my knowledge this exception is the only one.

B.   Initialize

Initialization is carried out once each time the routine is loaded. Even for overlay jobs the overhead of initialization is trivial. Although the use of a first time switch is not a part of ANS FORTRAN, it is known to be portable on all major North American computers including the stack-oriented Burroughs machine. Because first time switches are frequently used by programmers, it is unlikely that a compiler or computer would be tolerated if it did not allow the use of a first time switch. However, if the first time switch failed, the portable routines would still work, although they would initialize each time they were called. Initialization consists of two steps.

(1)   Calculate Legal Argument Bounds

Given several basic machine constants, it is possible to compute the legal argument bounds as illustrated in Section 5. These bounds must be conveniently available to the user in case he wishes to intercept illegal arguments. If the bound can be expressed simply in one line, then the user can easily incorporate the expression in his program. If, on the other hand, the argument bound is a non-linear function of the basic machine constants, then the bound should be calculated by a low level subroutine so that the user may obtain the bound. The calculated legal argument bounds will ordinarily contain over 99.9% of the range permitted by the hardware.

(2)   Determine Correct Number of Expansion Coefficients

This step could be omitted, because the routines
could be used on all computers with the same large
number of coefficients. But CDC routines usually
require about twice as many coefficients as IBM
routines. If the same large number of coefficients
were used on all machines, the IBM routines would
be 30% to 40% slower than necessary. Although this
difference is small, the proper number of coefficients
is easy to find and the calculation can be isolated in
a common low level routine. Perhaps a more important
reason for this optimization, though, is that users
will be discouraged from tampering with the routines
"to eliminate all that extra calculation."

C.   Check Input Argument for Validity

After argument bounds have been calculated, any input
argument error conditions can be detected. Two types of
error conditions must be distinguished. The first type is
fatal and no recovery is permitted. Fatal errors are gross
argument errors such as a non-positive argument passed to the
logarithm function or a large argument like 10^9 passed to the
exponential function. The second type of error condition is
fatal by default, but recovery is permitted. These errors
characteristically arise when the accuracy of the computed
function value is low, but possibly usable. For example,
an attempt to compute the cosine of 10^{10} on a CDC computer

would produce a result accurate only to four decimal places. Likewise, an attempt to compute the Gamma function of a number very close to a negative integer could yield an answer with no significance. Unfortunately, in such situations many routines give no warning or a warning only when the accuracy is less than two digits, about one-seventh precision on CDC machines. Routines in FNLIB issue warnings whenever the accuracy of a result is less than half the alleged machine precision. Programmers who know what they are doing may easily make these warnings recoverable errors. Programmers who do not know what they are doing will be forcibly warned. Inaccurate answers must not be returned without warning, and half precision seems a reasonable boundary between accurate and inaccurate.

All error conditions are handled portably and conveniently by Bell Laboratories utility {8}. Their error handler permits both fatal and non-fatal error conditions to be issued, and it allows sensible error messages to be printed rather than just a cryptic error number.

D. Reduce Argument Interval

Argument reduction for portable routines is no different than that for non-portable routines. The purpose of argument reduction is to obtain a sufficiently small interval for an efficient polynomial approximation to be used. For the elementary functions, special well-known results are generally used. For example, for trigonometric functions the argument is reduced to the interval 0 to $\pi/4$, then trigonometric identities are used to obtain the function value for the unreduced argument. For higher level special functions, semi-infinite intervals typically are reduced to just two basic intervals. For example, the sum of the exponential integral $E_1(x)$ and ln (x) behaves very much like a polynomial in the interval 0 to 1. For large arguments $x \cdot e^x \cdot E_1(x)$ behaves very much like a polynomial in the variable $1/x$.

E. Evaluate Auxiliary Chebyshev or Orthogonal Series

The approximations are evaluated by a simple low level routine with Clenshaw's recurrence. An auxiliary function such as $x \cdot e^x \cdot E_1(x)$ normally is approximated by a polynomial in an auxiliary independent variable closely related to the original variable. This auxiliary variable is linearly transformed to lie in the interval -1 to 1 so that expansions in unshifted Chebyshev (or other orthogonal) polynomials may always be used. For example, the auxiliary variable for $x \cdot e^x \cdot E_1(x)$ is $y = 1/x$, but for $1 \leq x \leq \infty$, $0 \leq y \leq 1$. A straightforward linear transformation yields a new independent variable in the required interval -1 to 1. Fortunately, these transformations are more easily programmed than explained.

F. Modify Auxiliary Series Value

The series value must be modified in order to correct it for form or interval. If an auxiliary function such as $x \cdot e^x \cdot E_1(x)$ has been evaluated, then it is necessary to multiply by e^{-x}/x in order to obtain the desired function value. And if Cosine x has been evaluated for an argument reduced to the interval 0 to $\pi/4$, then the auxiliary value must be modified to obtain the value for the required argument.

3.3 Summary

A recipe for the construction of portable special functions has been presented. Even though we cannot absolutely guarantee that no work must be done later for a new computer, we can almost provide a guarantee. This near guarantee is partially based on several years experience by a few groups that produce portable programs. Also, the increased emphasis on portability as evidenced, for example, by this conference will promote resistance to features in new compilers that might exclude the possibility of producing portable routines. Programmers who want to write portable programs need not be as paranoid as they once were.

Although the problem of execution speed has already been addressed in some detail, it should be noted that the modular algorithm proposed in this section is easily optimized. Much of the execution time required for each routine is consumed by a single low level routine, namely the Chebyshev series evaluator. If this routine is coded efficiently in assembly language, it may be possible to decrease the execution time of all routines. Of course, a hardware Chebyshev series evaluator could greatly improve the performance of all the routines. Even when some special functions are entirely hardwired, there still will be a number of routines that would benefit substantially from the availability of a hardware Chebyshev or orthogonal series evaluator.

4. Generalized Chebyshev Expansions

We show in this section that Chebyshev expansions are just a special case of a much more general expansion. Although mathematical details will appear elsewhere, the generalizations are easily described.

4.1   Chebyshev Series

Consider first as review the problem of constructing nearly
best absolute error polynomial approximations. We wish to find an
n^{th} order polynomial approximation $A_n(x)$ to the given function $F(x)$
such that the error $E(x) = F(x) - A_n(x)$ is nearly best in the sense
of Chebyshev, that is, the L_∞ norm. The interval of approximation
is taken to be $-1 \leq x \leq 1$ without loss of generality, because linear
transformation converts any other finite interval to $[-1,1]$. The
approximation $A_n(x)$, we know, is a truncated Chebyshev series, and we
consider the construction of this series in detail as a foundation for
generalization.

We first expand $F(x)$ in a series of orthogonal polynomials $T_i(x)$
such that

$$F(x) = f_0 T_0(x) + f_1 T_1(x) + \cdots.$$

The polynomials $T_i(x)$ obey the orthogonality condition

$$\int_{-1}^{+1} \frac{T_i(x) \, T_j(x)}{\sqrt{1-x^2}} \, \delta x = 0, \quad i \neq j,$$

and the monic orthogonal polynomials are easily constructed from this
defining equation (cf. Gautschi, {10}). Each polynomial is most conveniently normalized by taking its extreme absolute value to be unity.
Next we may find the coefficients f_i in the expansion from

$$f_i = \frac{\displaystyle\int_{-1}^{1} \frac{F(x)\, T_i(x)}{\sqrt{1-x^2}}\, \delta x}{\displaystyle\int_{-1}^{1} \frac{T_i^{\,2}(x)}{\sqrt{1-x^2}}\, \delta x} \; .$$

An approximation $A_n(x)$ is then obtained simply by truncating the expansion for $F(x)$ at n^{th} order. The error

$$E(x) = f_{n+1} T_{n+1}(x) + f_{n+2} T_{n+2}(x) + \cdots$$

is very nearly minimax if $|f_{n+1}| \gg |f_{n+2}|$,

because T_{n+1} is an equal ripple polynomial. An upper bound for the error is readily obtained from

$$|E| \leq |f_{n+1}| + |f_{n+2}| + \cdots$$

because $|T_i(x)| \leq 1$ for $-1 \leq x \leq 1$.

4.2 Chebyshev Series Generalized for Weighted Error

Now consider the generalization of Chebyshev expansions to (almost) arbitrary weight functions. We wish to find an nth order polynomial approximation $A_n(x)$ to the given function $F(x)$ with weight $W(x)$ such that the weighted error

$$E(x) = W(x) \cdot (F(x) - A_n(x))$$

is nearly best in the sense of Chebyshev. Again we expand $F(x)$ in a series of orthogonal polynomials $\phi_i(x)$ such that

$$F(x) = f_0 \phi_0(x) + f_1 \phi_1(x) + \cdots .$$

When we truncate this expansion at n^{th} order to obtain $A_n(x)$, we would like the weighted error

$$E(x) = f_{n+1} W(x)\, \phi_{n+1} + f_{n+2} W(x)\, \phi_{n+2}(x) + \cdots$$

to be nearly best. Thus, the product $W(x)\phi_i(x)$ is the analogue of the Chebyshev polynomial $T_i(x)$. And we see that the polynomials $\phi_i(x)$ should obey the orthogonality condition

$$\int_{-1}^{1} \frac{W^2(x)\phi_i(x)\phi_j(x)}{\sqrt{1-x^2}} \, \delta x = 0, \quad i \neq j.$$

Again, the monic orthogonal polynomials are easily found from this relation {10}. We normalize these polynomials so that the extreme absolute value of the product $W(x)\,\phi_i(x)$ is unity. Thus, the weighted error of the truncated series is again readily estimated by

$$|E| \leq |f_{n+1}| + |f_{n+2}| + \cdots.$$

The coefficients f_i are obtained from

$$f_i = \frac{\displaystyle\int_{-1}^{1} \frac{W^2(x) \, F(x) \, \phi_i(x)}{\sqrt{1-x^2}} \, \delta x}{\displaystyle\int_{-1}^{1} \frac{W^2(x) \, \phi_i^2(x)}{\sqrt{1-x^2}} \, \delta x}.$$

In evaluating the above integrals and the integrals for the recurrence coefficients for the orthogonal polynomials, Clenshaw-Curtis (i.e., automatic Gauss-Chebyshev) quadrature may be used {11, 2}. If the integrand is not easily integrated by Chebyshev quadrature, then the function $F(x)$ should not be approximated by a polynomial. The only constraint we place on $W(x)$ is the requirement that all the integrals should exist. In particular, the usual constraint that $W(x)$ be non-negative is unnecessary, because it always is squared. Note that the orthogonal expansion $A_n(x)$ may be efficiently evaluated by a generalization of Clenshaw's recurrence {6}.

These generalized Chebyshev series approximations can be made to appear arbitrarily better than conventional Chebyshev series approximations by choosing a weight function that deviates an arbitrarily large amount from a constant. For example, consider the problem of approximating e^x with weight e^{-10x} on the interval -1 to +1. The weighted error of the 4th order Chebyshev series approximation is approximately $2 \cdot 10^4$ times larger than the weighted error of the 4th order generalized Chebyshev series approximation, but the generalized approximation error is only $\sim 20\%$ larger than the true minimax approximation error. For more typical cases such as relative error approximation of $\mathrm{Erf}(x)/x$ for $0 \leq x \leq 1$, the difference between the Chebyshev and generalized Chebyshev expansions is approximately 50% because the weight deviates from a constant by only 50%.

4.3 Chebyshev Series Generalized for Arbitrary Basis

Generalization to (almost) arbitrary basis is straightforward. Instead of orthogonalizing the basis x^i, we orthogonalize the given basis functions with respect to the weight $\dfrac{W^2(x)}{\sqrt{1-x^2}}$. Naturally, we must require all the integrals to exist. In addition, however, we must require the basis set to be a Chebyshev set (as with generalized Gauss-Christoffel quadrature) in order to circumvent possible degeneracies. The integrals now cannot be evaluated efficiently by Gauss-Chebyshev quadrature, because the integrand multiplied by $\sqrt{1-x^2}$ ordinarily will not behave like a polynomial. It is therefore necessary to make use of special knowledge about the integrand or to use (expensive) adaptive quadrature.

4.4 Summary

These generalized expansions will not be used for the elementary special functions, because these functions can always be manipulated so that an auxiliary function can be efficiently approximated by a Chebyshev series. What really concerns us here is that Chebyshev series are just a special case of a more general expansion that may be used if necessary. Arguments that true minimax approximations should be used because they allow arbitrary weight and basis functions are not valid. If, indeed, an argument can be made for using minimax approximations on those grounds, then the variable-accuracy generalized Chebyshev expansions would be even more suitable.

Generalized Chebyshev series will be applied only to higher level functions and not to the relatively simple functions we are concerned with here. For example, the second order Vacuum Polarization Integral {9} is known to behave like a polynomial plus another polynomial times log x. Thus, an appropriate basis would be x^i and x^i log x, and an appropriate weight would be the reciprocal function (relative error). The use of a generalized Chebyshev series for the approximation would have obviated the separation of the function into the separate polynomial parts, unquestionably the most time consuming and boring aspect of the original approximation calculation.

5. An Example: The Gamma Function

Although a detailed discussion of a special function routine is inevitably as boring as the task of writing the routine, a specific example is needed to make the earlier philosophical and possibly obtuse discussion more concrete. We choose to illustrate how portable

function routines are written by considering the Gamma function.
Even though the Gamma function is a relatively elementary function,
it is one of the more difficult routines to write. Large argument
values may cause overflow and small arguments may cause underflow.
Moreover, the Gamma function is singular for non-positive integers,
and arguments too close to a negative integer will introduce considerable significance loss. The Gamma function, therefore, proves to be
a non-trivial example.

5.1 Legal Argument Bound Calculation

As promised in Section 3.2, we illustrate how to calculate
legal argument bounds in a portable manner. For large positive
arguments Stirling's approximation shows that the Gamma function
behaves like

$$\Gamma(x) \simeq e^{-x} x^{x-\frac{1}{2}} \sqrt{2\pi} \ (1 + 1/12 \ x + \cdots).$$

We want to find the values of x for which $\Gamma(x)$ is the largest
representable machine number, call it B. Approximately,

$$-x + (x - \tfrac{1}{2}) \ \ln x + \tfrac{1}{2} \ln 2\pi \simeq \ln B.$$

Given x_o, a guessed solution to the above equation, we may use
Newton iteration to refine the estimate. We easily find

$$x_{i+1} = x_i \left\{ 2 - \frac{(x_i - \tfrac{1}{2}) \ \ln x_i - x_i + \tfrac{1}{2} \ln 2\pi - \ln B}{x_i \ \ln x_i - \tfrac{1}{2}} \right\}.$$

Typically three or four iterations starting with $x_0 = \ln B$ produce
a solution accurate to an absolute error of 0.005. Because the
function is convex downwards, the Newton solution will be too large.
It is therefore necessary to subtract a small constant several times

larger than the estimated error of the solution. In a program the constant B can be obtained from a low level routine, e.g., Bell Laboratories' utility R1MACH {8}.

For large negative arguments, we find

$$\Gamma(-x) = \frac{-\pi \csc \pi x}{x \Gamma(x)}$$

$$\simeq - \sqrt{\frac{\pi}{2}} \csc \pi x \, e^x \, x^{x-\frac{1}{2}} \, (1 - 1/12 \, x + \cdots).$$

We want only that value of x such that underflow might occur. Obviously, because of the presence of csc πx, underflow will occur near a half integral value first. We write

$$|\Gamma(-x)| \geq \sqrt{\frac{\pi}{2}} \, e^x \, x^{-x-\frac{1}{2}} \, .$$

We wish to find the value of x such that $|\Gamma(-x)|$ is approximately the smallest representable machine number, call it S. Choose $x_0 = -\ln S$, and use

$$x_{i+1} = x_i \left[1 - \frac{(x_i + \frac{1}{2}) \ln x_i - x_i - \frac{1}{2} \ln \frac{\pi}{2} + \ln S}{x_i \ln x_i + \frac{1}{2}} \right]$$

in order to refine the estimate. The minimum legal argument is the negated result of the iteration to which a small constant must be added in order to insure that the limit really will not produce an underflow. Finally, if the reflection formula

$$\Gamma(-x) = - \frac{\pi \csc \pi x}{x \Gamma(x)}$$

is used to obtain negative argument values, we must insure that $x \Gamma(x)$ will not overflow. This insurance is obtained by choosing the lower bound to be the maximum of the earlier calculated lower bound and 1 minus the calculated upper bound.

We have now derived formulas that enable us to find two legal argument bounds portably given only the largest and smallest machine numbers. These limits, moreover, are very close to the true machine limits. On CDC machines, for example, the upper bound is near 200, and we find that limit to within 0.01%.

One final error condition deserves consideration. Argument values too close to a negative integer will cause substantial significance loss. Observe in the above reflection formula that csc πx will have a large relative error if x is close to an integer not zero. We choose to produce a recoverable error condition whenever the accuracy is less than half machine precision. Large negative arguments therefore cannot be as close to an integer as can small negative arguments, because a greater portion of the computer word is needed to represent the integer part of large negative arguments. We require that negative arguments satisfy the condition

$$\left| \frac{x - [x]}{x} \right| > \sqrt{\varepsilon} \, ,$$

where [x] is the nearest integer to x and where ε is the machine epsilon, a constant to be obtained from a low level utility routine.

All the above calculations and error conditions must be considered if one wishes to write reliable routines, whether they are portable or not. We have simply carried out the calculations for the general case rather than for many different special cases. And certainly no more effort is required to do the general case.

6.   Conclusion and Acknowledgements

Mercly saying that portable routines can be written will justifiably fail to convince many readers that such a possibility exists. Hopefully, the portable software already produced by a few groups combined with the discussion in this volume will hasten widespread acceptance of portable software. Nonetheless, the proof is in the pudding, and a portable special function library must exist before readers can assess the wisdom of the many choices made in Sections 2 and 3.

I am most grateful to D. D. Warner for many stimulating discussions about approximation methods and software portability. These discussions led to the results in this paper. I am also indebted to W. S. Brown and A. D. Hall for their willingness to share their viewpoints on portability. I have unashamedly incorporated several of their views in this paper. W. J. Cody prodded me to develop my own ideas beyond an embryonic stage, and I am grateful for his encouragement.

REFERENCES

1. Bailey, C. B. and Jones, R. E. 1975, ACM Transactions on Mathematical Software, 1, 196. "Usage and Argument Monitoring of Mathematical Library Routines."

2. Branders, M. and Piessens, R. 1975, J. Computational and Applied Math., 1, 55. "An Extension of Clenshaw-Curtis Quadrature."

3. Brown, W. S. and Hall, A. D. 1977, These Proceedings. "Fortran Portability Via Models and Tools."

4. Clenshaw, C. W., Miller, G. F., and Woodger, M. 1963, Numerische Mathematik, 4, 403. "Algorithms for Special Functions I."

5. Cody, W. J. 1975, ACM Transactions on Mathematical Software, 1, 13. "The FUNPACK Package of Special Function Subroutines."

6. Fox, L. and Parker, I. B. 1968, Chebyshev Polynomials in Numerical Analysis, Oxford U. Press, London.

7. Fox, P. A. 1977, These Proceedings. "PORT - A Portable Mathematical Subroutine Library."

8. Fox, P. A., Hall, A. D., and Schryer, N. L. 1975, Bell Labs Computing Science Tech. Report #37, "Basic Utilities for Portable FORTRAN Libraries."

9. Fullerton, L. W. and Rinker, G. A., Jr. 1976, Physical Review A, 13, 1283. "Accurate and Efficient Methods for the Evaluation of Vacuum-Polarization Potentials of Order Z."

10. Gautschi, W. 1968, Mathematics of Computation, 22, 251. "Construction of Gauss-Christoffel Quadrature Formulas."

11. Gentleman, W. M. 1972, Comm. ACM, 15, 353. "Algorithm 424. Clenshaw-Curtis Quadrature."

12. Rudsinski, L., McGirt, A. F., Melendez, K. J. and Fullerton, L. W. 1973, (unpublished).

13. Ryder, B. G. 1974, Software-Practice and Experience, 4, 359. "The PFORT Verifier."

14. Schonfelder, J. L. 1977, These Proceedings, "The Production and Testing of Special Function Software in the NAG Library."

15. Schonfelder, J. L. 1976, Software--Practice and Experience, 6, 71. "The Production of Special Function Routines for a Multi-Machine Library."

16. Schonfelder, J. L. 1974, Software for Numerical Mathematics (Ed. D. J. Evans), Academic Press, London and New York. "Special Functions in the NAG Library."

17. Waite, W. 1977, These Proceedings. "Intermediate Languages: Current Status."

THE IMPORTANCE OF STANDARDIZED INTERFACES
FOR PORTABLE STATISTICAL SOFTWARE

N. Victor and M. Sund

1.   Introduction

Statistical computations which are being performed daily all
over the world are innumerable because there is hardly any empirical
science which does not have to consider random processes of one sort
or another and, consequently, needs to take recourse to statistical
calculations. The amount of statistical software available is tre-
mendous, too. Schucany et al. {12} in their survey of statistical
software already list 55 packages. Francis and Sedransk {6} estimate
the number of packages existing in 1976 to be well over one hundred,
and no one will ever be able to count the number of individual programs
not contained in any package. This vast amount of software, however,
can by no means be accounted for by actual need as can be seen from
the fact that the packages all cover nearly the same range of statistical
methods. One may safely conclude that in the development of statistical
software portability aspects did not play a great role.

The quality of statistical computations depends heavily on the
quality and adequacy of the numerical algorithms they employ. In this
way anybody in statistical computing is like others in the position
of a user in need of numerical software of all kinds such as matrix
algebra, solving linear and non-linear systems of equations, computing
eigenvalues and eigenvectors, generating pseudo-random numbers, quadrature,
fitting curves and the like. Therefore, if you are in the business of

designing and producing portable statistical software you inevitably have to cope with all the problems a producer of portable numerical software is confronted with. We all know of course that these problems cannot be solved solely by the people who write codes. The list of portability requirements for numerical software is rather lengthy as can be seen from the topics covered in this workshop, a fact which resulted in the conviction on our side that we as statisticians should gladly leave this enormous task to the numerical people. We therefore want to confine ourselves to the discussion of the particular portability problems which arise in the development of statistical software.

Statistical analysis is a multistage process. There is the data checking and manipulation phase including looking into the data for admissible and missing values, correcting the data, replacing missing data by some reasonable values, grouping the data, transforming variables, selecting groups, cases and variables. There are preliminary computational steps to aid in directing further calculations using intermediate results, etc. All this is occasionally accompanied by a vast amount of output, both printed, plotted, and filed, done over a considerable time span. To offer then statistical software in the form of a numerical subprogram library, even if it includes routines which might be called statistical subprograms, and even if it is accompanied by detailed directions for its usage, is quite prohibitive. For, to perform the tasks sketched above, a considerable amount of additional programming on the side of the user is required, which can exceed the code of the numerical routines needed considerably. And this programming very often would have to be done by a person who is inexperienced and unsophisticated, both with respect to computing and statistics.

Therefore, many individuals and institutions who felt the same way developed so-called general purpose programs and program packages to avoid having to do the same type of programming over and over again. It is with these programs and program packages that the great bulk of the daily routine statistical computations is being done in computer centers, statistics departments and the like. But most of these program packages, besides very often having questionable numerical qualities, do not take into account the multistage character of statistical computing, "fostering," as Nelder {10} said, "single-stage procedures in a magic black box manner." In addition, these program packages more often than not cover a rather limited range of statistical procedures and they are sort of closed shops. That is to say, a user of a particular program has to accept the features of that program. He can neither drop nor modify features nor can he add new features, and he cannot combine the features of two or more programs in an easy way. These programs are often pretty much like a textbook which at the time of being distributed is already rather outdated in many respects. But if one wants to have up-to-date software which covers all the many areas of statistical data analysis, one must be able to have specialists in a given area contributing their expert knowledge to the software product. A form of subroutine pool is more flexible for incorporating program pieces developed by different people at different locations. Further, there are packages which have excellent data handling facilities and control languages which one would not like to miss.

To avoid the drawbacks of the two types of software while, at the same time, preserving their respective advantages requires a highly structured software product. If, in addition, one takes into account that statistical data analyses often require access to big data files of different structures, the necessity for the use of standardized data interfaces should be apparent. These are the main aspects of the portability problem which dominate in the statistical area.

2. Project DVM-107

The government of the Federal Republic of Germany sponsors projects to explore the feasibility of electronic data processing in the medical field. Within this framework, since 1974, it also supports our project, called DVM-107[+], which is aimed at the standardization and coordination of statistical software produced by the various projects. DVM-107 was initiated and, in the first phases, directed by the Department of Biomathematics of the University of Giessen, and was carried out in close cooperation with the Institut für Medizinische Datenverarbeitung (IMD) in Munich. Since 1976 responsibility for the project was transferred to IMD to guarantee continuity, this institute being a government-controlled scientific institute.

Two main goals of the project are (for a more detailed outline of the project see Victor et al. {13}):

1. To work out programming standards to facilitate the exchange of software produced by the various DVM projects;

2. To construct and collect portable statistical software in the form of a statistical software kit for the benefit of these projects.

[+]now: DVM-118

While working towards the realization of these goals we encountered among other difficulties three main problems. In what follows we want to portray these problems and describe our solutions to two of them.

The first problem to be solved was to seek agreement as to the programming language to be used, to define some standardized subset of the language and to get the programmers to accept these standards.

The next difficulty was to define a structure for the statistical program kit which allowed the processing of large data sets both for data manipulation and for statistical analysis, a structure which was to be situated somewhere between a numerical subprogram library and a closed statistical system at the same time keeping it portable and flexible. Since the way we defined this structure required the use of standardized data interfaces the definition of these interfaces became the most important aspect in our project.

The third problem we faced was to define a control language which would make the statistical program kit a statistical system with all the advantages of traditional systems but without their apparent disadvantages and without reducing the portability of the system as compared to that of the individual components. We did not do too much in this direction. On the one hand, it seems that most people responsible for doing the analyses in big projects prefer to compose the programs needed themselves using components available to them and to add their own products where necessary. On the other hand, the workload to link the modules is small as compared to that of programming these modules. In addition, one other project in the Federal Republic of Germany is to develop a portable control language system for controlling

FORTRAN-modules in the area of econometrics (Broadbend et al.{2}) and it
seems natural to utilize experiences and results obtained by that project.

3.   Language Standards for a Portable Code

The great variety of machine ranges used by the various DVM projects
made FORTRAN an obvious choice although there are certainly better
languages.   But even so we had, of course, to choose a subset acceptable
to the DVM machine ranges used.   After many meetings with other project
representatives and after having run a number of tests on various machines
we came up with DVM-FORTRAN-standards which are close to the ANS-standards
X3.9 taking into account the norms ANS X3.10 and ECMA-9 together with the
two ACM Clarification Reports.   This subset of FORTRAN (with the exception
that it contains direct access statements) is nearly identical with PFORT
described by Brown and Hall elsewhere in this volume.

More difficult than defining standards was enforcing these standards
on the side of the programmers.   According to our experience this cannot
be achieved without offering mechanical instruments which reduce the
effort of producing standardized code.   We published the standards in
German in the form of a manual (Friedrich {7}) but our most valuable aid
is a FORTRAN program which checks out the standards.   Incidentally,
this syntax-checker is itself coded according to the DVM-standards
and can, therefore, verify itself.   It has proven to be almost portable
(Schaumann {11}).   This program should be very similar to Ryder's
PFORT-Verifier also mentioned by Brown and Hall in this volume.   This
way the programmer does not have to learn the standards to the minutest
detail, is able to convert a given code to standardized code in a very
short time, and having done this a few times, turns more and more to code
in standardized form as he undergoes a learning process.

An additional aid is a programmers's guide containing for each non-standard statement of various FORTRAN dialects a translation into standard FORTRAN (Friedrich {8}). To write preprocessors which do this translation automatically is only a small additional step.

We are also working on recommendations for code editing, internal and external program documentation, output formatting, programming techniques, etc. For example, we recommend structured programming techniques. A group at IMD extended FORTRAN by statements which allow one to write structured code which is then converted by means of a preprocessor (Drechsler {3}) into regular FORTRAN. We also use a program to aid in editing the code like bringing statement numbers into sequence, indenting statements for better depiction of the logical structure of the code, etc.

4.   Structure of the Program Kit

As we stated before, a subroutine pool does not suffice for easing the task of a statistician confronted with large data sets, because the additional code needed for data management can exceed that of the statistical algorithms considerably. Nor are self-contained program systems the best solution, because they are not very much portable, they are not flexible enough, the range of methods realized by them can never cover all statistical areas, and, in many areas they are outdated. The best algorithms for solving special problems are usually found in programs written by experts in that field and are superior to corresponding package programs (e.g., for tabulation programs cf. Francis and Sedransk {6} and for discriminant analysis programs cf. Victor and Trampisch {14}).

For the purpose of combining the best aspects of the two approaches we chose to proceed in a rigorous top-down manner to obtain a clear separation of control language part, data handling part, and statistical computations. Where statistical computations are concerned, we decided to regard as our smallest program unit a clearly distinguishable step during the course of a statistical analysis, which allows no further subdivision from the statistical point of view. These pieces of code we want to call (statistical) segments as opposed to (numerical) subprograms.

For the purpose of not reducing portability when linking segments we need to define interfaces. The segments themselves do not present great new problems for achieving portability. If, for the purpose of linking the segments, we use standardized data interfaces which contain all the data to be transferred and the description of the data structure, and, if in the segments we use special I/O modules for access to these data interfaces, the overall product becomes very flexible in a plug-in/plug-out fashion. What we mainly have to do is the realization of these portable interfaces together with the corresponding I/O modules.

First we had to define the class of admissible structures for the interfaces. When doing this we had to try to keep this class as general as possible to keep the interfaces to a minimum in number. But at the same time we had to avoid too far-reaching generality which would result in very complex descriptions and representations of very simple data structures. So far we have managed to get away with two data interfaces: one, called S_0, to link initial data handling with the analysis and one, called S_1, to store intermediate results, vectors and matrices

computed from S_0. For the incorporation of statistical procedures we
have not yet considered, it will surely be necessary to define additional
interfaces, e.g., for nested and incomplete designs. Our program kit will
be constructed in the following way (cf. figure 1):

The basis is a pool of subprograms (numerical, statistical, I/O, etc.)
which are characterized by the fact that transfer of data is being
done via parameter lists only.

The essential units, however, are the segments, each of which taken
by itself is capable of yielding a meaningful statistical result.
As such it can use any number of subprograms and may be positioned
in various places in a sequence of computational steps. Each
segment uses one or more data interfaces as input and can use
one or more interfaces which are created or modified as output;
there are, of course, segments which do not create new interfaces.
Segments, therefore, are characterized by the interfaces they access
and by those they create. In figure 2 we present some examples
of types of segments.

Logically an interface is a class of data-structures including
a language which allows us to describe a particular member of
this class. To physically realize the interface in a self-
describing form it is necessary to furnish the programmer of a
segment with control statements (i.e., calls to subroutines)
which allow him

    - to create a physical interface on a file,

    - to write into this interface the information which

      defines the structure and to retrieve this information,

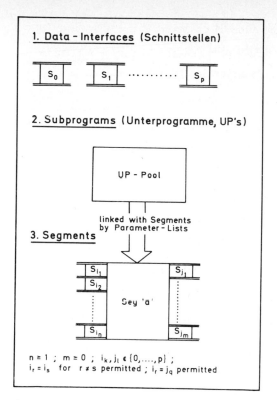

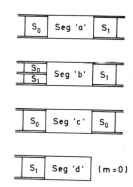

Fig. 2: Some common
types of
segments

Fig. 1: Structure of our statistical
kit

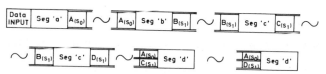

Fig. 3: An admissible sequence of segments

 - to write data elements into the interface according to
the structure defined above, and

 - to identify data elements according to the structure
and read them.

For a complete statistical computation several segments are linked sequentially (plugged into each other). A segment may be added to an already existing sequence of segments, if all interfaces needed by this segment have already been created. As the very first segment you need a segment which transforms the raw data into the structure of S_0. Physical interfaces are labeled any way you want, the name of the class of structures being added, e.g., $DATA_{(S_1)}$, $A_{(S_0)}$.

In addition, you need an indicator of when the content of the interface was created because a segment may just change the content of the interface. This way $(A_{(S_0)})_{t=3}$ means the data set of the physical interface A, which is of a structure S_0, after execution of the 3rd segment. The linking of segments can be seen in figure 3.

Some advantages of this concept can be summarized as follows:

 - Existing software can be used; e.g.,

o existing portable subroutine libraries can be incorporated
(e.g., for the numerical part we use the NAG library which is
mentioned at several places in this volume),

o existing good and portable programs for certain statistical
analyses can be incorporated into the kit after modifications
according to our specifications have been made, and

o existing data manipulation systems can be used after
incorporation of our data interface S_0.

- The workload for constructing a custom-made statistical analysis program is kept to a minimum. For frequently used standard analysis procedures you can save the sequence of segments needed as sort of a macro to be used by less sophisticated persons.
- The whole system is very flexible; e.g.,
   o each user can easily add the features he needs by writing his own segments,
   o data manipulation segments can be constructed by computer scientists and statistical analysis segments can be written by statisticians, and
   o the segments needed for certain areas of statistics can be constructed by experts who then can concentrate on the algorithms instead of having to deal with data handling and the logic of a large program.
- The portability problems are reduced; e.g.,
   o making the numerical subroutines portable can be left to the numerical people,
   o given portable numerical subroutines, the task of constructing portable segments is not too big,
   o portability is reduced to the task of writing portable modules for processing the interface control statements and this task has turned out to be relatively easy.
- The concept of using interfaces can also be used to link existing packages to make them more flexible.

The "expert concept" mentioned above, i.e., the concept of having programs for a certain area in statistics developed by persons who are accepted as experts in this area is a prerequisite of portability in the sense that the final product is not only portable but is also actually transported, i.e., accepted and used.

5. Interface S_o

Our most important interface is S_o by means of which segments doing statistical data handling and segments doing statistical calculations are linked. This interface should be described in more detail to point out its importance and meaning for achieving portability and flexibility.

We mentioned earlier that both data manipulation and statistical algorithms are equally important for analyzing a set of data, and that programs which cover only one of these two aspects are of little help to the person having to perform the analysis. Both tasks have to be performed in close connection with each other, for the results of one step can direct the other step. For the purpose of programming, however, it is desirable to separate the two tasks, since the methodology to be employed in both areas differs substantially: the treatment of data structures belongs to the field of the computer scientist, the statistical methods fall into the responsibility of the statistician.

The most convenient solution - the construction of a language which is appropriate for controlling both data manipulation and data analysis in an elegant way (cf. Gower's {9} suggestion for a statistical autocode) - is unrealistic. The other extreme,which is also unfeasible, is to fix a standardized data-format into which every user has to squeeze his data.

We tried with our concept of an interface for a certain class of data structures to stay somewhere in the middle, wanting

- to follow Gower's suggestions as closely as possible,
- to enable the user to do alternative passes through the data manipulation and the data analysis phase,
- to allow for complicated data structures during the data manipulation phase, and
- to keep the actual analysis modules free of complicated data descriptions.

The raw data a statistician is usually confronted with at first sight are very simply structured; they are almost always case oriented. But for each case there are a multitude of data elements of different types: there are elements which are used to structure the data, e.g., into groups like sex, bloodgroup, treatment groups. These variables are called "factors" in design terminology. Other values might be considered as realizations of random variables like measurements on weight and height. Some values might not be used at all for analysis or only after some transformations, e.g., as frequencies. Some might even be used as both types, as a random variable for one analysis (weight) and as a factor for another analysis (weight groups). This example demonstrates that data elements and information which describes the relationships among them are often not distinguishable.

The task of the data handling modules, therefore, is to map the raw data onto the internal data structure and to extract from the data structure a subset and put it into S_o in a form suitable for analysis. The data structures of S_o are method oriented and the class of structures is easy to describe. In this sense S_o separates the data manipulation and the data analysis sections.

In addition, the acceptance of one standardized interface could very well improve the flexibility of the classical systems, if only the systems producer would implement just one exit to this interface. To do this would require much less work than the efforts which are being made right now to link some existing systems pairwise (cf. Buhler {1}).

The class of data structures allowed in S_0 is general enough to cover uni- and multivariate data, simple data matrices and multifactor designs, measurements and counts; but, as mentioned before, for some designs S_0 is not suited very well, e.g., for nested or incomplete designs for which class of problems we will need an additional interface. The notation used for defining a particular data structure is very simple. For example, the expression

$$(F2; L2,3; D1; Z7,7,7,6,5;)$$

describes a univariate (dimension D=1) two-factorial (number of factors F=2) design, the factors having 2 and 3 levels (L=2,3), respectively. The cell frequencies (Z) are 7,7,..., 5 for the factor combinations 1-1, 1-2, ..., 2-3, respectively. If one had the same number n of observations in each cell, one could have written for short 'Zn;'. The specification D=0 indicates that the data elements are not measured data values but frequencies. By using an extended form of this notation to describe subsets of the data structure, we were able to define a set of powerful control statements (of course in the form of FORTRAN calls) to store and retrieve data structures and arbitrary subsets or data elements.

To implement this notation it is necessary to fix the mapping of data structures onto a physical file and for certain purposes this file should be made self-defining by writing all the information about the structure onto the same file. The physical representation of data structures of the class allowed in S_o on a file is given in figure 4. The seemingly complicated representation - which is fully needed for multifactor unbalanced MANOVA designs - reduces for the simpler designs to a very simple form. Figure 5 gives examples of some data structures.

We believe that we have found a reasonable way of implementing the interface in a portable manner. We implemented it almost entirely in standard FORTRAN. The subprograms for interpreting the notation, writing and reading into and from the file are part of the subprogram pool. There is only one contact point to the operating system: one file of fixed, user-specified record length has to be created which can hold all interfaces. File directory, index block and data block management, and internal data block chaining are exclusively done by FORTRAN routines. By using suitable buffers in most cases we get away with one physical access although we have to use an index-sequential file. The internal chaining of data blocks makes it possible to process the data sequentially, a fact which saves a lot of time, since this type of reading data is the one that is needed most in statistical computations. The increase in computer time needed is almost entirely dependent on the time-consuming read/write statements of FORTRAN. If one would replace these FORTRAN I/O-statements by routines written in assembler code, the savings in time would be considerable and this would perhaps justify the loss in portability. Tests we performed on different machine configurations

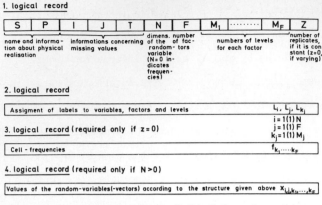

1. logical record

| S | P | I | J | T | N | F | M_1 | .... | M_F | Z |

name and information about physical realisation | informations concerning missing values | dimens. number of the random-variable ($N=0$ indicates frequencies) | number of factors | numbers of levels for each factor | number of replicates, if it is constant ($z=0$, if varying)

2. logical record

| Assigment of labels to variables, factors and levels | L_i, L_j, L_{k_j} |

3. logical record (required only if $z = 0$)

$i = 1(1) N$
$j = 1(1) F$
$k_j = 1(1) M_j$

| Cell - frequencies | $f_{k_1 \cdots k_F}$ |

4. logical record (required only if $N > 0$)

| Values of the random-variables(-vectors) according to the structure given above $x_{i,j,k_{ij}\cdots,k_F}$ |

Fig. 4: Representation of S_o-data-structures on files

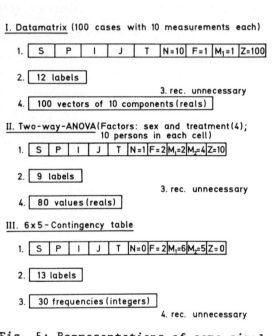

I. Datamatrix (100 cases with 10 measurements each)

1. | S | P | I | J | T | N=10 | F=1 | M_1=1 | Z=100 |

2. | 12 labels |

    3. rec. unnecessary

4. | 100 vectors of 10 components (reals) |

II. Two-way-ANOVA(Factors: sex and treatment(4); 10 persons in each cell)

1. | S | P | I | J | T | N=1 | F=2 | M_1=2 | M_2=4 | Z=10 |

2. | 9 labels |

    3. rec. unnecessary

4. | 80 values (reals) |

III. 6 x 5 - Contingency table

1. | S | P | I | J | T | N=0 | F=2 | M_1=6 | M_2=5 | Z=0 |

2. | 13 labels |

3. | 30 frequencies (integers) |

    4. rec. unnecessary

Fig. 5: Representations of some simple data-structures

(CDC, IBM and SIEMENS) indicated that it is not difficult to construct almost portable I/O programs for standardized interfaces.

Summary

We discuss portability aspects of statistical software, this topic being the main issue of a project sponsored by the government of the Federal Republic of Germany. From considerations relating mainly to the two most prominent forms of existing software - the subprogram library and the program package or system - we infer the structure of a flexible and more portable software product which incorporates the respective advantages of the two forms. We emphasize the use of standardized interfaces to link the various elements of the structure described and delineate one data interface which is to serve as the link between data handling and statistical algorithms.

REFERENCES

1. Buhler, R. (1976): The P-STAT System: Status in March 1976.
    9th Ann. Symp. on the Interface, April 1976, Harvard University.

2. Broadbend, J. E., Hauer, K., and Kindermann, G. (1975): Software-
    System zur Steuerung von FORTRAN-Moduln im Rahmen einer
    ökonometrischen Methodenbank. Tech. Report, Math. Beratungs- und
    Programmierungsdienst, Dortmund, 409 pp.

3. Drechsler, K. H. (1975): STRUCTRAN. Strukturiertes Programmieren
    in FORTRAN mittels Sprucherweiterung. Statist. Software Newsletter 2,
    23-25.

4. Ford, B., and Hague, S. J. (1973): The Organization of Numerical
    Algorithms Libraries. IMA-Conference on Software for Num. Mathematics,
    April 1973, Loughborough.

5. Francis, I., Sherman, S. P., and Heiberger, R. M. (1975): Languages
    and Programs for Tabulating Data from Surveys. 9th Ann. Symp. on
    the Interface, April 1976, Harvard University.

6. Francis, I., and Sedransk, J. (1976): Software Requirements for
    the Analysis of Surveys. 9th Internat. Biometric Conference,
    August 1976, Boston.

7. Friedrich, H. J. (1974): FORTRAN-Standards. Stat. Software Newsletter,
    Beiheft 1.

8. Friedrich, H. J. (1976): Programmiererhandbuch zur Übertragung
    von FORTRAN-Dialekten in Standard-FORTRAN. Stat. Software
    Newsletter, Beiheft 3.

9. Gower, J. C. (1969): Autocodes for the Statistician. In Milton,
    R. C. and Nelder, J. A. (eds.): Statistical Computation. Academic
    Press, New York.

10. Nelder, J. A. (1974): A User's Guide to the Evaluation of Statistical
    Packages and Systems. Int. Stat. Rev. 42, 291-298.

11. Schaumann, H. P. (1975): DVM-107-Syntax Checker. Stat. Software
    Newsletter, Beiheft 2.

12. Schucany, W. R., Shannon, Jr., and Minton, P. D. (1972): A Survey
    of Statistical Packages. Computing Surveys 4, 65-79.

13. Victor, N., Tanzer, F., Trampisch, H. J., Zentgraf, R., Schaumann, H. P.,
    and Strudthoff, W. (1974): A Project for Standardization and Coordination
    of Statistical Programs in the BRD. In: Bruckmann, G., Ferschl, F.,
    Schmetterer, L. (eds.): COMPSTAT 1974. Physica Verlag, Wien.

14. Victor, N., and Trampisch, H. J. (1975): Programme zur
    Diskriminanzanalyse. Stat. Software Newsletter 1, 48-58.

Note: "Statistical Software Newsletter" is a periodical for
        the benefit of the DVM-projects.

        "Beihefte" are the Technical Reports of Project
        DVM-107.

EXPLORING THE IMPACT OF PORTABILITY

The things which hurt, instruct.

— *Benjamin Franklin (1744)*

Portability, or the lack of it, has a profound economic effect on computing, felt most keenly by those managers who must deal with the many demands on an operating budget. The need to quantify portability problems for planning purposes is not matched by a supply of precise measuring tools; however various aspects of portability have been placed in perspective in certain practical situations. Two such studies are reported in this section.

J. Bentley and B. Ford extract a wealth of detailed information from the files of the NAG project. Records were kept of the nature of the changes as the NAG library was adapted to a number of machine ranges. A profile is presented in statistical form.

Ingemar Dahlstrand reviews a Swedish study of portability in technical and scientific computing. Broad-gauged in concept, this study examined language characteristics as well as tape labels and character codes. Dahlstrand summarizes the essential conclusions.

ON THE ENHANCEMENT OF PORTABILITY WITHIN THE NAG PROJECT
- A STATISTICAL SURVEY

J. Bentley and B. Ford

ABSTRACT

We present the results of a statistical survey
carried out on the contents of a numerical algorithms
library. The purpose of the investigation is to assist
in the development of a portable body of code.

INTRODUCTION

It is a brave man who suggests or indeed believes that the portability of numerical algorithms can be achieved with complete automation. In this paper we present the results of a statistical survey carried out on the contents of a numerical algorithms library; a library for which coding conventions are clearly defined, a library which at the time of the study was implemented on nine different machine ranges, a library which demonstrates man's instinct to "leave his mark."

The library in question is that of the Numerical Algorithms Group (NAG)* {2, 3, 4, 5} and the purpose of the investigation is to assist in the development of a portable {6} body of source code. Past experience has shown that the implementation of a routine on a machine other than that for which it was written requires a number of modifications, some of which are predictable, many of which are not. We aim to determine the modifications required in implementing the NAG library on different machine ranges and, using the results, suggest some solutions to our particular problem of portability.

We begin the paper by defining our objectives and then go on to describe the Master Library File System (MLFS) {7, 8, 9}, NAG's present solution to the storage of the vast amount of information resulting not only from the implementation of the library on different machine ranges but also from the updating of the library contents at regular intervals in time. It is the contents of these files which provide the subject matter of our investigation. In section 3 we present the categories for all modifications made within the routines and in section 4 detailed results of the number of occurrences of all such modifications.

---

*From March 18th, 1976; Numerical Algorithms Group Limited

Although the NAG library is implemented in Algol 60, FORTRAN and more recently in Algol 68, we have restricted this initial survey to the FORTRAN code but hope to extend the statistics to cover other languages in due course. With the exception of the error mechanism routine, the remaining library routines are all input/output free. This decision was taken at the outset of the project in order to improve the portability of the library. However, the test programs which support the library do by necessity include input/output and it is our intention at present to include these test programs in a future survey to be carried out at the next mark of the Library (Mark 5, May 1976).

1. OBJECTIVES

The contents of the Library are contributed at present by approximately 60 geographically dispersed individuals with conventions for the coding of routines being laid down in the NAG Reference Manual. The approach for the Contributed Library is basically one of language subset together with the more recent introduction of machine characteristics (unfortunately too recent to show up in the figures presented). The Library contents are updated approximately annually with each update being referred to as a new mark of the Library.

Implementation groups have been established for all machine ranges on which the Library is implemented. Again these groups are geographically dispersed and it is their responsibility to make any changes necessary to successfully run the contributed code on their particular machine. They are asked specifically to make the minimum number of changes consistent with accurate and efficient running. Unfortunately, as our results show, this has not always been done.

As the number of implementations steadily grows it becomes increasingly more important to develop a single transportable body of code. It is in the quest of this goal that we conduct this study. We wish primarily to determine the major changes made by the implementors to the contributed source code and to evaluate the nature of these changes. It is anticipated that the results will assist in proposing elements to be included in a new set of standards aimed at portable/transportable software and indeed will help determine whether a transportable library is feasible at this time.

2.   THE NAG MASTER LIBRARY FILE SYSTEM {7, 8, 9}

Implementing the NAG Library on different machine ranges has resulted not only in the need to hold a large body of software at the Central Office but also in vast maintenance problems not least of which is concerned with the updating of the library contents at regular intervals in time.

Considering the contributed code to be the base or predicted version of the library then, as shown by the results of Section 4, at most 21% of this code needed to be changed in order to implement the library on a machine other than that for which the code was initially written. Needless to say this experience relates only to the machine ranges covered in the survey. The MLFS, implemented in 1973 by Richardson and Hague {9}, takes advantage of this situation by creating a composite file with the implemented source codes held as variants to the contributed library.

The master library files holding the information are storage devices designed to hold software variants, the variation being recognised in two ways:

(i)   across and within machine ranges

(ii)   updates in time.

The body of software defined at a specific time is denoted by the term 'mark,' hence each update of the library contents results in a new mark of the library. For the purpose of this investigation we are interested in the consequences of the first variation at particular marks of the library.

The MLFS consists of a set of files in card image format, each file containing index records followed by control and text records, where the record is the basic unit of information. A variant therefore consists of a set of records, some of which may be particular to that variant, the rest being in common with other variants. The present MLFS can cater for up to 12 distinct implementation groups. Within each group there can be a tree structure where the nodes of the tree are referred to as subgroups. Changes introduced by a member of the group tree affect all members of its sub-tree. Between group trees, however, changes made by one group can only affect another if the first group is considered responsible for the maintenance of the library item involved.

MLFSRUN is a general purpose program developed to operate on the master library files. The program can be driven in six different modes, each performing a specific function on the MLFS. Details of these modes can be found in {10}. The program was employed in the accumulation of statistics since, when driven in mode *COMPARE, the output consists

of a clearly documented comparison between different implementation/mark
pairs which could then be examined by eye.

3.   CATEGORIES FOR CLASSIFICATION OF CHANGES

Although details of the nature of the changes made to the contributed
code could be determined only by careful examination of the contents of
the files, a certain amount of statistical information was obtained
with the aid of a program, developed to scan the master files keeping
account of the following:

for a particular mark

for the complete library

for each chapter

for each routine

(a)   The total number of lines in the contributed source text

(b)   The total number of lines of the original source text which
      had been replaced

(c)   The number of lines of source text replaced by each implementation

(d)   The total number of replacement lines of source text

(e)   The number of replacement lines for each implementation

(f)   The ratio of (d) to (b)

(g)   For each implementation, the ratio of (e) to (c)

A summary of this information is given in the next section.

Before the start of what became affectionately known as the
"eyeball search" a list was created containing categories for predictable
changes.   These included changes in precision, machine constants and
comments.   New entries were created as new categories were realised.
Having compiled the complete list we were able to classify all changes into
three general categories as given overleaf.   The effect of language standards,
present in our predicted list of general categories, was found to be redundant.

3. CATEGORIES FOR CLASSIFICATION OF CHANGES (contd)

A. Arithmetic/derived numerical characteristics	Comments
A1 Change to double precision	A11 to A14 represent the consequences of changing a routine from single to double precision whereas the A2 subgroups represent the effects when a routine is implemented on a machine which has a longer word-length than that for which it was originally written. This latter case is clearly demonstrated when an ICL 1906A routine is implemented on the CDC 7600. Both A1 and A2 are dependent upon the word-length of the machine and more directly upon NSIG, the number of significant decimal digits.
A11 Declaration of double precision variables.	
A12 DATA statements to set up double precision constants.	
A13 Conversion of intrinsic and basic external functions.	
A14 Statements changed due to effects of A11, A12.	
A2 Extension of basic precision	
A21 Changing the number of digits in constants etc. in order to give the maximum that the machine can handle.	
A22 Changing the number of terms in a series when more significant digits can be achieved.	
A23 Changing the maximum value allowed for a "variable" when more accuracy is possible.	
A3 Precision Criteria	The entries in this category are also dependent upon the word-length of the machine.
A31 The value of EPS, the criterion of relative precision.	
A32 The value of MAXREAL, the largest positive single length real.	
A33 The value of MINREAL, the smallest positive single length real.	
A34 The value of MAXINT, the largest integer I such that I and -I can be represented.	
A4 Commenting out implanted precision information not required for a particular implementation.	This represents the effect of including in the original code all precision information required for the different implementations and commenting out that which is not relevant.

3. <u>CATEGORIES FOR CLASSIFICATION OF CHANGES</u> (contd)

B. <u>Programming Style</u>	Comments

B1 Trivial Changes
   B11 Blank lines inserted.
   B12 Spaces inserted/deleted.
	Not only is B11 a trivial change but it does result in non-ANSI code and hence such changes should not have been made.

B2 Comments
   B21 Comments added.
   B22 Comments removed.
   B23 Comments changed.
	Although relevant comments are clearly desirable, many of the changes made within this category were unnecessary.

B3 Improvements in Efficiency
   B31 Multiplication replaced by (repeated) addition.
   B32 Changes made to save arithmetic, in particular in the implementation of loops.
   B33 Reducing the number of array accesses.
   B34 Redefinition of tests for negligible quantities, in order to avoid overflow caused by division by small reals.
	Many of these changes were probably unnecessary, in particular those in B31, B32 and B33, but may account for the correction of what was initially inefficient coding.

B4 Changes to ANSI FORTRAN, in particular the removal of mixed mode arithmetic, the elimination of zero labels and the omission of blank lines.	In accordance with NAG conventions such changes should not have been necessary.
B5 Changing variable names/alphanumeric constant names.	Again these changes were on the whole unnecessary.

3. CATEGORIES FOR CLASSIFICATION OF CHANGES (contd)

B. Programming Style (contd)	Comments
B6 The implementation of complex arithmetic.	For example, some implementors choose to employ intrinsic functions for the extraction of real and imaginary parts of complex variables whilst others use the EQUIVALENCE facility to map the two 'words' of a complex variable on to two real variables or array elements.
B7 The use of local dialects.	Examples for this category include the use of quotes for alphanumeric data and also changes to include mixed mode arithmetic.
B8 Change to simple parameters.	For example, replacing a logical expression in a parameter list by a logical variable with the same value.
B9 Correction of bad coding.	
B10 Changes made for no apparent reason.	
C. Compiler Operation	Comments
C1 The inclusion of array bound checks.	For many implementations such checks are made automatically.

3. CATEGORIES FOR CLASSIFICATION OF CHANGES (contd)

C. Compiler Operation (contd)	Comments

C2 Implementation of DO loops
 	
    C21 Whether the range is always executed at least once.
 	
    C22 Values permitted for control variables.
	For some implementations the initial value of the loop variable may take any value, for others it must be positive, hence C22.

C3 Omitting unnecessary declarations, in particular the EXTERNAL declaration.	
C4 Compiler Restrictions.	This category is directly relevant to the 4100 NL29 compiler which, amongst others, imposes the following restrictions. (i) The class of a statement must be recognisable from the first line. (ii) The maximum number of subroutine parameters is restricted to 31.

C5 Character Representation
 	
    C51 Effects of NCHAR, the number of characters per integer word.
 	
    C52 Justification of non-filled fields.
 	
    C53 Types allowed for non-numeric information.
	In the case of the NAG library, these entries influence the way in which Hollerith strings are represented.

C6 Continuation Codes.	Although there are NAG conventions to cover this category certain implementations do vary from the standard.

4. RESULTS

We begin this section by presenting a summary of the information obtained using the program as described in the previous section. Results are given at Marks 3 and 4 of the Library and the implementation groups included are as follows:

B   IBM 360/370

C1   CDC 7600

E   ICL 4100

F   ICL SYSTEM 4

G   NAG CENTRAL OFFICE

where B, C1, etc., denotes the identifier for the implementation group. The code for the NAG Central Office, which also acts as the 1906A implementation group, is taken as the 'base' version from which all modifications are made. The codes for the chapters of the Mark 4 Library, used throughout this section, are as follows:

A01   Machine Dependent Constants

A02   Complex Arithmetic

C02   Zeros of Polynomials

C05   Roots of One or More Transcendental Equations

C06   Summation of Series

D01   Quadrature

D02   Ordinary Differential Equations

E01   Interpolation

E02   Curve and Surface Fitting

E04   Minimizing or Maximizing a Function

F01   Matrix Operations, Including Inversion

F02   Eigenvalues and Eigenvectors

F03   Determinants

F04   Simultaneous Linear Equations

F05   Orthogonalisation

G01   Simple Calculations on Statistical Data

G02   Correlation and Regression Analysis

G04   Analysis of Variance

G05   Random Number Generators

H   Operations Research

M01   Sorting

P01   Error Trapping

S   Approximations of Special Functions

4. <u>RESULTS</u> (contd)

CHAPTER	LINES IN CONTRIB. VERSION	REPLACED LINES					REPLACEMENT LINES				
		TOTAL	IMPLEMENTATION GROUPS				TOTAL	IMPLEMENTATION GROUPS			
			B	C1	E	F		B	C1	E	F
A01	402	46	–	–	–	46	32	–	–	–	32
A02	69	46	46	–	–	46	106	53	–	–	53
C02	395	125	123	33	–	123	557	236	83	–	238
C05	649	100	98	–	2	99	311	150	3	7	151
C06	521	49	49	–	4	49	290	137	8	8	137
D01	1249	586	571	541	13	573	1786	601	547	37	601
D02	1511	293	266	33	24	266	698	308	51	31	308
E01	176	36	22	4		25	106	41	7	–	58
E02	483	65	63	2	1	65	178	81	6	7	84
E04	2916	451	387	4	5	448	1079	453	25	7	594
F01	3128	812	789	36	1	663	2087	1092	116	1	878
F02	2359	754	727	57	3	657	2127	1024	161	4	938
F03	1187	205	199	12	1	201	601	286	28	1	286
F04	698	183	180	14	–	180	541	254	33	–	254
G02	86	15	9	–	6	9	79	35	3	6	35
G04	177	102	–	1	102	–	197	40	8	109	40
G05	363	55	42	10	19	42	298	132	17	17	132
H	1165	205	104	59	143	73	491	181	69	145	96
H01	1697	120	26	92	–	4	267	102	159	–	6
P01	20	20	3	6	20	–	52	13	25	14	–
S	1001	569	500	458	12	515	2054	761	501	22	770
TOTAL	20252	4837	4204	1362	356	4084	13937	5980	1850	416	5691

TABLE 1. <u>Mark 3 Statistics</u>

The structure of the MLFS is such that all replacements are recorded for
each implementation. Hence, although a replacement line may be common to
all implementations other than the base version it will be counted separately
for each implementation. Consequently the total replacement lines in
Tables 1 and 2 should be treated with caution.

4. <u>RESULTS</u> (contd)

CHAPTER	LINES IN CONTRIB. VERSION	REPLACED LINES						REPLACEMENT LINES				
		TOTAL	IMPLEMENTATION GROUPS				TOTAL	IMPLEMENTATION GROUPS			
			B	C1	E	F		B	C1	E	F
A01	402	60	2	–	–	60	52	1	–	–	51
A02	69	46	46	–	–	46	106	53	–	–	53
C02	879	246	240	35	26	197	901	406	97	52	346
C05	651	104	103	1	3	103	309	150	3	6	150
C06	1111	99	99	–	4	99	361	174	9	8	170
D01	1249	683	673	529	13	573	1760	587	546	26	601
D02	1211	245	229	30	15	229	558	252	43	11	252
E01	177	49	35	4	–	43	111	45	11	–	57
E02	485	71	67	3	1	70	182	83	7	5	87
E04	5953	753	699	15	31	718	1754	820	63	30	841
F01	3173	1071	828	60	5	988	2607	1113	195	1	1098
F02	2468	878	819	72	3	855	2524	1078	214	153	1079
F03	1209	255	249	13	1	251	710	330	38	11	331
F04	780	219	210	27	–	210	1295	594	90	16	595
F05	57	28	28	–	–	28	37	18	1	–	18
G01	732	167	141	4	2	166	376	172	10	3	191
G02	3328	537	525	18	43	502	1217	584	62	11	560
G04	178	103	1	2	101	1	220	48	11	113	48
G05	370	89	79	1	19	77	315	137	29	14	135
H	1341	250	155	62	138	154	683	242	73	140	228
M01	1756	125	30	93	29	29	321	90	160	–	71
P01	20	20	10	10	20	–	73	8	51	14	–
S	1628	937	795	646	59	848	2973	1104	642	71	1156
TOTAL	29227	7035	6063	1625	513	6247	19445	8089	2355	685	8118

TABLE 2.   <u>Mark 4 Statistics</u>

4. RESULTS (contd)

From Tables 1 and 2 we summarise certain aspects of the results in Tables 3 and 4.

TABLE 3. Summary of Statistics

CONTRIBUTED LIBRARY		UNION OF REPLACED LINES		CDC 7600		IBM 360/370		ICL SYSTEM 4		ICL 4100		
No. of LINES OF CODE	%	No. of REPLACED LINES	%	No. of REPLACED LINES	%	No. of REPLACED LINES	%	No. of REPLACED LINES	%	No. of REPLACED LINES	%	
20252	100	4837	23.9	1362	6.7	4204	20.8	4084	20.2	356	1.8	MARK 3
29227	100	7035	24.2	1625	5.6	6063	20.7	6247	21.3	513	1.8	MARK 4

TABLE 4. Summary of Line Replacement

IMPLEMENTATION GROUP	MARK OF LIBRARY	REPLACED LINES	REPLACEMENT LINES	% INCREASE
CDC 7600	3	1362	1850	35
	4	1625	2355	45
IBM 360/370	3	4204	5980	42
	4	6063	8089	34
ICL SYSTEM 4	3	4084	5691	39
	4	6247	8118	30
ICL 4100	3	356	416	17
	4	513	685	34

Taking the ICL 1906A, whose real arithmetic uses a 48 bit word with a mantissa of 37 bits as the base version, then from Table 3 we may observe the following: For the ICL 4100, whose real arithmetic also uses a 48 bit word with a 37 bit mantissa, 98% of the library code remains unchanged. For the CDC 7600, using a 60 bit word with 48 bit mantissa, 93% of the code is unchanged but for the byte machines, whose real double precision arithmetic use a 64 bit word with a 56 bit mantissa, approximately 79% of the code remains unchanged. We may therefore draw the tentative conclusion that the variation in the percentage of changes to the contributed library at Mark 3 and Mark 4 appears to be directly dependent upon the floating point arithmetic features of the configuration.

Before establishing the nature of these changes it may be worth noting the significance of Table 4. For all implementations the number of replacement lines exceeds the replaced lines, on average by 35%. Although the more intensive search reveals that this figure may be somewhat misleading (for example the program has no way of distinguishing between genuine replacement lines and lines of original code which are commented out by other implementations) it does warrant consideration if one wishes to develop automatic precision conversion aids.

In presenting the results of the 'eyeball search' we attempt to show the progression of the library from pre-mark 3 up to mark 4. For this purpose we give results for the pre-mark 3 Library, for the mark 3 releases and re-issues and for the mark 4 releases and re-issues, where a re-issue is a new routine which is introduced to replace a withdrawn routine and which adopts the same name. It is interesting to observe how styles have changed as hopefully the problems of portability have become more widely appreciated within the project.

4. RESULTS (contd)

TABLE 5. Summary of Replaced Line Information

CATEGORY	PRE-MARK 3 % OF TOTAL	MARK 3 % OF TOTAL*	MARK 4 % OF TOTAL*
A1	51	57	73
A2	12	33	7.6
A3	0.5	0.6	1
A4	-	-	4
B1	16	2	1.7
B2	12	5	8.8
B3	2	1	-
B4	1	0.2	1.4
B5	2	-	-
B6	0.3	1	-
B7	0.3	-	-
B8	-	-	0.5
B9	0.4	-	-
B10	1	-	-
C1	1	-	0.3
C2	0.6	-	0.1
C3	0.07	0.2	1
C4	-	-	0.7
C5	0.07	-	-
C6	0.04	0.2	-

* Releases + re-issues

Table 5 shows a summary of the categorised changes made to the contributed code. This information is given by implementation in Table 6 and by chapter in Table 7. An obvious but nevertheless important point to note is that the results obtained are directly dependent upon the nature of the algorithms covered by the survey. Although the pre-mark 3 results represent an overall picture of the library, the later additions tend to be more limited in scope.

In particular the smaller number of routines added at Mark 3 (53 as opposed to 88 at Mark 4) were in specific areas of the Library; for example the quadrature and ordinary differential equations chapters. Since both these chapters involve a large number of constant coefficients which need to be represented to full machine precision, this would account for the high percentage of changes falling into category A2. Category A1, however, is always representative since the routines are always implemented in double precision on the IBM and System 4 machines, and the necessary changes are seen to be of a uniform level throughout the Library. The problems of extending the basic precision of a routine will always be present unless a 64 bit word is used as the base library version.

Man's desire to leave his mark is evident from the high but fortunately decreasing introduction of cosmetic formatting into the base version. Of comparable magnitude to these trivial changes is the problem of comments. Whilst we may reduce the number of such changes by advocating the use of machine independent comments, the problem, in particular that of commented out implanted information, remains fundamental to portability exercises employing software transformers and formatters.

It is noteworthy that the implementation of DO loops, which were thought to be a significant problem in earlier portability discussions, are hardly evident in the statistics. Unfortunately the effects of different library mechanisms for subroutine collection for user programs continue to be an issue.

The information summarised in Tables 5, 6 and 7 is available in full for each routine in the library, for each chapter and for each implementation. This detailed information was considered too extensive for the scope of this paper. It is interesting, however, to use the chapter information in conjunction with the results of Tables 1 and 2. For example, we see from Table 1 that almost 60% of the code in the Special Functions chapter had to be replaced. We can establish that 90% of these changes were in the category of precision accounting for 26% of all changes made within that category. Similarly we can establish that over 60% of the high proportion of changes made in the G04 chapter were in the category B12, "spaces inserted/deleted"!

TABLE 6. Replaced Line Information for each Machine Implementation (Expressed as a Percentage of the Total Number of Lines in the Base Version)

CATEGORY	PRE-MARK 3					MARK 3*					MARK 4*				
	IBM	CDC	4100	SYS 4	TOTAL CHANGED LINES	IBM	CDC	4100	SYS 4	TOTAL CHANGED LINES	IBM	CDC	4100	SYS 4	TOTAL CHANGED LINES
A1	20.6	-	-	19.5	20.7	15.7	-	-	15.7	15.7	16.2	-	-	15.7	16.6
A2	-	4.9	-	-	4.9	-	9.1	-	-	9.1	-	1.7	-	-	1.7
A3	0.2	0.2	-	0.2	0.2	0.1	0.06	-	0.1	0.2	0.2	0.2	-	0.2	0.2
A4	-	-	-	-	-	-	-	-	-	-	0.9	0.8	-	0.9	0.9
SUBTOTAL	20.8	5.1	-	19.7	25.8	15.8	9.16	-	15.8	25.0	17.3	2.7	-	16.8	19.4
B1	4.8	0.01	1.8	4.5	6.6	0.1	-	0.5	0.1	0.7	0.4	-	0.2	0.4	0.4
B2	4.7	-	0.1	3.6	4.8	1.2	0.04	0.1	1.2	1.4	1.8	-	-	1.7	2.0
B3	0.4	0.2	0.01	0.8	0.8	0.06	-	0.2	0.06	0.3	-	0.2	-	-	0.2
B4	0.5	-	0.2	0.5	0.5	0.06	-	-	0.06	0.06	0.3	-	0.3	0.3	0.3
B5	0.7	0.5	-	0.6	0.8	0.2	-	-	0.2	0.2	-	-	-	-	-
B6	0.2	0.02	0.04	-	0.2	-	-	-	-	-	-	-	-	-	-
B7	0.1	-	-	0.1	0.1	-	-	-	-	-	0.1	-	-	0.1	0.1
B8	0.1	-	-	0.1	0.1	-	-	-	-	-	-	-	-	-	-
B9	-	-	-	0.2	0.2	-	-	-	-	-	-	-	-	-	-
B10	0.3	-	0.01	0.3	0.3	-	-	-	-	-	-	-	-	-	-
SUBTOTAL	11.8	0.73	2.16	10.7	14.2	1.62	0.04	0.8	1.62	2.66	2.6	0.2	0.5	2.5	2.8
C1	0.4	-	-	0.3	0.4	-	-	-	-	-	0.07	-	-	0.07	0.07
C2	0.2	-	-	0.06	0.2	-	-	-	-	-	0.02	-	-	0.02	0.02
C3	-	-	-	-	-	-	-	0.1	-	0.1	-	-	0.2	-	0.2
C4	-	-	0.03	0.02	0.03	-	-	-	-	-	-	-	0.15	-	0.15
C5	0.02	-	0.01	-	0.03	-	-	-	-	-	-	-	-	-	-
C6	-	-	0.01	-	0.01	-	-	-	-	-	-	-	-	-	-
SUBTOTAL	0.62	-	0.05	0.38	0.67	-	-	0.1	-	0.1	0.09	-	0.35	0.09	0.44
TOTAL	33.22	5.83	2.21	30.78	40.67	17.42	9.2	0.9	17.42	27.76	19.99	2.9	0.85	19.39	22.64

* Releases + re-issues

4. RESULTS (contd)

TABLE 7. Replaced Line Information for each Chapter (Summary) (Expressed as
a percentage of the total number of lines in the base version)

CHAPTER	PRE-MARK 3 CATEGORY			MARK 3* CATEGORY			MARK 4* CATEGORY		
	A	B	C	A	B	C	A	B	C
A02	0.3	0.1	-	-	-	-	-	-	-
C02	1.4	0.4	-	-	-	-	1.8	0.2	-
C05	0.4	0.1	0.04	1.0	0.07	-	-	-	-
C06	0.5	0.2	-	0.3	0.2	-	0.3	0.07	0.02
D01	-	-	-	16.7	0.7	-	-	-	-
D02	0.5	0.1	0.03	4.0	0.3	0.1	-	-	-
E01	0.07	0.03	-	0.01	-	-	-	-	-
E02	0.6	0.07	-	-	-	-	-	-	-
E04	2.7	1.1	0.02	0.8	-	-	4.7	0.3	0.2
F01	2.3	3.2	0.2	0.5	0.9	-	-	-	-
F02	2.5	3.6	-	0.3	0.1	-	-	-	-
F03	0.7	0.9	-	0.3	0.01	-	-	-	-
F04	0.7	1.0	-	0.04	0.06	-	0.1	0.1	-
F05	-	-	-	-	-	-	0.3	0.02	-
G01	-	-	0.01	-	-	-	2.1	0.2	0.07
G02	0.08	0.07	0.1	-	-	-	5.5	1.1	0.04
G04	0.02	0.8	0.2	-	0.06	-	-	-	-
G05	0.4	0.2	0.03	-	-	-	-	-	-
H	0.4	1.6	-	0.2	0.1	-	0.2	0.6	-
M01	0.2	0.4	0.1	0.06	0.06	-	-	-	-
S	12.3	0.3	-	0.7	0.01	-	4.4	0.2	0.1
TOTAL	26.07	14.17	0.73	24.91	2.57	0.1	19.4	2.79	0.43

* Releases + re-issues

TABLE 8. Overall Summary

GENERAL CATEGORY	PRE-MARK 3	MARK 3*	MARK 4*
	%	%	%
Arithmetic/derived numerical characteristics	63	90	86
Programming Style	35	9.5	12
Compiler Operation	2	0.5	2

* Releases + re-issues

CONCLUDING REMARKS

The results confirm our initial beliefs that of fundamental importance in the portability of numerical software are the arithmetic and derived numerical characteristics of a machine. Possibly of more far reaching consequence, however, is the realisation that we may never expect to predict all of the changes all of the time and hence a corrective step will invariably be necessary. We can without doubt deal with the problems of precision, indeed this can be done automatically with the aid of tools such as APT[1]; we can avoid the problems of such features as array bound checks and DO loops by programming them into the original code; we can with consummate ease eliminate the trivial, unnecessary changes, but until the ideal machine with the standard compiler becomes a real proposition, we shall always have problems of a different nature.

---

[1]APT is an automatic precision transformer developed at the Central Office by J. L. Siemieniuch and colleagues {1}. All changes deemed necessary by categories A1, A2 and A3 of this survey are automatically dealt with by the transformer.

There appears to be little problem with language dialects, once the principle of using language subsets is accepted, and it is reassuring to see how programming styles have changed to accommodate the new trend towards portability.

It had been our opinion that too little quantitive information was available to sustain generally informed discussion of the portability of numerical software and we therefore hope that the results of this study will help reveal some relevant facts. We believe that the machines included cover a reasonably broad spectrum of computing configurations and look forward to repeating the exercise at Mark 5.

REFERENCES

1. Du Croz, J. J., Hague, S. J., and Siemieniuch, J. L., (1976), "Aids to Portability Within the NAG Project." This Proceedings.

2. Ford, B., "Developing a Numerical Algorithms Library," (1972), I.M.A. Bulletin, Volume 8, pp. 332-336.

3. Ford, B., and Hague, S. J., (1974), "Software for Numerical Mathematics," Ed. Evans D. J., Chapter 20, London, Academic Press.

4. NAG Annual Report, (1974), Oxford, Numerical Algorithms Group Ltd.

5. NAG Annual Report, (1975), Oxford, Numerical Algorithms Group Ltd.

6. Ford, B., "Portability of Numerical Software in the NAG Project," (1976), CREST ITG Software Portability Course, Kent.

7. Prentice, J. A., Ford, B., and Hague, S. J., (1974), "Software for Numerical Mathematics," Ed. Evans D. J., Chapter 22, London, Academic Press.

8. Hague, S. J., and Ford, B., (1976), "Portability - Prediction and Correction," Software P and E, Volume 6, pp. 61-69.

9. Richardson, M. G., and Hague, S. J., (1976), "The Design and Implementation of the NAG Master Library File System," Software P and E, In Press.

10. NAG Reference Manual, (1976), Oxford, Numerical Algorithms Group Ltd.

A STUDY OF PORTABILITY IN TECHNICAL AND SCIENTIFIC COMPUTING

Ingemar Dahlstrand

ABSTRACT

The author reviews a portability study funded
by the Swedish Board for Technical Development,
covering portability not only for Fortran and
similar languages but also for data and command
languages. Fortran is shown not to be a portable
language; even the proposed new standard leaves
some deficiencies. The most important improvements
would be:

- to make it possible to define range and precision
  for numeric quantities;

- to define a collating sequence for string handling
  based on the international standard character code;

- to introduce overflow control or, more generally,
  means for ensuring computational security.

Data portability can be achieved by using the
standard character code in conjunction with standard
labels. In command languages, no standard exists,
but international committee work is in progress
and the prospects for technical success are good.

This presentation will mainly be a review of a study of portability in technical and scientific computing which was carried out in 1974/75 on a grant from the Swedish Board of Technical Development. The study was motivated by a strong feeling that present-day lack of portability is a serious drawback to the user. This is especially so in technical and scientific computing where most users are amateurs and the standardization level is lower than in administrative ADP. The study recognized the need for a total approach, i.e., not only would programming languages proper, like Fortran, but also command languages and data have to be portable. In this sense the level of ambition of the study was high. On the other hand, a ceiling to ambition was set by the fact that the project was envisaged as a Swedish - possibly Scandinavian - effort with limited implementation resources and impact on manufacturers. If and when portability were to become a reality through international effort, this particular project would be discontinued. This level of ambition called for implementation through pre-compilers rather than writing wholly new software (or persuading manufacturers to do so). The study investigated and showed the feasibility of this approach. Altogether, the study came to cover the following topics:

- Meaning of and advantages of portability
- Character codes
- Fortran
- Tape labels
- Command languages
- Machine independent intermediate languages
- Algol

In each case the existence of, quality of and adherence to standards were investigated, the problems discussed and solutions outlined.

The next stage of the study is now to plan and budget for an implementation phase.

What is Portability?

Portability is defined as the ability to run computer programs unchanged on different machines and get essentially the same results. Obstacles to portability are differences of memory size and type, differences of I/O equipment, differences in word length and differences of implementation of programming languages. Two points worth noting:

1. Reproduction of results exactly is usually not called for. The user must be assumed to accept rounding errors, minor differences in system printouts, etc.

2. Lack of portability is legitimate when it is due to different capacity or type of equipment. Each machine may cover only part of the universe of permissible programs; in many cases the benefit of portable languages is being able to send your run to a better suited machine. But a program that runs on one machine should either run and give essentially the same result on another machine, or it should be rejected for capacity reasons; it must not run and give quite a different result!

The study indicates that it should be possible to obtain 100% portability for Fortran programs and a high portability for command language sequences as well. The advantages of such a level of

portability are obvious for program exchange, education and shifting
one's load from one computer to another in many situations. The cost
of program porting (and personnel re-training) actually carried out is
estimated to be $0.75 - $1.00 Million in Sweden alone - still considering
only technical and scientific computing. But looking at the program
porting actually carried out today tends to grossly underestimate the
advantages of portability; it is somewhat like measuring the value of
a standard railway gauge by referring to actual re-loading costs in
the year 1850!

After these general points let us look at particular aspects of
how to achieve portability.

Character Representation

International standard (ISO 646, also called "ASCII code") exists
and is quite adequate. Adherence is uneven, but improving - we all
know this process is agonizingly slow, due to manufacturer and user
conservatism and the heavy costs involved in conversion. The study
recommends that the standard should be used for data exchange, but -
as we shall see later - other character representations may be used
locally without loss of program portability if the character code
and the string handling fulfill some reasonable conditions.

Fortran

Standard for Fortran (ISO 1539) exists since several years and
is well adhered to. When the standard was adopted many specifications
were intentionally left out, because it would have been too difficult
to agree on them.

Portability Deficiencies in Fortran

A.   No definition of word length and precision.

B.   Rudimentary and non-standard string handling.

C.   Special constructs undefined (e.g., DO loop with zero iteration count, computed GO TO out of range, SAVE of local variables in subroutines).

D.   No overflow control.

E.   End-of-file and error sensing in I/O undefined.

F.   No direct access I/O.

G.   No special I/O (plotters, paper tape).

Consequently, ANS Fortran does not really define a portable language. Subsets of it are portable but rather too restricted for all-round use.

Draft Proposed New Fortran

It is interesting to see how the new Fortran standard now being proposed would treat the portability problem. String handling is expanded and a new data type, CHARACTER, introduced. End-of-file and error handling is introduced, as is direct access I/O (the latter, however, in a manner deviating from de facto standard). Most of the special constructs are now defined. On the other hand, the word length problem remains - though eased by the introduction of generic functions - and there is still no collating sequence for strings and no overflow control. Let me elaborate a bit on each of these points.

Possible Methods of Handling Word Length and Precision

Word length varies greatly between Fortran implementations. Basically there are three ways of handling this problem.

1.   Standard meanings of the types, e.g., let REAL and INTEGER have 32 bits and DOUBLE PRECISION 64 bits word length.

2.   Environmental enquiries: functions allowing the
     programmer to ascertain the local word length and choose
     a path in his program accordingly.

3.   Type definition: language features allowing the programmer
     to define the word length desired.

In my opinion, the first method is not viable because it would
be inefficient on those machines which did not happen to have exactly
the word length defined as standard, and such a standard would not be
adhered to. The second method puts too great a burden on the user to
write alternative program sequences. It is, however, useful in certain
cases, e.g., when you just want to know if the word length is sufficient,
or if you want to modify the tolerance or the step in an iteration to
make the best use of the machine at hand. The third method, allowing
the user to define word length, is the one to be recommended in the
majority of cases. The ordinary user should, of course, have the
option of leaving out these word length definitions and get the
default definitions of the local implementation if he does not need
portability.

Suggestions for Type Definition

An example of how type definitions might look is the following:

INTEGER = INTEGER 32 BIT, FORMAT CODE I

REAL*4 = REAL 21/11 BIT, FORMAT CODE E, SIZE = INTEGER

REAL*8 = REAL 53/11 BIT, FORMAT CODE D, SIZE = 2 * INTEGER

The example defines three types INTEGER, REAL*4 and REAL*8,
which may then be used in type statements in the program. Each type
definition gives the word length, split up in the case of real

quantities between a mantissa and an exponent, the format code associated with the type, and possible relations between the storage space used for the types. The REAL 21/11 BIT part of the definition of REAL*4 would mean a mantissa of 21 bits and an exponent of 11 bits to the base of 2 (this is roughly equivalent to a 24-bit mantissa and an 8-bit exponent to the base of 16).

A type definition could be seen as a directive to the precompiler. The precompiler would consider the demand for range and precision (and a possible SIZE directive) and implement the type as single precision on some machines, double precision on other machines and in some cases give a warning that even double precision would not be enough. No loss of efficiency would result, unless the demand for range and precision was unnecessarily high in the first place.

On a higher level of ambition the precompiler might implement multiple integer and real arithmetic. This would give a much slower program but would of course only be used when the demand for range and precision was such that the program would otherwise have to be run on another machine.

As seen from the above, a demand for precision and range, say REAL*5 = REAL 32/8 BIT, would normally be interpreted to mean that the mantissa and exponent should hold at least - but not necessarily exactly - 32 and 8 bits, respectively. In special circumstances, however, namely when testing a program for use on many different machines, one would like to know that 32/8 bits is really enough, i.e., the implementation should destroy precision above 32 bits and give overflow if the exponent exceeded 8 bits. (This might be signaled by adding the

attribute STRICTLY to the type definition.) This would of course slow
down the arithmetic a lot, but is to be regarded as a special testing
feature - the alternative being to run the program on different machines
to be sure it worked everywhere.

String Collating

Let us first make one thing clear: There is no way around using
a standard collating sequence if you want portability at all for string
handling programs.

Variations of word length and other language deviations usually
have the effect that a program that works on one machine either does
not work on another machine or seems to work but gives a nonsense result;
but there is usually no discussion about what is a right and wrong result
once the difference is detected.

Not so for string collating. In this case what is wrong on one
machine (say, "A" > "1") is right on another, and if there is no
agreed collating sequence there is no way of deciding which.

The argument advanced against a standard collating sequence is
that it is too expensive because it calls for using the standard code.
This argument seems wrong to me on two counts:

1.   In the long run, it will be even more expensive <u>not</u> to
     use standard code.

2.   In the short run it is possible to use the standard collating
     sequence without actually storing the strings in standard
     code. Consider the following example, collating two Russian
     text strings:

     САМЫЙ
     САБАК

Without knowing any Russian, we see that the strings are equal in their first and second positions, but differ in their third. All we have to do to compare the strings is to take the "М" and the "Б" and see which comes first in the Russian collating sequence.

In a similar manner we may compare strings stored in, say the EBCDIC code, converting only one character from each string into ISO code.

When porting data the whole text obviously would have to be converted, but that is already the position today so no efficiency is lost.

Overflow Control

Overflow control must be improved so that overflow can always be sensed by the programmer, and if unexpected overflow occurs this must at least give a warning printout. The situation today may be illustrated by the following example (written in an easily understandable Fortran-like language):

```
INTEGER K, L
K = 4000000
L = K * K
PRINT L
END
```

Try running this on four different machines:

- Machine A (24 bits word length) signals a compile-time error (the constant 4000000 is too large for this machine).
- Machine B (32 bits) gives a run-time abort (overflow).
- Machine C (36 bits) gives a wrong result without warning (integer overflow is not trapped).
- Machine D (48 bits) gives the right result - believe it or not!

In a more mature branch of technology this sort of thing would be regarded as slightly scandalous; and let us realize that we cannot go on forever basing computations that result in moon rockets and bridges on such loose foundations.

Given the hardware deficiencies of present-day machines, it would be quite expensive to achieve a reasonable level of overflow control or, more generally, computational security. It will have to be phased in over a period of 10 years or so.

I shall not go further into the details of Fortran portability but end up with a few comments on other aspects of portability.

Data Exchange

The existing standard, ISO R/1001 or "ASCII labels," defines rules for labelling and block structure on magnetic tapes. It is generally adhered to, but gives data portability only if combined with the standard character code, which is often not the case.

Command Languages

In this area, there is no standard or even proposed standard. Command languages are very dissimilar, though basic features tend to be present in all. The field has been surveyed by several committees and an IFIP Working Conference, and there is a consensus that defining and implementing a machine independent command language is no more (and no less) difficult than was defining and implementing languages like Fortran and Algol. Machine independent command languages have in fact been proposed. Some of these, usually based on precompilation to existing languages or on local operating systems, are in limited use. The study discusses a number of questions arising in connection with precompilation:

- Can interactive processing take place in a precompilation environment?
- Division of functions between Fortran and the command language. Since this division is different on different machines, portability is in general possible only when a job is completely defined (i.e., programs are in source form and there is full information about files).
- Need for surrounding the job with local commands.
- Problems with memory files, especially local naming conventions and how to give machine independent information about record sizes. Some of these problems in fact prevent complete portability for command languages.
- I/O files.
- Compilation, linking and loading.

The main recommendation in this area is to continue following international committee work before undertaking any implementation.

The study ends with a brief discussion of machine independent intermediate languages and Algol, in both cases concluding that work in these areas is of great potential long-range interest, but should have second priority in the short run to making Fortran portable.